THE NAVY TIMES BOOK OF SUBMARINES

A POLITICAL, SOCIAL, AND MILITARY HISTORY

Brayton Harris
edited by Walter J. Boyne

Berkley Books, New York

A Berkley Book
Published by The Berkley Publishing Group
A division of Penguin Putnam Inc.
375 Hudson Street
New York, New York 10014

Cover photos courtesy of PNI/Check Six and U.S. Navy
Cover design by Steven Ferlauto
Interior text design by Sherman/Durlon Design

PRINTING HISTORY
Berkley hardcover edition / December 1997
Berkley trade paperback edition / August 2001

Berkley trade paperback ISBN: 0-425-17838-2

Visit our website at
www.penguinputnam.com

The Library of Congress has catalogued the Berkley hardcover edition as follows:

Harris, Brayton, 1932–
The Navy times book of submarines : a political, social, and military history / Brayton Harris : edited by Walter J. Boyne.—1st ed.
p. cm.
Includes bibliographical references and index.
ISBN 0-425-15777-6
1. Submarines (Ships) I. Boyne, Walter J., 1929– . II. Title.
V857.H37 1997
359.9'3—dc21 97-38078
CIP

PRINTED IN THE UNITED STATES OF AMERICA

10 9 8 7 6 5 4 3 2 1

Contents

List of Illustrations and Photographs

DUST JACKET: World War I French submarine, late-war German coastal boat *UB-133* and early model, *UB-24*, photographed in 1918. *Credit: U. S. Navy*

LIST OF ILLUSTRATIONS

PHOTOGRAPHS

Foreword

This beguiling review of the history of the submarine imparts a tremendous amount of new information, and at the same time, puts to the lie many of the oft-told myths about this fascinating subject.

In his careful reconstruction of the history of the men and the machines that dared to go beneath the surface of the water to enter combat, Brayton Harris combines his extraordinary sense of humor with intensive research to present an engaging, informative, and sometimes startling book. Readers will be amazed at the number of things long accepted as gospel that are in fact completely in error, including hoary quotations from Prime Minister William Pitt, Admiral David Farragut, and Admiral Hyman Rickover. But Harris does far more than debunk; his careful analyses of the political and economic situations prevailing in the various countries during each period of submarine development are thought-provoking, and have application even today. During its entire history, the submarine's greatest value has always been to weaker naval powers; it was a great leveler from the beginning, and even today, many of the world's smaller countries depend upon submarines to be an equalizer to the stronger navies of potential enemies.

The present doomsday power of nuclear submarines is daunting, and it is almost a relief to go back in history to the time when the power of humankind to damage itself was limited and tinged with human foible. Perhaps the most surprising aspect of submarine warfare is that it was conceived so early, had a centuries-long gestation period, and then was born with many a bang during World War I.

During the crucial first forty years of the 20th century, the submarine was handicapped by the narrow boundaries of conventional naval thinking in every seafaring country. Thus, after having almost decided the outcome of the war at sea during World War I with a handful of submarines, German Naval leadership failed to prepare the submarine force that it might have for World War II, instead building virtually useless so-called "pocket battleships" and initiating a heavy battleship program. The British Navy, for its part, after having suffered enormous losses to submarines in the 1914–1918 war, assumed that the problem was solved by underwater listening devices, and entered World War II almost completely unprepared for Admiral Dönitz and his wolf-pack tactics.

Harris records many instances of heroism in the conduct of operations, just as he properly notes the horrors of submarine warfare, extending his sympathetic treatment to the victims of the submarines and to the submariners when they in turn became victims, as they so often did. When he brings his narrative forward into the present, one can only be struck how, despite the many casualties, all previous submarine warfare was of such a limited, measurable danger. Despite the many thousands of sunken ships, the damage done was trivial when compared to the incalculable destructive power lurking in the ballistic-missile-laden vertical tubes of U.S. and foreign nuclear submarines. At the earlier level ships might be lost—at the current level of nuclear submarine power, entire countries could be exterminated within an hour. Paradoxically, these potent submarines have—to date—produced no casualties. One can only hope that this will remain true for the future.

Walter J. Boyne
Ashburn, Virginia

Introduction

The most complete bibliography of submarine books I have found—and I found it posted on the World Wide Web!—lists more than 425 titles. Most are out of print but still available, lurking in the better libraries of the world. Many—most—of the more recent books concentrate on the wars of the 20th century, often with a focus on "exploits." Exploits make good drama, but not necessarily good history.

When recent authors cover the early years, more than three hundred years of invention and trial and frustration and growth, they do so *de minimus*: a few pages, some times only a few paragraphs, apparently borrowed with minor changes from another author's brief treatment; copies of copies, losing detail, leading to errors of fact or interpretation. Of course, in dealing with information that is perhaps four hundred years old, some error is inevitable. Documents from the distant past are fragments, and "eyewitness" accounts often turn out to be someone else's account of the eyewitness account. Copies of copies . . .

As a result, much of what we think we know about the history of submarines may not be true. Some samples?

- One author describes America's first submariner, the Revolutionary War pioneer David Bushnell, as a "sturdy Maine native"—he was sickly, and from Connecticut.

- A book written in 1994 states with authority that "Bushnell returned to the charge in the Anglo-American War of 1812." In 1812, Bushnell—whose family had not heard from him for twenty-five years—was in his seventies and living under an assumed name in Georgia.

- Another book, vintage 1995, presumes that two submarines, designed for Japan by John P. Holland after his retirement from the Electric Boat Company in 1904, were "smaller" and "inferior to the earlier *Holland* types" delivered by Electric Boat. They were twenty feet longer, and twice as fast.

Well, those are minor quibbles, three amidst hundreds; in the history of the submarine boat, let one significant error stand for all. This, from a popular history of the World War I British submarine force:

- Almost a hundred years earlier, Pitt had watched Robert Fulton's *Nautilus* nose her way under the waters of Walmer Roads before attacking and sinking the Danish brig *Dorothea*.

There are only two things wrong with this: Pitt was not present at the demonstration, and Fulton never had his "diving boat" *Nautilus* in England; he built only one (many reports to the contrary), which he demonstrated for the French. In France. And which he scrapped in France. There is no evidence that *Nautilus* was ever used to sink anything, anywhere.

He did indeed sink *Dorothea* off the coast of England, but did so with carefully placed explosive charges that he called "torpedoes." To place the charges, he used boats. Rowboats. A short time later, Fulton gave the Earl of St. Vincent (who likewise had not been present at the demonstration) a briefing. The earl's response has ever since been enshrined as the ultimate pithy saying about submarines:

> Pitt was the greatest fool who ever lived to encourage a mode of war which they who commanded the seas did not want, and which, if successful, would deprive them of it.

As pithy sayings go, it would be first-rate *if* these were the Earl of St. Vincent's words. However, the statement is Robert Fulton's *report* of what the earl said. As delivered, the quotation is suspiciously tinged with Fulton's views of his own transcendent importance.

As pithy sayings go, it would have been even better had the earl been talking about submarine boats. He was not. The conversation centered on Fulton's schemes for delivering torpedoes by surface boats or rafts, not by submarine.

Some of the historical confusion arises from misunderstood terminology. Over the years, "submarine" has been applied variously to any weapon of war used underwater. Sometimes it is specific, as in the submarine boat, the submarine bomb, the submarine battery, the submarine torpedo. Sometimes it is just "submarine" and could mean any of the above. Sometimes it is not really a *submarine* but is, instead, what is loosely called a *submersible*—meaning an "almost" submarine, a boat designed to run just awash, with only a small portion above the surface.

Another confusion: When is it a "torpedo," when a "mine"? The term "torpedo" as applied to underwater explosives is taken from the electric ray fish, genus *torpaedae,* folk name "cramp fish," which delivers a numbing shock to its victim. The words *torpor* and *torpid* are derived from the same source. The first use in underwater warfare is attributed to Robert Fulton, circa 1800; examination of apparent earlier attributions reveal the hand of the modern translator. "Torpedo" originally meant any underwater explosive mine—as in the order given by Admiral David Farragut at the Civil War Battle of Mobile Bay. A ship had fallen out of position in the attack column; Farragut asked, "What's the trouble?" The answer came back, "Torpedoes," and Farragut exploded, "Damn the torpedoes! Four bells, Captain Drayton. Go ahead!" Since about 1866 the name has stuck to the "automobile," or self-propelled, torpedo originally perfected by Robert Whitehead. Towed, moored, or floating maritime weapons subsequently have been called "mines." *That* term is derived from the military practice of undermining fortifications and planting explosives in the tunnel. I'll try to be clear and consistent, but won't fiddle with the style, spelling or grammar of contemporary quotations. Damn the torpedoes, indeed!

This present work is neither a catalog nor an encyclopaedia; it is, rather, an excursion through the years, an exploration of the steps and stumbles along the way toward the development of the submarine boat. I hope to establish a cohesive record, which does not seem, at this time, to exist. This book is necessarily written from an American perspective—but not, I hope, an American bias. This book is also written from the perspective—but not, I hope, the bias—of a career Naval officer.

Brayton Harris

Administrative Remarks

The following are offered to refresh the reader's memory on basic submarine physics before meeting an actual submarine boat:

- The weight of ocean water imparts a pressure of about half a pound per square inch for every foot of depth. This is approximately one atmosphere (14.7 pounds per square inch) for every thirty-three feet below the surface. Water itself is basically incompressible.
- Any boat can be made to sink, by making it heavier than the amount of water it displaces (i.e., negative buoyancy). A boat at negative buoyancy will keep sinking (and also losing volume, which progressively increases the rate of sinking) until something is done to add positive buoyancy. Creating a boat that safely can sink, and rise again, requires a controlled method to either change the volume of the boat with some sort of extending/contracting structure, or change the weight of the boat.
- The usual method for changing weight is by adding or removing water. (A well-meaning British naval engineer, around the time of World War I, is reported to have asked a commanding officer, "Where do you get

the water?") On early boats, the water was removed by a "forcing" pump, operated by hand or foot; the later and common solution is to introduce compressed air into the tank at a pressure greater than the outside water pressure. When pumps are used—for example, for waste removal—the pump must be able to overcome outside pressure at whatever depth it is being operated.

- Weight and volume in submarines must carefully be matched; in a small submarine, balance can be upset by the simple measure of raising a periscope, which increases volume without changing weight. As depth and pressure increases, hull volume decreases and weight must be reduced to match; a typical late-model World War II submarine might pump out fifty gallons of ballast for every one hundred feet of depth.

- A modern submarine can hover at a selected depth by constantly adding and expelling a small amount of ballast; this was not so easily done in earlier boats. Any submarine can rapidly return to the surface by quickly getting weight out of the boat. Some early designs had "drop keels" of lead or iron or stone which could be released in an emergency; another approach was a ballast tank which could quickly be blown by air at high pressure.

- Until the middle of World War I—until, that is, someone developed a weapon which could damage a submerged boat—there was little reason to design a boat that could dive any deeper than needed to get out of sight. Depth was dangerous, and depth control was inexact. Some early designers insisted that a submarine boat should always operate on a level keel, with depth controlled by ballast and "haul-down screws," to avoid uncontrolled sinking. Others created "diving" boats, designed always to run at slightly positive buoyancy, using a combination of forward momentum and diving planes—horizontal rudders—to drive the boat down or up. If a level-keel boat became unstable and tipped down (or up), it could be in trouble; a diving boat was designed to operate at an angle.

- If the air pressure inside the hull exceeds the water pressure outside, water will not come in through any opening below the internal water level. This is the principle of the diving bell: a large, strong tub, literally bell-shaped, with the bottom open to the sea and weighted so as to sink evenly when suspended from the surface. This traps air inside—the same effect as when you take an empty drinking glass and push it down, inverted, into a pail of water. Once the pressure is equalized, a sufficient reservoir of air remains to permit a diver to breathe for some time. This principle was to be adapted to a number of submarine boats in the 19th and early 20th century, to provide underwater access for combat or salvage divers.

- An unprotected diver can reach exceptional depths—greater than most submarines—because the human body is largely water and solid tissue; pressure at depth effects only air passages and pockets. Free divers, like the sponge divers of Greece or the pearl divers of Japan, take a big breath of air at the surface. This air becomes increasingly compressed with depth, and is uncompressed as the diver returns toward the surface. The depth which the diver can reach is basically controlled by the length of *time* which that individual can go without a fresh breath, not the pressure. In 1969, U.S. Navy Petty Officer Robert Croft set a world-record for open-ocean breathhold diving: 240 feet. He took a deep breath, and let a hand-held weight pull him down to his limit—which included allowing for the time it would take to haul himself, hand-over-hand up a tethered line, back to the surface. In the years since, other divers have pushed that frontier to somewhat more than 400 feet by using flotation devices to aid in speedy return to the surface. The maximum recorded pressure in which an unprotected human has worked: a simulated ocean depth of 2,300 feet in a hyperbaric chamber.

Prologue

There is a tradition that, in 332 B.C., Alexander the Great went underwater in some sort of glass globe. "He entered into a glass case," according to one account written too long after the event to be credible, "covered with asses' skins, that had a door that could be made fast closed with chains and a ring. Upon descending he took with him such food as was necessary, and two friends for company." That has the ring of myth—the great potentate launched upon a perilous journey, provided with food and companions. The journey is preserved in several works of art, each different, none showing companions.

Eighteen centuries later, Leonardo da Vinci (1452–1519) claimed to have developed some sort of military diving system, but refused to reveal his secret "on account of the evil nature of men who practice assassination at the bottom of the sea."

The myth of Alexander celebrates man's eternal fascination with the submarine world; Leonardo's dark caution foreshadows the topic we are about to address. He, no doubt, was referring to combat swimmers, perhaps operating from diving bells, but in the five hundred years since Leonardo, men created ever more effective methods to "practice assassination at the bottom of the sea." Whether or not these men were of evil nature has rested, in each generation, with the arbiters of law and morality. International Law of the Sea, while taught in naval officers' schools as an absolute, is little more than an abstract principle. International Law "works" only insofar as the rules are agreed to and followed by all players, and compliance disappears when survival of the state is at issue. To the victor goes the right of postwar adjudication.

Universal fascination and transient morality are joined by another thread in this narrative: Call it professional intransigence. Toward the end of the 19th century, Alfred Thayer Mahan—the premier naval historian and theoretician of his day—wrote, "An improvement of weapons is due to the energy of one or two men, while changes in tactics have to overcome the inertia of a conservative class . . . a great evil." (*The Influence of Sea Power upon History 1660–1783*)

By "conservative class," he meant naval officers.

But—why not? The naval professional—any professional—understands the tools of his trade; he grew up with them, knows how and when and to what purpose to use them. He manages groups of men who are trained in their maintenance and operation. He can create broad plans, whether for business or war, understanding each element and knowing the role of each element in the whole. This may be conservative, but it makes sense.

In addition, in the days before instant (or reasonably timely) communications, naval officers were selected, trained, and certified for independent operation; they would often be weeks, if not months—at times, years—removed from "guidance" from headquarters. Those men were expected to make decisions on the spot. Doctrine was created to help with those decisions, and any officer who felt he must deviate from doctrine was prepared to stand accountable. The type of man attracted to the profession was, of course, the type of man who excelled in such environment: often a man with a large ego, convinced of his own invincibility, secure in the rules of his game.

Thus, in a conservative profession dominated by strong-willed leaders, imagine the reception given to the uninvited amateur who came with a wondrous new device and promised great changes *if only* the professional would give him a chance. More to the point, imagine the reception given the outsider who claimed that his device would thenceforth render all navies obsolete or irrelevant!

Such, indeed, was the general proposition offered by the inventors of submarine weapons. We don't have to look very far to find ludicrous posturing, blind faith, and "vision" which does not even reach the other side of the room—on both sides. But we also find a number of prescient naval officers and politicians who helped move the inventors' dreams to the battlefield. The history of the past could be read as a text for today: Human activities may change, but human nature does not.

1 1580–1696—Of Innkeepers, Alchemists, and Mathematicians

". . . it is not hard to imagine what would be the usefulness of this bold invention in time of war, if in this manner (a thing which I have repeatedly heard Drebbel assert) enemy ships lying safely at anchor could be secretly attacked and sunk unexpectedly. . . ."

—*Constantyn Huygens, 1631*

The submarine boat has been, with minor and largely uninteresting exceptions, a delivery system for weapons of war. The first verifiable proposal for a submarine boat was contained in the book *Inventions or Devices. Very necessary for all Generalles and Captains, or Leaders of men, as wel by Sea as by Land,* first published as a series of articles circa 1580. The author was William Bourne (1535–1583), a popularizer of science and technology who created a nautical almanac (to aid in navigation) and translated material published elsewhere for the benefit of English readers. Bourne was not of gentle birth, had not attended the finer schools, and earned his primary living as an innkeeper. He introduced himself to his audience as "a poore gunner" (having served in that capacity in the Navy) who wished only to advance the cause of knowledge. "To ye gentell reder: I praye you hold me excused, I being alltogether ignorante, lacking the capacitye bothe of knowledge and experience have taken upon me . . . to open any science."

Bourne offered a lucid description of "why" a ship floats (by displacing its weight of water), and described the mechanism by which

> It is possible to make a Ship or Boate that may goe under the water unto the bottome, and so to come up again at your pleasure. Any magnitude of body that is in the water, if that the quantity is bignesse, having alwaies but one weight, may be made bigger or lesser, then it shall swimme when you would, and sinke when you list; and for to make anything doe so, then the jointes or places that doo make the thing bigger or lesser, must bee of leather; and in the inside to have skrewes to winde it in and also out againe; and for to have it sink, they must winde the thing in to make it lesse, and then it sinketh unto the bottome: and to have it swimme, then to wind the sides out again, to make the thing bigger, and it will swim according to the body of the thing in the water.

Note: Bourne is describing a *principle,* and provided no plan or drawing to illustrate the scheme. Later writers tried to turn that principle into a vessel, with perhaps more imagination than skill. The interpretation which seems to have been accepted for more than three hundred years suggests the use of paired port and starboard ballast compartments open to the sea, but with movable inner leather-lined walls; pull the wall back (by turning a large screw) and water would flow in, the effective volume of the hull would be reduced and the boat would sink. Turn the screw the other way to squeeze the water back out, and the boat would rise.

However, reread the text, not the interpretation. To my mind, it is a clear description of something quite different, a telescopic device with a watertight leather joint. The "bigness" is made "greater or lesser" (without changing the weight) by opening or collapsing the structure through the pushing or pulling action of a pair of screws. Further, no one has ever described an actual submarine boat built in accordance with the common interpretation—but submarines using telescopic hulls were put to practical use in England in 1729 (see page 23) and in France in 1863 (see page 110).

* * *

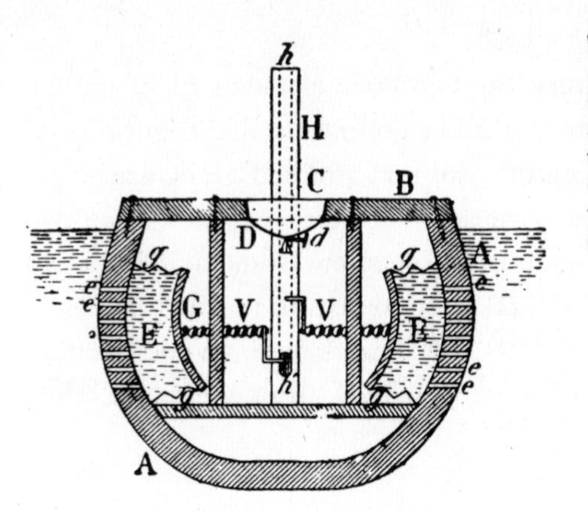

What purports to be William Bourne's scheme, circa 1580—leather-wrapped pads which can be screwed in toward the centerline to create a flooded chamber, and screwed out to expel the water. However, as Bourne wrote of expanding and contracting *structures,* not flooding chambers, this may not be accurate. (Credit: Mary Evans Picture Library)

Credit for the first use of maritime explosives—and thus the first steps toward the submarine *weapon*—is generally given to an Italian military engineer named Frederico Gianibelli, circa 1585. During a siege of Antwerp, the Spanish had built a half-mile-long bridge across the Scheldt River, about half resting on piles, the rest floating on pontoons chained together. Gianibelli's solution to breaking the siege—or at least, the bridge: an exploding ship.

He loaded seven thousand pounds of gunpowder in the hold covered by six feet of flat stones—atop which was another thick layer of rocks, cannon-balls, and other rubble. As a result, a two-hundred-foot gap was blown in the bridge and about eight hundred Spaniards could share the honor of being the first to be killed by what Gianibelli called his "floating marine volcanoes." The Spanish called them "the hell-burners of Antwerp"; other names, applied to similar devices over the next several hundred years, included "infernal machines," "fire ships," "bomb ships," and "mines."

Gianibelli's next venture was in the service of the English Queen Elizabeth. Threatened by the Spanish Armada, July 1588, the Royal Navy set fire ships—purportedly under Gianibelli's direction—to drift down on the enemy fleet, then in anchorage at Calais. The Spanish sailors had heard of "the hell-burners of Antwerp," and panicked, cutting anchor cables, some ships running aground, some ships chased out into the North Sea by the English fleet, there to be left to the weather.

Over time, attempts were made to package explosives in watertight "bombes" or "carcasses," but with varying degrees of success; it was to be more than two hundred years before the maritime version of the mine, as we know it today, achieved any notable result.

* * *

Thus, by the end of the 16th century, the two basic elements of submarine warfare had started on a journey which would culminate in the "frightfulness" of the two world wars of the 20th century. But they still had a long way to go. In the 335 years between Bourne and world war, there were more than two hundred proposals for submarines, some quite ingenious, some so ill informed that it was well that the "inventor" lacked the resources to pursue the dream. Almost fifty were turned into actual boats. We'll examine some and mention a few, but basically follow the trail of those which led, from one to the other, to the present day.

Bourne's scheme included no apparent provision for locomotion; the boat would sink or rise, but not move about on its own. But Bourne's writings moved about quite a bit, and were still widely read when an expatriate Dutch inventor turned Bourne's theory in to practice about forty years after his death. Cornelius Van Drebbel (1572–1633) not only built what may have been the world's first submarine boat, but himself left a trail in the public record still being followed at the dawn of the 20th century.

Drebbel was trained in his native Alkmaar, Holland, as a glassmaker, engraver, and alchemist. Arriving in England in 1603, just as James I became king, Drebbel earned a living as the court inventor—an ill-defined position but one which paid well enough for him to support a profligate wife and family—and also served as sometime companion and tutor to the king's children.

A self-promoting braggart and sycophant, Drebbel once wrote to his patron, James I, "I have always burned, and still burn, with a great desire to serve your gracious majesty and to divert you with my inventions." In his lifetime, Drebbel earned an amazingly variegated reputation, inspiring by turns, enmity, awe, derision, praise, scorn, adulation; but he did build the first publicly demonstrated submarine, seems to have discovered oxygen (he called it the "quintessence" in air that was necessary for life) years before the chemical element was discovered, and for his reward, ended his days in penury as an innkeeper.

In 1623, Drebbel created what was probably the world's first working submarine boat. There are enough reports of Drebbel's work to verify this success, some of which may even have been written by people who actually saw the boat. One of Drebbel's boats—there may have been several—made a submerged journey down the Thames River, perhaps at a depth of fifteen feet, from Westminster Bridge to Greenwich. Of this voyage, the *Chronicle of Alkmaar,* celebrating the accomplishments of hometown heroes and published in 1645, said,

> . . . [Drebbel] built a ship which could be rowed and navigated under water from Westminster to Greenwich, the distance of two Dutch miles: even five or six miles, or as far as one pleased. In this boat a person could see under the surface of the water and without candle-light, as much as he needed to read in the bible or any other book.

Robert Boyle, the professional chemist who among other contributions established "Boyle's Law," was fascinated by Drebbel. In 1662, he wrote:

> That deservedly famous Mechanician and Chymist, Cornelius Drebbel, who among other strange things he performed, is affirmed by more than just a few creditable persons to having contrived for the late learned King James a vessel to go under water; of which trial was made in the Thames with admirable success, the vessel carrying twelve rowers beside passengers; one of which is yet alive, and related to an excellent mathematician that informed me.

Since James I was Drebbel's employer and patron, the oft-repeated story that the king was present at one of the demonstrations is probably true. The sometimes-repeated story that the king went for an underwater ride in the boat should be regarded as a charming fantasy.

All contemporary writers agree that propulsion was provided by oars run out through watertight sleeves, and most note that a compass was fitted to aid in navigation. However, the method Drebbel used to put his boat under the water is not so easily determined. John Wilkins, Bishop of Chester, wrote an article "Concerning the Possibility of Framing an Ark for Submarine Navigation," published as a chapter in his 1648 treatise *Mathematical Magick: or the Wonders That May Be Performed by Mechanical Geometry.* The bishop offered this prescription for Drebbel's submarine:

> Let there be certain leather bags of several bigness . . . and strong to keep out the water . . . answerable to these, let there be divers windows, or open places in the frame of the ship, round the sides of which one

end of these bags may be fixed, the other end coming within the ship being to open and shut like a purse.

Simon Lake—a latter-day submarine pioneer—gave a similar explanation in his 1918 book *The Submarine in Peace and War:* goatskin bags with the neck fitted to hull openings through which water could freely flow in or out. A set of planks, manipulated by a Chinese windlass, could be used to squeeze the bags up against the hull to force water out, or when the pressure was relaxed, allow water to flow back in.

Lake cites no source, but his version is close to a general submarine scheme put forth by an Italian priest, Giovanni Alfonso Borelli, in 1680. Borelli described how fish controlled swimming depth with a bladder, and suggested using goatskin bladders which could be squeezed empty by levers. Borelli's work of 1680 may have been indebted to Bishop Wilkins's description of 1648, although his text did not say so.

Borelli's work included an engraving: fish in various configurations, a diving bell, a swimmer with what appears to be a buoyancy tank with variable displacement, and what has become the best-known and least understood early picture purporting to be a submarine. In fact, the "Borelli" engraving has appeared in at least three different versions in (at least) four different books published since 1982—not always attributed to Borelli—and as explored in

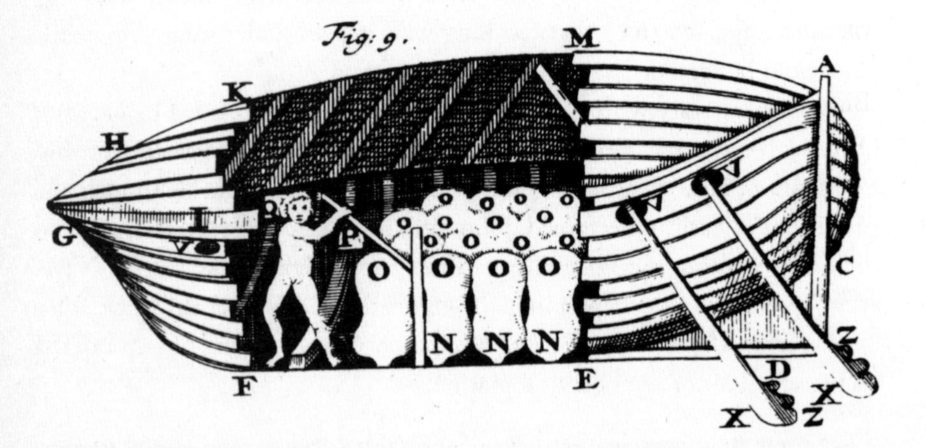

Borelli's scheme of 1680. This may have been an original idea or may have been Borelli's interpretation of Bishop Wilkens's 1648 description of Drebbel's effort, but cannot plausibly be linked with any submarine boat actually built—published opinions to the contrary notwithstanding. See page 20.

the next chapter, the goatskin-ballast-bag system has been "installed" in at least two submarines after Drebbel.

It is not likely, however, that it was ever used in any successful underwater effort. No one, in 1678 or later, seems to have asked the hard question: How could such a flimsy system have worked? The whole scheme smacks of the sort of arrant nonsense dreamed up by uninformed writers to explain something they didn't understand. Would not the water pressure on the inside and the pressure from the levers on the outside too easily have burst the stitching necessary to close up the bags?

It is just as likely—more likely—that Drebbel calculated the buoyancy of his boat so that it would run on the surface, just awash, and be driven under the water by forward momentum, as were many later pioneering boats and most modern submarines.

Bourne had written that his submarine boat might be useful in exploring the ocean's bottom; Drebbel had more warlike intentions. In his *Autobiography* (1631), Drebbel's friend Constantyn Huygens wrote;

> . . . it is not hard to imagine what would be the usefulness of this bold invention in time of war, if in this manner (a thing which I have repeatedly heard Drebbel assert) enemy ships lying safely at anchor could be secretly attacked and sunk unexpectedly by means of a battering ram—an instrument of which hideous use is made now-a-days in the capturing of gates and bridges of towns.

Bishop Wilkins echoed Huygens: ". . . a man may thus go to any coast in the world without being discovered or prevented in his journey. . . . It may be of very great advantage against a navy of enemies, who by this means may be undermined in the water and blown up."

Huygens's "battering ram" was an explosive at the end of a stick, the *petar* or *petard* used by 16th century military engineers in breaking down barriers and immortalized by William Shakespeare's *Hamlet*: "'tis sport to have the engineer/Hoist with his own petar." That is, carried aloft by the mischance of his own weapon.

Shakespeare's contemporary and sometime rival Ben Jonson directly celebrated Cornelius Drebbel's invention in his 1625 play *The Staple of News,* Act II, Scene 1:

THOMAS: They write here, one Cornelius-son hath made the Hollanders an invisible eel, to swim the haven at Dunkirk and sink all the shipping there.

P. JUNIOR: But how is't done?

CYMBAL: I'll show you, sir. It is an automa runs under water with a snug nose, and has a nimble tail, made like an auger, with which tail she wriggles betwixt the costs [ribs] of a ship, and sinks it straight.

P. JUNIOR: A most brave device, to murder their flat bottoms.

At about the same time that Drebbel's boat was being rowed down the Thames, his patron had a falling out with Parliament. James I was a true believer in the God-given right of kings to do whatever they choose, and this king did not choose to grant Parliament authority over his choices. Parliament felt otherwise, and put limits on the king's power to, among other things, reward his favorites from the public purse. The "Statute of Monopolies" also established the rights of an inventor to patent protection for his work.

Drebbel, having been one of those perhaps too generously rewarded by the king, was probably a target of the legislation. His reputation seems to have plummeted quickly; James I died in 1625, to be succeeded by his son Charles I (who later was to lose his head because he thought even less of Parliaments than his father, but that is a whole different story). In the funeral procession, where position and rank were synonymous, "Drebbel the Engineere walked between Baston le Peer the dauncer, under officers of the Mynte, Actors and Comedians."

He continued on in some sort of employment with Charles I for the next year, and then in 1626 was hired by the government to manufacture a submarine boat and underwater weapons. There was a war on—there always was a war somewhere in Europe. The French coastal town of La Rochelle—an enclave of revolutionary Protestantism—was under siege by the Catholic French, and the British were determined to break through. It is apparent that Drebbel's experiments (and salesmanship) had caught the attention of nonroyal authorities, and he was hired by the Master of Ordnance to manufacture a quantity of "divers water mines, water-petards" and "two boates to conduct them under water." He was later told to produce the "water engines" but not the "boates," and spent about a year on the project.

Subsequently, there were three separate attempts to break the siege of La Rochelle; the "water mines," although taken along, appear not to have been used in the first two. Having invested time and money in these weapons, and a bit perturbed that the on-scene military commander had elected not to use them, the government hired Drebbel to go along on the third expedition at the exceptional salary of £150 a month. However, the mines again were not used (or had no useful effect), true reason unknown; Drebbel blamed the "fear and cowardice" of the expedition's leaders; the government held Drebbel accountable and canceled his contract.

Drebbel became a not-very-successful brewer and innkeeper, and died about five years later (1633). Drebbel's son Jacob and son-in-law Johannes Kuffler saw promise in the munitions business, and they and other family members spent most of the rest of the century in a search for customers. In 1661 First Lord of the Admiralty Samuel Pepys agreed to consider the "engine to blow up ships," and the family petitioned the king for a reward of £10,000. This seems to have disappeared into that dark hole which all bureaucracies reserve for outlandish petitions. In 1689, a nephew lamented that Oliver Cromwell—who for a time had replaced Charles I in the hearts of his countrymen—had promised a great sum for the invention to sink ships, but had died before the arrangements could be made. As late as 1695, one of Drebbel's daughters was still looking for a sponsor.

Of great significance—and also of significant mystery—Drebbel claimed to have discovered a method for "freshening the air" inside the boat.

And this was Boyle's main interest:

> Drebbel conceived, that it is not the whole body of the air but a certain quintessence, or spiritous part of it that makes it fit for respiration; which being spent, the grosser body, or carcasse of the Air, if I may so call it, is unable to cherish the vital flame residing in the heart. Besides the mechanical contrivance of his vessel, he had a chymical liquor, which he accounted the chief secret of his submarine navigation. For when from time to time he perceived that the finer and purer part of the air was consumed or over-clogged by the respiration and steames of those that went with his ship, he would, by unstopping a vessel full of this

liquor, speedily restore to the troubled air such a proportion of the vital parts, as would make it again for a good while fit for respiration.

It was not until 1768 that the French chemist Lavoisier verified that four-fifths of the air was the "grosser body . . . unable to cherish the vital flames," which we now know to be nitrogen, and one-fifth was the quintessence "fit for respiration" which Lavoisier called oxygen.

Boyle noted that "Drebbell would never disclose the liquor unto any, nor so much as tell the matter whereof he had made it, to above one person, who himself assured me what it was." That one person was most likely son-in-law Joannes Kuffler, with whom Boyle discussed Drebbel's work; if Boyle in fact was given the secret, he kept it secret.

Writing in 1604—almost twenty years before he demonstrated his submarine—Drebbel suggested that "saltpeter [potassium nitrate], broken up by the power of fire, thus changed into something of the nature of air." No later investigator was able to verify that this was the key to his "secret," although today we know that oxygen can chemically be produced through any number of methods. In 1996, oxygen generators figured prominently in the crash of an airliner in the Florida Everglades.

Drebbel's work was the catalyst for a flurry of submarine activity—or at the least, for a spate of theorizing about submarine activity. Most prominent among the theorists was the French priest Marin Mersenne (1558–1648). Mersenne was a bit like Bourne—not an original thinker, but a compiler of the works of others; a contemporary called him "a good thief." In 1634, before knowing of Drebbel's success, Mersenne had theorized that submarine navigation was possible; ten years later, he was citing Drebbel's work and adding his own improvements. A submarine boat, he suggested, should be made of copper, cylindrical in shape with pointed ends (streamlined, for easy movement through the water; cylindrical the better to resist pressure, and double-ended so it could easily reverse course without having to turn around). The boat should have an escape hatch, some method for viewing above the water, strong pumps to move air in and out, wheels for running along the ocean bottom, and underwater cannon for encounters with an enemy.

Along with his collaborator, Father Georges Fournier, Mersenne recommended the training and employment of combat divers. "It is a matter of utmost importance," they wrote, "that every general commanding so much as an army and every governor commanding a port should have experienced swimmers at hand to give him advice, and to carry out a selection of schemes and undertakings intended to sink an enemy fleet or preserve his own. . . ."

They were but two of the dozens of writers who proposed schemes and provided sketches for combat or salvage divers, most of which would not have worked. Even Leonardo da Vinci was uncharacteristically naive when it came to ship construction and the lack of leverage a diver would enjoy underwater, when he suggested that a combat diver could sink a ship by prying apart the planks.

Many—most—of the schemes called for the diver to be connected to the surface by tubes or hoses. None would ever have worked. Even with a rigid tube which would not collapse, the diver would be physically unable to overcome the pressure of the water squeezing in on his chest. At only fifteen feet, that pressure would be 7.5 pounds per square inch, thousands of pounds over the chest of an average diver. The only way a diver can breathe at any depth below a few inches is by having the air also provided at the pressure measured at his chest, and it was not possible to provide breathable air from the surface that way until efficient pumps had been developed, around the end of the 18th century.

One actual inventor who gave some attention to the subject, although his major field was horology (he devised the spiral clock spring), was Abbé Jean de Haute-Feuille. Writing in 1681, he had it right: "The lungs of a man are similar to a bellows, and in order for him to inflate them he must raise that column of water which is pressing down on him." Abbé Jean distrusted Drebbel's claim to a special liquor *("une essence volatile")*, and thought instead that Drebbel employed some sort of bellows-and-pipe arrangement to cycle surface air in and out. However, none of Drebbel's "witnessess" ever mentioned breathing tubes.

There followed some years of bits and pieces contributed by a dozen or so inventors—submarine boats or diving bells, they are difficult to differentiate at this time—none of which amounted to anything more than a footnote in history; in fact, lacking details, they can't be given even that much respectability. They become one-line mentions in a list:

1640: John Barrie, France, government-financed submarine for salvage operations.

In passing, however, we will include mention of the 1653 "Rotterdam Boat" because, while unsuccessful, it may have been the first underwater vessel specifically built to attack an enemy. A Frenchman named de Son obtained the support of the Belgian government to build a seventy-two-foot boat for use against the English. The boat was designed to run awash, to approach an enemy ship unnoticed and sink her by ramming; in the manner of Mersenne, his boat was double-ended, each end reinforced.

De Son had great hopes. Cross the English Channel and back! Sink a hundred ships in a day! The boat was to be powered by a spring-driven clockwork motor, which would turn a paddle wheel in a central well. This arrangement seemed to work before the boat was launched, but once it was in the water, the motor did not have enough strength to even budge the paddle wheel. De Son's boat, quite literally, went nowhere.

The next submarine along the historic time line also was designed to attack a specific enemy, but unlike de Son's abortive effort, this one worked. Well, it was not actually one, but two, the first more or less a prototype for the second. And although the "impoved" version was never quite finished and never sent into combat, the innovations demonstrated by Denis Papin (1647–1712) became an important part of the public record.

Papin not only followed in Drebbel's footsteps, he was associated with some of Drebbel's most ardent admirers. He started his working life as a laboratory assistant to Christian Huygens—the better-known son of Drebbel's

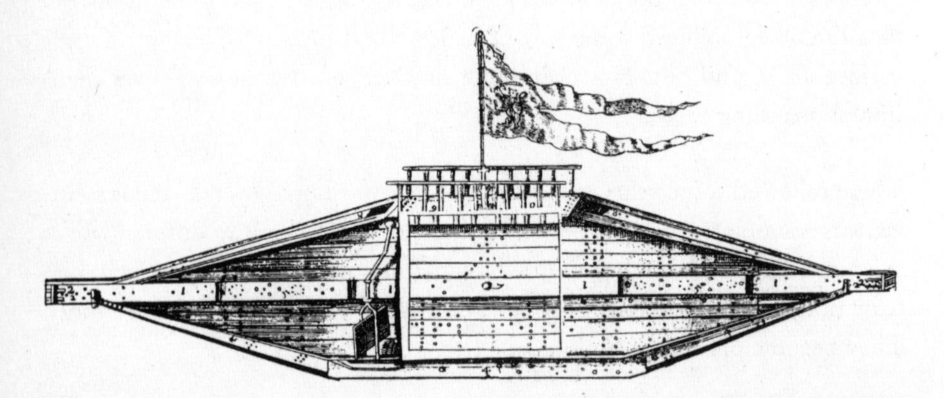

De Son's "Rotterdam Boat," 1653—a double-ended semi-submersible ram that was built, and launched—and went nowhere.

friend Constantyn—and then worked for Robert Boyle. He developed quite a reputation as a scientist: Mersenne called him *"le petit Archimede."*

Papin's first submarine activity involved attracting fish with a lighted candle, which he discussed in a paper, "How to Preserve a Flame Under Water." This work—which involved a diving bell—soon led to an assignment from the local prince, Charles, Landgrave of Hesse-Castle. As Papin was to explain;

> Drebbel's submarine had made such a noise in the world, so many Authors had spoken of it, and it seemed that one ought to expect from it so many uses of such great consequence, that S. A. S. CHARLES Landgrave of Hesse has not disdained to have the invention perfected.

Perfected, that is, as a machine or war and perfected, of course, by Professor Papin (he held the Chair of Mathematics at Marburg University). The prototype was made of metal, a tin box reinforced with iron, equipped with a barometer to measure depth (by pressure) and a detachable ballast for quick recovery. The box was about large enough for one operator, and water was admitted or discharged by juggling a valve and working a pump to increase internal air pressure—a controlled diving bell, if you will. Submerged, the boat was moved by oars; there was a folding mast and sail for running on the surface. To meet the offensive purpose, the box had certain holes through which the operator could "touch enemy vessels and ruin them in sundry ways." This boat was lost in an accident.

The second boat, built of wood, was slightly larger and would accommodate a two-man crew. An operator could "touch enemy vessels" by lying down in a six-foot-long copper cylinder that was jutting out from the hull and equipped with a viewing port and a pair of armholes (with tight-fitting sleeves) at the outer end. The cylinder was only 1.5 feet in diameter, which seems a bit confining. Perhaps Papin's associates were small.

Papin described his efforts in a paper, and included schematic diagrams of the boats. They are a bit fuzzy on details. In truth, they do not seem to depict any kind of boats at all, looking more like awkward steam kettles. But, that's not too surprising—Papin is the inventor of the pressure cooker and the safety valve which makes the pressure cooker safe. His engraver might simply have modified an earlier illustration.

Papin demonstrated what may well have been Drebbel's method:

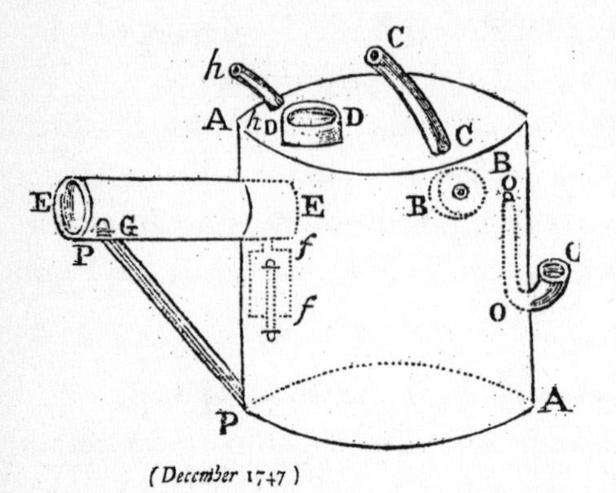

A diagram of Professor Papin's second submarine (as published in *Gentleman's Magazine* in December 1747). Some explanation: DD is the entry port, CC is a breathing tube, OO is a pressure tube for measuring depth, EE is a chamber in which one operator would lie (held up by prop PP) and G is the port through which he would extend his arm to "touch enemy vessels and ruin them in sundry ways." The fact that this diagram looks more like a steam kettle than a boat *may* be linked to Papin's parallel invention of the pressure cooker. It may also—no proof at hand—be a joke, or Papin's attempt at secrecy.

> . . . to prevent the ship from being quite sunk by letting in too much water, two men ought always to be trying to depress it by the help of oars, and when they find it can be done without much stress, the cock is immediately to be shut, by which means the ship will for any space of time be kept lighter than water, and yet may, by means of the oars, be depressed more and more at discretion. The oars are to come thro' lateral holes, which are most exactly closed by leather bound around them, as, we are told, was also practis'd in *Drebel's* ship. When we think fit to emerge, or raise the ship, the thing is easily effected, partly by the help of oars, and partly by expelling the water by a pump contrived for that purpose.

Air, for breathing, was moved in and out with a pump via tubes leading to the surface. This, of course, was an expedient, not a solution, and Papin tried to discover Drebbel's secret. He sought assistance from the Baron von Leibnitz—a friend of Huygens and Boyle, best known today as the inventor of differential calculus. Leibnitz questioned Boyle, one of Drebbel's daughters and her husband, and discovered nothing.

Around 1696, Papin was ready to demonstrate his almost-finished second boat for the Landgrave. His patron was impressed, but too impatient to be on with other business: "It was then the season to take the field." The submarine was never completed.

2 1729–1773—Of Carpenters, Adventurers, and Wagon Makers

He made his boat in two parts, and joined them in the middle very tight, with leather, that no water could get in. . . . he had a screw to each side of the boat, which, when within it, he could manage himself, and which, by means of the leather which joined the parts of the boat, contracted them to that degree that the boat would sink. . . . and then, by extending it with his screws, he rais'd it to the surface again without any assistance. He said, that tho', at last, the air began to be thick, he could bear it very well.

—*Letter from Samuel Ley, to the Editors of* Gentleman's Magazine, *July 17, 1749*

The December 1747 issue of London's *Gentleman's Magazine and Historical Chronicle* carried a brief note about an explosive antiship weapon to be delivered underwater. The author did not explain the method of delivery. The editors offered a solution of their own by including a condensed version of Papin's description of his second boat, along with the schematic drawing. Thus was launched a leisurely exchange of letters with readers, echoes of which continue to the present day.

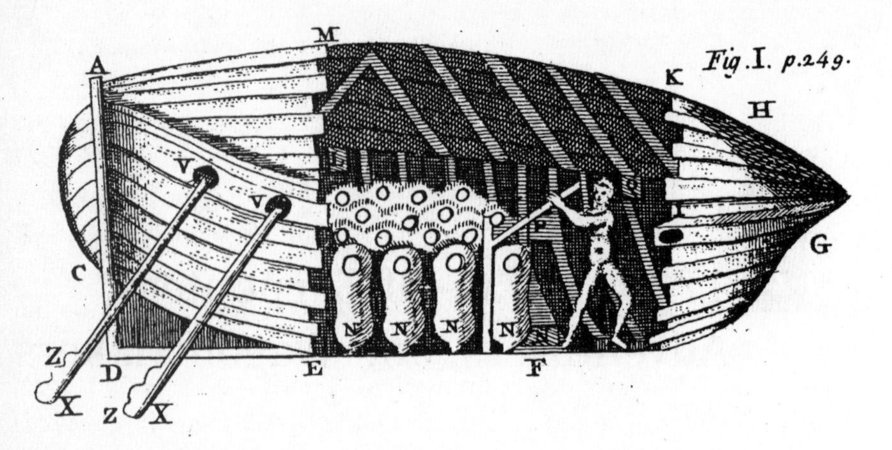

The reversed "Borelli" as published in *Gentleman's Magazine,* June 1749. At some time in the past, some author assumed this was an illustration of the Symons's boat described in the July issue of the magazine—and as such it has been known, ever since. It is not.

The first letter was published in the June 1749 edition of the magazine, and suggested a method for controlling the buoyancy of Papin's boat.

> [Editor]
> The description of the curious diving-vessel in your Mag. for Dec. 1747, p. 581, left us at an uncertainty about the method of pumping out the water, so as to raise or lower the vessel; but I was pleased to find it very clearly shown by M. Marriott, and doubt not of your giving the draught and description a place, for the satisfaction of your other inquisitive readers, who might be in the same uncertainty with myself. Yours, T.M.

The text of "M. Marriott's" work followed. It was pure Borelli. The letter writer also included the illustration from Borelli's book, but as reproduced in *Gentleman's Magazine,* the illustration was reversed, left to right. This has been treated by some writers as a mystery of special but unknown significance. The image was reversed for a very simple and easily explained reason—if, that is, you understand the printing techniques of the day. Until late in the 19th century, the only practical method for printing an illustration in a mass-produced book was with a copper, steel, or woodcut engraving. This is a simple process, still employed by schoolchildren today who use a bit of more-easily-carved linoleum mounted on a wood base. With a sharp knife or scoop, you cut away the parts of the picture you want to be white and leave those which, when smeared with ink and pressed into a piece of paper, will leave an imprint. The printed image will always be the mirror image of the block.

The most expedient method to make a copy block, for printing *almost* the same illustration, is to paste an already printed sheet on the surface of a fresh block, and cut away all of the "white" spaces. Of course, the new block will be a mirror image of the original block, and the resulting prints will be reversed.

In the following month's issue (July 1749), another writer described the virtues of an English submarine boat built by Nathaniel Symons, carpenter. This is a boat about which some considerable nonsense has been printed over the years since, perpetuated in at least one book published as recently as 1995. Some writer at some time in the past linked the suggested scheme for Drebbel's 1620 boat (the Borelli engraving of 1680 as reproduced in *Gentleman's Magazine* in 1747) with the "diving boat" demonstrated in the River Dart by Symons (circa 1729). They are not, should not, *cannot* be linked. They certainly were never linked in *Gentleman's Magazine.*

Here's some of the nonsense:

> Many other inventors [after Drebbel] produced designs but none offered a true advance until the Abbé Borelli propounded a scheme for destroying the buoyancy by filling a number of leather bottles and returning to full buoyancy by expelling the water from these bottles by hand. Nevertheless no construction took place until, in 1747, an Englishman by the name of Symon built a boat. In this design he employed the idea of controlling the buoyancy by a number of leather bottles on the Borelli principle. Thus a practical step nearer the ballast tank had been taken.
>
> —*Historic Submarines*

> It was now 1729, and Abbé Borelli's underwater theories enjoyed a reawakening. Apparently a carpenter from Devonshire came into possession of his plans and built a vessel according to their specifications. A picture of this boat appeared in *The Gentleman's Magazine* for April of that year. Following this public presentation, some very ungentlemanly accusations were bandied about as to who stole what plans from whom. . . . Regardless of the uncertainties involved, it is an established fact that the boat made several descents in the River Dart before large

> crowds. . . . *The Gentleman's Magazine* of London carried this drawing, in reverse, under the claimed ownership and invention of a man signing himself only as "M.T." This touched off a battle of claims among a large group, each claiming to be the originator.
>
> —*Oceanographic Submersibles*

Well, it was "T.M" and not "M.T." who sent in the drawing, and it was in 1747, not 1729 (the magazine did not even begin publication until 1731). "T.M." made no claim for himself; the "battle of claims" came eighteen months after this letter, and the "large group" appears only to have been one man, Jonathan Lethbridge (see below), who made no such claim concerning either this drawing or the Symons boat.

One more bit:

> It was not until 1747 that Nathaniel Symons, a joiner, built a "Borelli" in the River Dart in Devon. His design drawing is a mirror image of the Italian prelate's, obviously copied in every detail. According to a contemporary issue of the *Gentleman's Magazine,* Symons conducted successful trials in the Thames, staying under water for forty-five minutes. What became of his work is unknown, but it seems that when the ballast bottles were full there was not enough room for the oarsmen to move. . . . Yet Bourne's concept of the ballast tank had at least been put to the test and not found wanting.
>
> —*Stealth at Sea*

This version takes a lot more out of *Gentleman's Magazine* than the editor ever put in. 1747? Joiner? Design drawing? Trials in the Thames? Ballast bottles? Oarsmen?

Do writers ever *read* anymore? To keep the record straight, here is the complete description of the Symons boat, as published as a letter to the editor in the July 1749 edition of *Gentleman's Magazine* and accompanied by *no* drawing, design or otherwise:

> [Editor]
> In your *Mag.* for *June p.* 249, you have given us a description of a diving ship, &c, which, according to my notion of it, is far inferior to one made some years since, by one *Nathanial Symons,* of the parish of *Harberton,* near *Totness, Devon,* a common house-carpenter: Though I did not see

> it, I shall trouble you with such a description as my memory will permit, after so long a time as twenty years, and which I had from the inventer himself.
>
> He made his boat in two parts, and joined them in the middle very tight, with leather, that no water could get in; he made a false door in the side, which, when he was in, shut very tight; and tho' his going in admitted a small quantity of water, it was no inconvenience; after this outer door was shut, he opened the inner one to get into his boat. There was more than four-score weight of lead to the bottom of his boat, but this I presume must be according to the dimensions; yet he had a screw to each side of the boat, which, when within it, he could manage himself, and which, by means of the leather which joined the parts of the boat, contracted them to that degree that the boat would sink.

That describes a sort of a telescopic or accordion device, whereby the operator could crank up on a pair of screws to reduce volume to sink and restore it to rise. This has nothing in common with the goatskin bags depicted in the purported "design drawing," but to my mind is markedly similar to even the most casual reading of Bourne's basic scheme.

The letter continued;

> . . . He went into the middle of the river *Dart,* entered his boat by himself, in sight of hundreds of spectators, sunk his boat himself, and tarry'd three quarters of an hour at the bottom; and then, by extending it with his screws, he rais'd it to the surface again without any assistance. He said, that tho', at last, the air began to be thick, he could bear it very well.
>
> . . . And tho' a great number of gentlemen of worth were present at shewing his boat, he told me he received but one crown piece from the all—I think he has been dead for some years.
>
> —*Samuel Ley; July 17, 1749*

From this, we might deduce that the house-carpenter had thought to create a public entertainment, but that his hopes for a quick profit were disappointed in the event. *One* event, in whatever river.

A small controversy did arise when Samuel Ley added that "This same person invented the famous diving engine for taking up the wrecks" but was "deprived . . . both of the honour and the profit" by a cousin, identified as "L——e."

In the September issue, John Lethbridge answered that charge: "As I am the first inventer of a diving engine in *England* without communication of air from above, I, therefore, presume Mr *Ley* means me, under the title of *Symons's* cousin L——e, (to which kindred I have not the least pretension)."

Lethbridge insisted that the design—for the diving engine, not the submarine boat—was his, and gave a lengthy explantion of his work in developing "a machine to recover wrecks lost in the sea."

"Necessity is the parent of invention," he wrote, "and being, in the year 1715, quite reduc'd, and having a large family, my thoughts turned upon some extraordinary method, to retrieve my misfortunes; and was persuaded that it might be practicable to recover wrecks lost in the sea. . . ."

Lethbridge's one-man diving apparatus was a leather-covered weighted barrel, six feet long, two and a half feet in diameter at the head, and about eighteen inches diameter at the foot. It was "hoop'd with iron hoops without and within, to guard against pressure" and had two armholes and a glass viewing port (à la Papin's second boat). The device was slung from a surface ship positioned over the work area, and could be maneuvered over a twelve-foot square, the crew on the surface following instructions sent up by a signal line. "I have stayed," Lethbridge noted, "many times, 34 minutes. I have been ten fathom deep maby a hundred times, and have been 12 fathom, but with great difficulty."

Lethbridge described his experience with Mr. Symons—for, as we see, Mr. Symons was not a stranger to him—who had come down "to see my engine, which he liked so well, that he desired to adventure with me." The two men worked on some wrecks near Plymouth, without success. "Some time after this," Lethbridge continued, "Mr *Symons* reported, behind my back, but I declare, never to my face, that he was the inventer of my engine; but, I protest, I never saw a diving engine, before I saw my own, nor did I ever see Mr *Symons* diving boat (as Mr *Ley* calls it) nor ever saw him dive in an engine in my life, of all which I am ready to make affadavit."

Lethbridge also noted the interest and support he had gained from the English astronomer Edmund Halley (who predicted the periodic return of the comet which bears his name). In 1690, Halley had developed a method of replenishing the atmosphere in a diving bell with weighted barrels of air, sent down from the surface. There was a vent hole in the bottom of each; the air inside the barrel would be compressed and forced out through a hose in

Edmund Halley's diving bell, 1690, in which underwater salvors could work in a dry environment. The weighted barrel replenished the air supply.

the top, which was led into the diving bell. A valve released stale air from the top of the bell as fresh air flowed in from the barrel. In an early demonstration, Halley and two other people spent more than one hundred minutes at a depth of sixty feet in the Thames River. Some years later, a sixty-five-year-old Halley sat at a depth of sixty-six feet for more than four hours.

Lethbridge had indeed created a practical diving machine, and turned it into a profitable business. While the average man is fascinated by tales of sunken treasure, the day-to-day financial rewards for Lethbridge and his contemporaries came from more mundane but surprisingly valuable scrap. The brass which could be recovered just from the ordnance of a hundred-gun ship of the line was worth more than $50,000, and thousands of shipwrecks littered the ocean bottom just around the British Isles.

There is one more early submarine worth noting: In 1773, a wagon maker named J. Day built a submarine with a detachable ballast—large stones, hung outside by ringbolts which could be released from inside. His first test, in thirty feet of water, was successful enough to attract the interest of a professional gambler, who suggested that they team up and take bets on Day's ability to remain submerged for a specified length of time (twelve hours by one account, twenty-four by another). The gambler supported the cost of a larger boat and an attending sloop; following a successful shallow-water test, they shifted to deeper water for the big event. Then, while the gambler collected bets and sailors hooted in derision from HMS *Orpheus*, anchored nearby, Day tried to submerge—and couldn't.

He had not allowed for (or didn't understand) the change in buoyancy as he moved from the relatively fresh water at the head of the river-fed bay, to the salt water further out. The density of fresh water is 62.4 pounds per square foot. Salt water, with a density of sixty-four pounds per cubic foot, has a greater buoyant force than fresh water; also, cold water—salt or fresh—is denser than warm. He figured out soon enough that he needed to add more ballast; his associates, riding alongside in the sloop, hung more stones from the boat, which quickly sank out of sight. On the way to the bottom at 132 feet, the hull would have collapsed before Day could even begin to release ballast.

Back on the surface, we presume that the gamblers counted the minutes, the hours—and finally sailed away. We find no record of the final disposition of the money laid down on the bet.

3 1775–1795—Of Yankee Tinkerers and Expatriots

I then thought, and still think, that it was an effort of genius; but that a combination of too many things were requisite, to expect much success from the enterprise against an enemy, who are always on guard.

—*George Washington to Thomas Jefferson, 1785*

A thirty-three-year-old Connecticut native, graduate of Yale University, Class of 1775, built the first submarine to make an actual attack on an enemy warship. Well, actually, Bushnell's *Turtle* made the world's first, second, and third submarine attacks, although none were successful. The story of *Turtle* is well known, included in grade-school histories of the American Revolution. However, writers often give Bushnell credit for more than he accomplished:

> It was the first truly practical submarine in the history of all mankind, and the first submarine ever designed for offensive warfare. In addition, Bushnell put into action the first mines ever used in underwater warfare.
>
> —*Submarine Fighter of the American Revolution: The Story of David Bushnell* (1963)

Drebbel, Gianibelli, de Son, and Papin could each take exception to some part of that claim. Drebbel could challenge it all. Neither, as other recent writers have asserted, was Bushnell the "Father of Submarine Warfare," nor would he appear to be a man whose "submarine boat may justly be considered an entirely original conception." Bushnell may—or may not—have had access to those issues of *Gentleman's Magazine* which described the work of Papin, and others, but there are enough similarities to Papin's first boat to beg the question of originality: valves to let water in, and pumps to force water out of the hull to control buoyancy; a barometer for depth measurement; tubes to the surface, with a pump for drawing in or exhausting air.

That caveat aside, Bushnell made a definite contribution to submarine warfare—although he did not start out to invent a submarine boat. His primary interest was underwater explosives, which may have started as a school experiment, and he created *Turtle* as a method to deliver an explosive to a target.

There is no contemporary drawing of Bushnell's boat; the most widely used illustration is based on an inaccurate drawing used in a lecture delivered at the U.S. Navy Torpedo Station in Newport, Rhode Island, in 1875 by Lieutenant Francis Morgan Barber, USN.

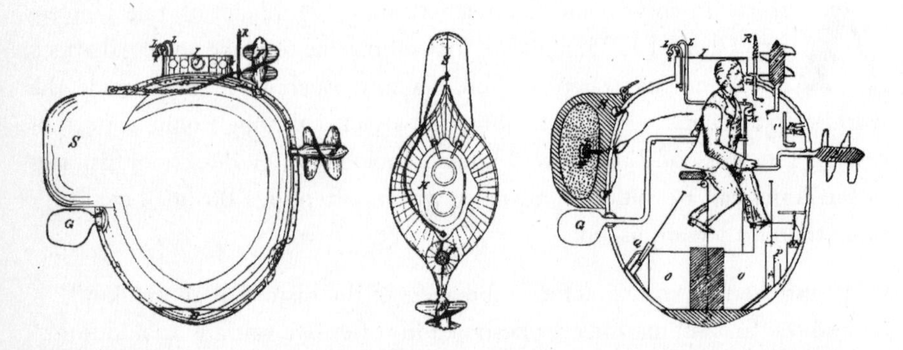

Bushnell's *Turtle* as illustrated in a lecture given by Lieutenant Frances M. Barber, USN, at the U.S. Navy Torpedo Station, Newport, Rhode Island, 1875. The lecture was seminal in the later development of the submarine. Barber created this illustration from the best information he could gather, and noted, "If not exactly accurate, it will at least give a tolerably correct idea of its general appearance." There are several important errors. One, it shows ballast tanks when there were none; it shows an Archimedes Screw (helical) for locomotion instead of the propeller like the "arms of a wind mill" or a "pair of oars" described by Bushnell and others. It also shows—but this we may forgive—the operator dressed in a rather foppish late 19th-century outfit.

However, Bushnell and his contemporaries left abundant descriptions of *Turtle,* many of which had been scattered in various archives and largely unnoticed until assembled in *Naval Documents of the American Revolution, 1775–1783,* published by the Office of Naval History and released in ten volumes beginning in 1964. Much of this material was not considered—nor even known to exist—by earlier writers on the subject.

The record starts with a letter of August 7, 1775, from Dr. Benjamin Gale to Benjamin Franklin:

> Your Congress doubtless have had intimation of the Invention of a new machine for the Destruction of Enemy ships, but I sit down to Give you an account of that Machine and what Experiments have already been made with it, what I relate you may Intirely rely upon to be fact. . . . The Person who invented it, is a student of Yale Colledge, and is Graduated this Year—Lives within five Mile of me.

Turtle was shaped like a flattened egg, pointed end down, but took its name from a supposed resemblance to a sea turtle floating head-up—as Dr. Gale noted in a letter to Silas Deane of the Continental Congress, November 9, 1775: "The Body, when standing upright in the position in which it is navigated, has the nearest resemblance to the two upper shells of a Tortoise joined together. . . ."

Respiration was accommodated through a pair of brass tubes, equipped with float valves to keep water from entering if the water level reached their tops; the valves re-opened when the water level dropped. The tubes were also fitted with internal check valves as proof should the float valves fail.

Bushnell clearly understood the effects of pressure—his boat had internal bracing to keep the sides from collapsing. He understood both dynamic and static ballast: A foot-operated valve let water into the hull, which could be expelled by two foot-operated pumps, and a droppable two-hundred-pound lead weight under the keel provided for an emergency ascent. For locomotion,

> . . . he has a pair of oars fixed like the two opposite arms of a wind mill, with which he can row forward, and turning them the opposite way, row the machine backward; another pair fixed upon the same model, with which he can row the machine round, either to the right or left, and a third, by which he can row the machine either up or down; all of which are turn'd by foot, like a spinning wheel.

Barber's illustration showed Archimedes helical screws for forward and vertical locomotion, rather than Dr. Gale's "arms of a wind mill" (which Bushnell called a "pair of oars"). Among other features, as later described by Bushnell himself: The rudder could do double duty as a scull, for moving forward; all controls were arranged, "as convenient as could be devised: so that everything could be found in the dark." Depth was measured by a barometer, an eighteen-by-one-inch glass tube, closed at the top and connected to the outside through a brass tube; a phosphorescent cork of foxfire floated inside. "When the Vessel descended," Bushnell wrote, "the water rose in the water-gage, condensing the air within. . . . by the light of the phosphorous, the ascent of the water in the gage was rendered visible, and the depth of the Vessel under water ascertained by a graduated line."

The weapon was a 150-pound keg of gunpowder, with a clockwork attached to a gun lock, as detonator. Bushnell's scheme: Use tidal flow to drift *Turtle* into the vicinity of a target; submerge to an awash condition; crank heartily to come alongside; crank the down-haul and perhaps let in some additional water to get under the hull; use a top-mounted augur bit to drill enough into the hull to attach a tether for the keg; release the drill from the boat, leaving it in the hull, and pull away; this would tug at the lanyard and start the timer; release the keg, which, being slightly buoyant, would rise up against the hull; crank *very* heartily to clear the ship and return to the awash condition and clear out before the explosion.

Dr. Gale didn't think that an underwater explosion would have much effect upon the stout hull of a warship, "the water giving way before." Bushnell satisfied his concern with one demonstration, which "produced a very great effect . . . to the astonishment of the spectators."

Warfare in the 18th century was a seasonal thing; armies went into winter quarters and navies took shelter in icebound harbors. In December, Bushnell discovered that *Turtle* was allergic to cold weather. "He proposes going in the night," Dr. Gale reported, "on account of safety. He always depends on foxwood, which gives light in the dark, to fix on the points of the needle of his compass, and in his barometer . . . but he now finds that the frost wholly destroys that quality in that wood."

Bushnell tried using a candle, "but that destroys the air so fast he cannot remain under water long enough." He was forced to wait until spring.

* * *

Bushnell had at first refused government support—"He Builds it on his own Acct," Dr. Gale had affirmed—and when urged to seek assistance, was offered "so Inconsiderable a sum" by the Connecticut government that he elected to continue on at his own risk. However, by February 1776, working at "his own risk" had put him at risk, and he went back to the Connecticut Council of Safety. The Council proved "agreeable" to have him proceed, "with expectation of proper public notice and reward," and the next day voted a sum of sixty pounds to be used "according to instructions from this Board."

Dr. Gale had suggested that Bushnell deserved: "If he succeeds, a stipend for life, and if he fails, a reasonable compensation for time and expense is his due from the public." He received neither. A few months later, Bushnell did meet with General George Washington, who appears to have offered assistance, although in what form or amount is today unknown.

Meanwhile, we should not assume that the British had remained ignorant of this "turtle" being hatched within their midst. New York Governor William Tryon, a loyalist, reported to the Admiralty in November 1775:

> The great news of the day with us is now to Destroy the Navy, a certain Mr. Bushnel has compleated his Machine, and has been missing four weeks, returned this day week.
>
> It is conjectur'd that an attempt was made on the *Asia*, but proved unsuccessful—Return'd to New Haven in order to get a Pump of new Construction which will soon be completed,—When you may expect to see the Ships in Smoke.

The source of Governor Tryon's information was a valet in the service of a member of the Continental Congress. The treachery was discovered, but the spy escaped in January 1776.

Eventually, *Turtle* and the weather both seemed ready, but Bushnell then had a setback in selecting and training an operator. As he later explained;

> In the first essays with the submarine Vessel, I took care to prove its strength to sustain the great pressures of the incumbent water when sunk deep, before I trusted any person to descend much below the surface: and I never suffered any person to go under water without having a strong piece of rigging made fast to it, until I found him well acquainted with the operations necessary for his safety. After that I made him descend and continue at particular depths, without rising or sinking, row

> by the compass, approach a Vessel, go under her, and fix the Woodscrew . . . into her Bottom &c. until I thought him sufficiently expert to put my design into execution.
>
> I found agreeably to my expectation that it required many trials to make a person of common ingenuity, a skilful operator. The first I employed was very ingenious and made himself master of the business, but was taken sick in the campaign of 1776 at N. York, before he had the opportunity to make use of his skill, and never recovered his health sufficiently afterwards.

The "first I employed" was Bushnell's brother, Ezra. Bushnell next found an army volunteer, Sergeant Ezra Lee, who went through the training regimen and finally, toward the end of summer, seemed ready.

The British fleet was concentrated in New York harbor—almost three hundred ships from which Bushnell and Lee could choose a target. *Turtle* was towed into rough position by a pair of rowboats, and then, aided by the tidal flow, headed for the fleet anchorage. Sergeant Lee picked his target, submerged, went under the hull—and could not get the drill to bite. Believing that he had struck some iron strapping or fastening, Lee tried to shift to another position, but then—as Bushnell later explained—"not being well skilled in the management of the Vessel . . . he lost the Ship. After seeking her in vain, for sometime, he rowed some distance and rose to the surface of the water, but found daylight had advanced so far, that he durst not renew the attempt."

At some point, Lee was spotted by the British (who had been warned of such an attack) and pursued by at least one boat. Then, according to Bushnell, the sergeant cast off the powder magazine "as he imagined it retarded him, in the swell." That started the clockwork timer; the magazine exploded an hour later. The blast was described as spectacular. The British increased their vigilance. The ships moved to anchorage further out in the bay.

The oft-repeated story that Sergeant Lee attacked the sixty-four-gun *Eagle,* flagship of Admiral Howe, only to be frustrated when the drill could not penetrate the copper sheathing on her bottom, is not accurate.

The next morning, Lee told Captain Samuel Richards that the drill failed to take hold because it hit a bolt or iron brace; to his diary, Richards "judged it just as probable that the point was prevented from entering the ship by the copper sheathing."

Richards's speculation long has been treated as the fact; if it were true, however, the ship would not have been *Eagle.* According to British Naval records, her bottom was not copper-plated (to prevent underwater attack by a more natural enemy, the toredo worm) until 1782. Thus, Captain Richards's speculation was off the mark, but does not disqualify *Eagle* as the intended victim. We might judge that Lee's failure to carry off the attack was just as likely the result of carbon-dioxide poisoning and fatigue. Bushnell's estimate of "30 minutes" seems a bit long for a man working at full capacity—and in some reasonable degree of panic, to boot.

One recent author believes that *Eagle* was not even at New York on the night of the attack, and cites unidentified "contemporary British records." However, as reproduced in *Naval Records,* the log of *Eagle* for September 7, 1776, put the ship squarely in New York harbor, "Moored off of Bedlows Island." *Bedloe's* Island (now Liberty Island) is the site today of the Statue of Liberty.

Whatever the target, an entry in the log of *Asia* verifies that there was that night an attack by somebody on some ship: "At 1/2 past 10 sent 4 boats to the Assistce of the Advanced Guard p Signal."

Lee made at least one more attack, on an unidentified frigate anchored upriver from the city—which he described in an 1815 letter to General David Humphreys:

> I now made another attempt upon a new plan—my intention was to have gone under the ship's stern, and screwed on the magazine close to the water's edge, but I was discovered by the Watch and was obliged to abandon this scheme, then shutting my doors, I dove under her, but my cork in the tube,(by which I ascertained my depth) got obstructed, and deceived me, and I descended too deep and did not track the ship and then I left her.

On October 6th, the British sank the boat which carried *Turtle;* she was recovered, but not further used. Bushnell was ill, worried about money, and convinced that his "operator" needed a great deal more training and practice.

"I waited," he would later write, "for a more favourable opportunity, which never arrived."

Bushnell continued to experiment with floating mines, launched from a whaleboat: in August 1777, he made an attempt to sink the frigate *Cerberus* near New London, but the mine missed the warship and sank a schooner anchored beyond. However, it would be stretching things to credit this as the "first" sinking of a ship by a mine in wartime. Four sailors spotted the device and pulled it aboard out of curiosity. It detonated while they were looking it over, killing three and blowing the fourth overboard.

In December 1777, Bushnell staged a full-scale assault with floating mines on ships in the Delaware River. He was misinformed about (or misjudged) the tide and the weather; by the time the mines had drifted down to the British anchorage—it took a week—the ships had moved. Just as in New London, one mine destroyed "a boat, with several persons in it, who imprudently handled it too freely."

Startled, British troops stood on the river banks and fired away at any bit of floating trash. This event was "celebrated" in a satirical poem soon making the rounds, attributed to Declaration of Independence-signer Francis Hopkinson, the "Battle of the Kegs."

Gallants attend, and hear a friend
Troll forth harmonious ditty,
Strange things I'll tell that once befell
In Philadelphia City.

'Twas early day, as poets say,
Just as the sun was rising.
A soldier stood on a log of wood
And saw a thing surprising.

As in amaze he stood to gaze,
The truth can't be denied, sir,
He spied a score of kegs or more
Come floating down the tide, sir.

These kegs, I'm told, the rebels hold
Packed up like pickled herring,

And they're coming down to attack the town,
In this new way of ferrying.

Therefore prepare for bloody war,
The kegs must all be routed,
Or surely we despised shall be
And British valor doubted.

The royal band now ready stand
All ranged in dead array, sir,
With stomach stout to see it out
And make a bloody day, sir.

The cannon roar from shore to shore,
The small arms make a rattle,
Since wars began, I'm sure no man
E'er saw so strange a battle.

The kegs, 'tis said, though strongly made,
Of rebel staves and hoops, sir,
Could not oppose their powerful foes,
The conquering British troops, sir.

Bushnell was captured by the British, was not recognized as anyone of consequence, and was released. He served out the balance of the war as captain in an Army engineer unit, based at West Point, and was mustered out, with five years' pay, in 1783.

In 1785, Thomas Jefferson—then Ambassador to France—sent a note to George Washington. He wrote;

> . . . be so kind as to communicate to me what you can recollect of Bushnel's experiments in submarine navigation during the late war, and whether you think his method capable of being used successfully for the destruction of vessels of war. . . .

Washington responded with his personal view of the failure of *Turtle*—reproduced below—and sent the request along to several other people, at least one of whom forwarded Jefferson's letter to Bushnell. He responded, two years later, with this apology: "I was seized with a severe illness, which disabled me from writing, & though I attempted it several times, obliged me to desist." Bushnell went on to describe his experiments in great detail—touched on above—and defended his passion for secrecy:

> I have ever carefully concealed my Principles & Experiments, as much as the nature of the subject allowed, from all but my chosen Friends, being persuaded that it was the most prudent course, whether the event should prove fortunate or otherwise, although by the concealment I never fostered any great expectations of profit, or even of a compensation for my time & expences; the loss of which has been exceedingly detrimental to me.

Just about the time he wrote this letter to Jefferson, in 1787, Bushnell dropped out of sight—for the rest of his life. His family heard nothing more until his death in 1826, when he was found to have been living in Georgia, at least since 1795, working first as a teacher, then as a medical doctor under the assumed name of Dr. Bush.

There is a strong presumption that he went to France in the interim; his brother later reported that Bushnell had "been receiving letters from France" and that he had talked about making the voyage. At least one most likely came from Jefferson—Bushnell had asked for an acknowledgment of his own letter, to be sent in care of the president of Yale, "if it were not too great a favour, to hear that this finds a safe conveyance to your excellency." Jefferson, appointed Secretary of State, left France three years later, in 1790.

Other letters may have been sent by Joel Barlow, a fellow Yale alumnus also living in Paris. The evidence here is a bit more circumstantial but nonetheless compelling: Barlow's brother-in-law was Abraham Baldwin, Yale Class of 1772, and for some years the professor of divinity at the college. Baldwin moved to Georgia in 1784. He became a member of the Continental Congress, and later served in both the House of Representatives and the Senate. In 1795, the same year Barlow left Paris to serve for a time as U.S. Consul to Algiers, Bushnell surfaced in Georgia, where, as the pseudononymous Dr. Bush, he now shared a house with Baldwin, until Baldwin's death some years later.

A connection? Likely. Even more so when we find that Barlow was an associate of Robert Fulton in the development of the next submarine boat, three years later. In Paris. Next chapter.

As a fitting close to *this* chapter, we offer a longer extract from the letter with which we began: George Washington's answer to Thomas Jefferson, September 26, 1785:

> I then thought, and still think, that it was an effort of genius; but that a combination of too many things were requisite, to expect much success from the enterprise against an enemy, who are always on guard.—That he had a machine which was so contrived as to carry a man under water at any depth he chose, and for a considerable time & distance, with an apparatus charged with Powder which he could fasten to a ships bottom or side & give fire to in any given time (sufft. for him to retire) by means whereof a ship could be blown up, or sunk, are facts which I believe admit of little doubt—but then, where it was to operate against an enemy, it is no easy matter to get a person hardy enough to encounter the variety of dangers to which he must be exposed. 1 from the novelty 2 from the difficulty of conducting the machine, and governing it under water on acct. of the Currents &ca. 3 the consequent uncertainty of hitting the object of destination, without rising frequently above water for fresh observation, wch., when near the Vessel, would expose the adventure to discovery, & almost to certain death. To these causes I have always ascribed the non-performance of his plan, as he wanted nothing that I could furnish to secure the success of it—This to the best of my recollection is a true state of the case.

4 1795–1801—Of Painters and Politicians

Considering the great importance of diminishing the power of the British Fleets, I have contemplated the Construction of a Mechanical Nautulus. A Machine which flatters me with much hope of being Able to Annihilate their Navy . . .

—*Robert Fulton to the French government, December 1797*

About Robert Fulton, a contemporary, but not a friend, wrote: ". . . a man of very slender abilities though possessing much self confidence and consummate impudence." Fulton made outrageous claims to have found a method not only to eliminate maritime war but the English aristocracy as well, which certainly counts as "self confidence and impudence," but from a simple list of just some of his accomplishments, we must argue that his abilities were far from slender:

He built the world's first commercially viable steamboat, and helped turn it into a moneymaking monopoly.

He built the world's first steam-powered warship.

He invented an underwater gunpowder-launched torpedo; in fact, it was Robert Fulton who borrowed the name of a variety of electric stingray, genus *torpaedae,* and applied it to the general family of underwater weapons.

He was the first person to actually sink a ship by hitting it with a mine (drifting on the surface).

He took Bushnell's submarine to a logical conclusion within the technological limitations of the day, and in so doing, built and tested the first boat to bear what is arguably the best-known submarine name, *Nautilus.*

Robert Fulton was a compulsive inventor, hopping from one idea to another like a drop of water on a hot griddle. No sooner was he satisfied with one scheme than he was off on two others. Improvements in canals. Cast-iron bridges. A machine for sawing marble. A machine for preparing flax. He spent much of his energy on weapons of war, but for profit, not patriotism.

Profit seems to have been the motive behind everything he did, and much of what he did in fact turned a profit. Paradoxically, he kept his secrets close but broadcast his schemes widely—as a result, of course, the "secrets" did not long remain secret.

In truth, he left behind such a rich trail of correspondence and government reports that, except for a few bare patches, the record seems complete—and completely fascinating, a look into the posturing of a total mercenary and the always cautious and often bemused response of governments in three countries. He offered, in turn, to blow up ships and help the French against the British, the British against the French, the Americans against the British, the British against the Americans. The Dutch, who did not seem at the time to have been at war with anyone, may have been tossed an offer on general principles—although if Fulton had any "principles," they revolved around money, not scruple.

Robert Fulton, born in 1765 in Little Britain Township (long since renamed "Fulton"), Pennsylvania learned the gunsmith's trade, then trained as a portrait painter, and in 1786, with the encouragment of Benjamin Franklin, went to England to improve his technique under the watchful eye of Benjamin West. However, he was not a very good portrait painter, and at a time when commissioned portraits were not much in demand, he would not have made a living in the visual arts. But he took up the study of civil engineering, canals and bridges and the like, which gave him a modest income for a time; in the summer of 1797, Fulton went to France, theoretically on his way home to America, to apply for French patents on his English inventions.

In Paris, he joined forces with fellow expatriate Joel Barlow. Barlow was a writer and poet by training and inclination, but had moved to France as part of an Ohio-French real estate venture which failed. He kicked around Europe and licked his wounds for a few years, searching out opportunities, and became a wealthy man. Barlow and his wife (née Ruth Baldwin) took the younger Fulton (by ten or so years) under their wing and into their home. He stayed for seven years.

Soon after Fulton's arrival in Paris, he and Barlow tried to develop a self-propelled torpedo: a device which would move off in a given direction, for a given distance, and then explode. They did not succeed, and almost killed themselves in the attempt. From whence sprang this interest in underwater warfare? No one knows, but the Bushnell influence may well have been at work here. We've noted the close connection between Bushnell and Barlow's brother-in-law, and Barlow himself was a Yale freshman the same year that the senior Bushnell was working on *Turtle.*

If Barlow was one source of the mysterious "letters from France" which may have triggered Bushnell's apparent move to Paris, we can speculate: Might Fulton have met Bushnell? Barlow and his wife were living in London, 1790–92, as was Fulton; the record shows that Fulton made a three-month visit to Paris in 1790. There is no record, however, that he knew the Barlows at that time, or of any meeting with Bushnell, who was back in the United States living with Baldwin before Fulton's 1797 return to Paris (and verifiable friendship with the Barlows).

Did Fulton even know about Bushnell? Yes. Fulton once admitted to familiarity with Bushnell's work, although he did so more in the vein of offering criticism than of giving credit.

Toward the end of the year, not long after the failure of the torpedo, Fulton sent an unsolicited proposal to the French government—now at war with England:

> Considering the great importance of diminishing the power of the British Fleets, I have contemplated the Construction of a Mechanical Nautulus. A Machine which flatters me with much hope of being Able to Annihilate their Navy . . .

He did not describe his invention, believing it to be too complex for the average mind; he would, however, be pleased to explain the details to "a good engineer . . . [like] general Bonaparte." His newly formed Nautilus Company would build and operate the device at his own expense, he wrote, and would expect to be compensated by a payment for each British ship destroyed. He proposed a sliding scale based on size, a typical reward being 400,000 French francs for a thirty-gun frigate; any British ships and cargoes captured would become his property; if the war should end before his machine could be brought into service, he would be reimbursed for all of his out-of-pocket expenses.

He and his crew, he suggested, should be given naval commissions for protection. As he noted, in the climate created by the work of Gianibelli, "fire ships and other unusual means of destroying Navies are Considered Contrary to the Laws of War, And persons taken in Such enterprise are Liable to Suffer death."

The Minister of Marine was interested in the scheme, but made a counteroffer: The fee schedule was reduced by half, and compensation for an early end to the war would be granted only if the weapon so terrorized the English that they surrendered before he could actually sink any ships. The request for naval commissions was flatly denied—it was not possible

> . . . to grant commissions to men who made use of such means to destroy the enemy's forces and, even so, that such commissions could be any guarantee to them. For the reprisals with which the French Government could threaten the English Cabinet would be useless, since there existed in England three times more French prisoners than English prisoners in France.

A contract incorporating the changes was prepared under date of February 4, 1798. The contract was never signed—most likely because Fulton insisted on the commissions. Fulton changed his tactics, and for the time, instead of knocking on the doors of the ministry, he sought support from prominent citizens. He occupied his time with the study of other projects: canals, steam engines, propellers.

By July, a new Minister of Marine was in office and Fulton re-submitted "Nautilus" along with this stirring call to arms:

> The destruction of the English Navy will ensure the independence of the seas and France, the nation which has most natural resources and population, will alone and without a rival hold the balance of power in Europe.

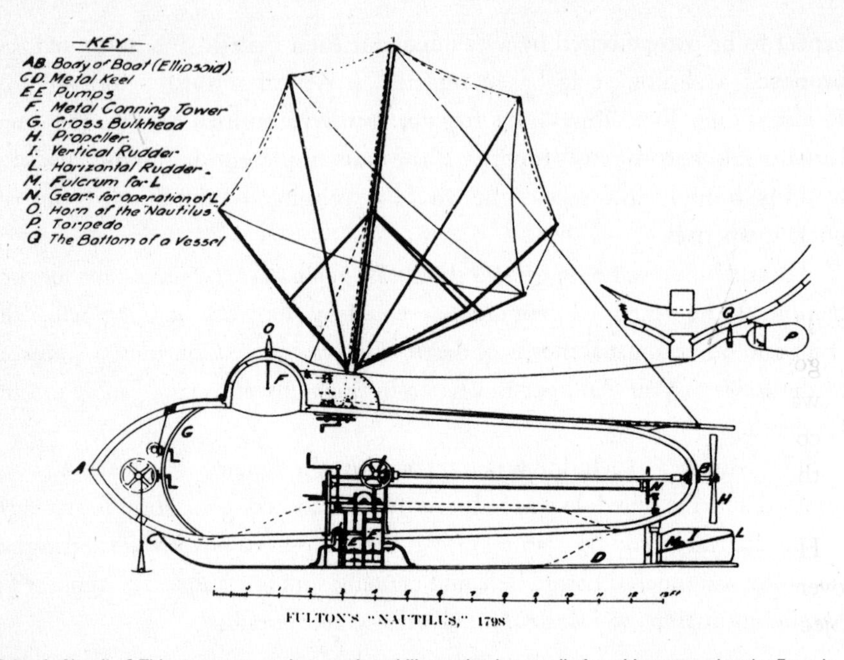

Fulton's *Nautilus*? This most commonly reproduced illustration is actually from his proposal to the French Ministry of Marine, two years before *Nautilus* was built. This is probably a fair, but not definitive, representation. Fulton added a deck, six feet by twenty, and most likely made other changes—perhaps substantial—in the finished boat. Three things support this supposition: one, a commissioner who reviewed the drawing told Fulton that the sail rig was top-heavy; two, Fulton declined to submit a final drawing, even when asked to do so by Napoleon in 1801; three, drawings of the "improved" boat, offered to Britain in 1804 and left behind in England when Fulton returned to the United States in 1806 show a markedly different design. (See photos 1, 1a.)

This drawing shows the "Horn of the *Nautilus*"—Fulton's version of Bushnell's drill. The "horn"—a spike—would be driven into an enemy hull by a hammer blow; a rather lengthy line, attached at one end to *Nautilus* and at the other to a detachable mine, ran through a ring at the bottom of the spike. As *Nautilus* moved away, the mine would be pulled into the hull.

The new Minister, Admiral Bruix, was intrigued; he asked for more details, and appointed a special, professionally astute eight-man commission to evaluate the proposal. It is from the report of this commission, issued September 5, 1798, that the picture of this first "Nautilus" emerges.

And rather than a chambered mollusk, we see a modifed *Turtle.* In general size, method of submerged locomotion, system of ballast and pumps, it was Bushnell's craft with minor changes—an extended tail to accommodate an enlarged crew compartment, a larger propeller (four feet in diameter), horizontal wings rather than a propeller to guide vertical movement, and a mast and sail for surface operation. Even the weapon and method of attachment to an enemy hull were Bushnell's, with only some improvement in the point of the drill, which Fulton called "the horn of the *Nautilus.*"

The special commission was impressed, and recommended that the government authorize construction of a prototype. Encouraged, Fulton raised his fee to 500,000 French francs for the first English ship, but offered to turn that sum immediately into building "a fleet of Nautili." In a letter to a high government official, he re-emphasized the significance of it all:

> The enormous commerce of England, no less than its monstrous government, depends upon its military marine. Should some vessels of war be destroyed by means so novel, so hidden and so incalculable the confidence of the seamen will vanish and the fleet rendered useless from the moment of the first terror. . . .

He also addressed what he believed to be one of the concerns within the government which earlier had blocked adoption of his scheme—strong moral objections to clandestine warfare:

> If at first glance the means I propose seem revolting, it is only because they are extraordinary. They are anything but inhuman; it is certainly the most peaceful and least bloody mode that the philosopher could imagine to overturn the system of plunder and of perpetual war, which has always vexed the maritime nations.

Nothing happened.

The recommendations of the Special Commission were forwarded to that black hole that most governments reserve for reports from Special Commissions. Fulton's efforts had not gone unnoticed, however, by the English. "I hear from France," wrote one friend to another, "that Mr. Fulton has not yet gone to America . . . besides which the Government and he are amusing each other (I think, however, to little purpose) on his new invention of the submarine boat."

Fulton continued much as before, working on various projects—some of which earned him a living. He obtained a patent for a rope-making machine, which he sold, and another patent for a panorama—a large, nay, *huge* painting depicting some scene from history, stretched across a stage between two posts, to be unrolled from one and taken up on the other. Viewers paid a small admission for the privilege of watching. Fulton sold the rights to a fellow

American and painted the massive scene: *The Burning of Moscow.* This was not, however, the burning associated with Napoleon, as that was not to happen until 1812; Moscow apparently burned often enough to excite the public curiosity. (The site of the Panorama today is in an arcade in Paris known as the *Passage des Panoramas,* near the Bourse—although the Michelin Tourist Guide misnames the famous inventor-painter as "Henry" Fulton.)

In July 1799, another new Minister of Marine (the third) took office, another proposal was submitted by Citizen Fulton, another commission appointed, and another favorable report issued. "The inventor is no charlatan," the commissioners emphasized. "He proposes to captain his engine himself and thus gives us his head as a hostage for his success."

Nothing happened.

Out of frustration over dealing—or, rather, *not* dealing—with the French, Fulton may have slipped across the border to make much the same offer to the Dutch. He did indeed drop out of sight for a few months in 1799, and although hard proof of a Dutch initiative has never been found, many of his biographers accept the possibility as certain.

Back in Paris by early October, Fulton made another approach to the Minister of Marine. However, a month later, with some progress being made, Napoleon seized full control of the government and yet *another* Minister of Marine was appointed. Lo! Minister Pierre Alexandre Laurent Forfait had been a member of the favorably disposed special commission of 1798. In fact, M. Forfait himself is reported to have devised a submarine scheme of his own a few years earlier, but details are lacking.

Perhaps on the strength of this change, certainly before any formal arrangements were made and on his own initiative, Fulton began construction of *Nautilus.*

The submarine was launched July 24, 1800, and began trials on the 29th. These were held on the River Seine, according to an eyewitness, just opposite the army barracks at Les Invalides (now the site of Napoleon's tomb). This was some distance from the builder's yard, where the trials would more logically have taken place; the "eyewitness" may have confused *Nautilus* with the public demonstration of Fulton's steamboat three years later—which newspaper accounts certifiably put opposite Les Invalides.

Be that as it may, on July 30th, a jubilant Fulton made the following report:

> Citizen Minister:
> Yesterday I tryed my experiments with the Nautilus in water 25 feet deep and have succeded to Render the sinking and Rising easy and famelior. . . . [now] having succeded to sail like a common boat and plunge under water when I think proper to avoid an enemy—it may be sufficient at present to render an operation against the enemy successful. . . .

He expressed his readiness to head for the coast—and, in fact, did so the next day with *Nautilus* under tow. Within two weeks, he had completed all the trials he felt necessary, including one dive to fifteen feet, with two other companions and a lighted candle, for a duration of one hour, two minutes. He was pleased to verify that "the compass acts in the same way below water as on the surface" and that he was able to dive "by means of lateral wings" and maintain "the boat at a desired level." On the surface, "using wings like the sails of a windmill for propulsion"—i.e., a four-bladed screw propeller driven by two men operating hand-cranks—*Nautilus* went 192 yards in four minutes, which was about twice as fast as the same two men using oars could row the boat over the same distance. At one point Fulton spent six hours sitting underwater, ventilating through a tube to the surface.

He also toyed on the surface with a tethered mine. He found that, using the current as if it were the wind, he could "fly" a floating mine much like a kite, guiding it to a target from a considerable distance.

He now was ready, he wrote, to make an attack upon the English ships, but for the legal reasons already advanced, was unwilling to proceed unless he and his crew were given the protection of commissions in the French Navy.

The Minister of Marine was sympathetic, but once again was unwilling to comply: "It cannot be disguised that the Nautilus is a machine not yet in use and that it infringes in several points the laws of war."

Fulton sent a note to Barlow, back in Paris, and asked him to visit the Minister of Marine to make an appeal in his behalf. For reasons unknown (especially to Barlow himself) the visit was successful; the commissions were dispatched a day later by mail. According to the report of a British merchant captain (verified from no other source), Fulton was made a rear admiral, Nathaniel Sargeant, captain of a man-of-war, and associate M. Fleuret was commissioned lieutenant. Barlow found the experience, at the least, amusing. He wrote to Fulton: "Your old idea, that these fellows are to be considered parts of the machine, and that you must have as much patience with them as with a piece of wood or brass, is an excellent maxim. . . ."

Thus, given the veneer of authority and with *Nautilus* on station, armed and ready, Fulton went forth to do battle. However, neither the weather nor the English were willing to cooperate that first week of September. The weather turned sour, and when he targeted two English ships, "both times, whether by accident or design, they set sail and were quickly at a distance."

It wasn't by accident. The captain of one of the ships, HMS *L'Oiseau,* sent a thank-you note to the Admiralty on September 21st, acknowledging a letter to him of the 14th "giving an account of Mr. Fulton's Plan respecting the possibility of destroying the ships on this station," adding, "I shall be very much on my guard."

Fulton was unfazed by what, after all, was the failure of his mission. After spending another thirty-five days watching winter close in and vainly hoping for the return of an enemy ship, he headed back to Paris. He was as full of confidence as ever. "Navigation under water is an operation whose possibility is proved," he wrote, and then he went on to introduce a thought which would before long would cause him some great difficulty with M. Forfait: "and it can be said that a new series of ideas have just been born as to the means for preventing naval wars or rather of hindering them in the future; it is a genus which only demands for its development the encouragement and support of all friends of science, of justice and of society."

Fulton now proposed that the government advance him money to build a larger *Nautilus,* provide two small boats for support, outfit her with twenty mines (which he called "bombs," not yet having invented the term "torpedo") and cover the costs of sea trials. He offered to give up command of the submarine, staying on only to supervise construction, and suggested that his crew—Captain Sargeant, Lieutenant Fleuret, and a new member, Citizen Guillaume—be retained as government employees.

Yet another commission was assigned to study Fulton's proposal, and recommended that Napoleon authorize a test: If Fulton could blow up an old hulk, the government should finance his progress. "We do not doubt his success," they wrote, "especially if the operation is conducted by the inventor himself who combines with great erudition in the mechanic arts an excellent courage and other moral qualities necessary for such an enterprise."

It was just at this time that Fulton was invited to his one and only meeting with Napoleon Bonaparte; there is no record of the conversation.

However, Minister of the Marine Forfait advised Napoleon that getting a worthless old hulk into position for the proposed test would cost *something* and, once sunk, would cost considerably more to clear out of the way. If a ship was to be destroyed, he suggested, let it be an enemy ship. Of course, it being winter, that test would necessarily be delayed.

Minister Forfait also keyed on what Fulton had written a few weeks earlier about a "New series of ideas." Fulton had been eager to explain; ". . . there has come to me a crowd of ideas infinitely more simple than the means I have employed hitherto and in an enterprise so new and without precendent one ought to expect that new ideas should present themselves, tending to simplify the execution of the great object in view." Fulton's frustrations in trying to approach the enemy's ships had shown him that it would be difficult if not impossible for a submerged *Nautilus* ever to come up against an enemy hull long enough to set the drill into place. He had been working, therefore, on his alternative methods for delivering the weapons.

In a memo to Napoleon on December 4th, Forfait wrote: "I have always been the most ardent defender of the plunging boat, and it is with pain that I see it abandoned; for it is abandoned in the new system since it plays only a secondary part. . . ." The plunging boat (a literal rendering of the French *plongeur*, "diver") now was relegated to a transport role, merely "carrying some destructive machines and the men who must direct them in a manner new and independent of the boat."

Nonetheless, Fulton was advised in February that his proposal had been approved and that 10,000 French francs had been deposited to his credit. Fulton's proposed bounty schedule was accepted: 400,000 francs for a ship of thirty guns or more, sliding down to 60,000 francs for a ten-gun sloop. The government realized that verification of a sinking might be difficult, but assumed that public announcements from the English government would provide the necessary corroboration.

Meanwhile, the harsh winter weather had not been kind to *Nautilus* and the boat was not ready for service. *Nautilus* had been built to test a theory, not for longevity; iron was used in fastenings and fittings when copper or brass should have been the materials of choice. A refit was started in May and was completed by July, whereupon Fulton had *Nautilus* loaded aboard a heavy wagon and hauled to the seaport of Brest.

There he continued to prove the theory but, curiously, avoided putting *Nautilus* into the combat which was the basis for government support. He conducted diving tests to twenty-five feet (which he felt was the practical limit for his hull). He had calculated that, with a capacity of 212 cubic feet of air, the boat had "sufficient oxygen to nourish four men and two small candles three hours."

A chemist had suggested that he could scrub out the carbon dioxide by passing the air through a bed of lime—or, as an alternative, take bottled oxygen down with him. Fulton decided that the equipment required for either of these would take up too much space in the boat. However, by adding a tank holding two hundred cubic feet of compressed air, he doubled endurance—supporting four men with no candles for six hours.

He installed a glass port, 3/4 inch thick, so that he could dispense with the candle, at least during daylight hours, and reported that at twenty-five feet, "I had sufficient light to count the minutes on the watch." He tested the boat's sailing ability—adequate—and found that it only took about two minutes to rig for "the operation of Plunging."

But having proved the concept, he abandoned the submarine boat and took up another of the "rush of ideas." This was a small surface boat, a pinnace, propelled by a hand-cranked screw. Fulton the Inventor could not resist making an elaborate series of comparative tests: With "24 of the best seamen of the fleet," he tried several variations on propeller size and blade pitch, and each at two different cranking ratios. He had hoped to achieve twelve knots, but never got above four—which seems about the limit for hand-cranked propulsion—although he found that some combinations were easier on the crew. The mechanism also proved noisy, and the boat could be heard more than four hundred yards away. A French official wondered why Fulton didn't just use canoes and paddles, or even better, why not just "fly" the floating mines guided by a line, like a kite, as he had already shown to be effective?

Fulton then turned his attention to what he called the Bomb Submarine. "It is this Bomb which is the Engine of destruction, the plunging boat is used only for the purpose of conveying the Bomb to where it may be used to advantage." The Bomb was, essentially, a copper container filled with gunpowder and fitted with a gun lock. On August 11th, he finally had the opportunity to test this weapon on a forty-foot sloop provided by the Navy, which was "torn into atoms" by the explosion. There is no evidence that *Nautilus* was involved; there is some evidence that Fulton used the pinnace and his primitive "wire-guided" torpedo.

Fulton made another attempt against the English, but the continued state of alert now included lookouts posted at the mastheads: The ships would move away at the slightest sign of unwanted approach. He definitely used the pinnace, not *Nautilus,* for this attack; Cafarelli, the Maritime Prefect at Brest, complained to his superiors in Paris that "Mr. Fulton, not making use of the plunging boat, which by its invisibility would assure success of the operation, does not respond to the expectations of the Government."

Fulton himself complained to the same authorities, as if they were to blame for the failure, that "for lack of a good plunging boat I have been unable to do anything this summer against the enemy."

At this point, he regarded his submarine experiments as complete. He took *Nautilus* apart and sold the metal parts for scrap, although he didn't tell anyone in the government until learning at the end of September that Napoleon wanted to see the boat. He was sorry, he then wrote, that he didn't know of Bonaparte's interest, but did not apologize for breaking up the boat. He also had a request from Napoleon to submit his drawings to a Committee of Engineers, a request which he refused for two reasons: the first, a fear that the details might "pass from one to another til the enemy obtained information; the Second is that I consider this invention as my private property, the Perfectionment of which will give to France incalculable advantage over her most Powerful and Active enemy, and which invention, I conceive, ought to Secure to me an ample Independence, that consequently the Government Should Stipulate certain terms with me before I proceed to further explination."

The government suggested that, at the least, his plans should be put in safekeeping. Fulton affirmed that he had "placed correct Drawings of the Machine and every movement with their descriptions in the hands of a friend; so that any engineer capable of constructing a Steam engine, could make the plunging boat and Carcasses or Bombs."

A week later, on the same day (October 1, 1801) that a preliminary peace agreement was signed with England, yet *another* Minister of Marine took office—an admiral, described by Fulton's biographer Dickinson as "one of the old school, and consequently bitterly opposed to the new method of warfare; in this he only voiced the prejudices of his time."

"Go, sir," the admiral is reported to have told Fulton, "your invention is fine for the Algerians or corsairs, but be advised that France has not yet abandoned the Ocean."

Peace, and this change of ministers, seem to finally have put an end to Fulton's underwater warfare in France. This did not prevent Fulton from continuing a not-so-subtle campaign. For example, a laudatory item in an 1802 issue of the French *Naval Chronicle* almost certainly was written, or heavily dictated, by Citizen Fulton. The inventor, the article noted,

> . . . not only remained a full hour under water with three of his companions, but held his boat parallel to the horizon at any given depth. He proved the compass points as correctly under water as on the surface, and while under water the boat made way at the rate of half a league per hour by means contrived for that purpose. Mr. Fulton has already added to his boat a machine by means of which he blew up a large boat in the port of Brest: and if by future experiments the same effects could be produced on frigates or ships of the line, what will become of maritime wars, and where will sailors be found to man ships of war, when it is a physical certainty that they may every moment be blown up into the air by means of a Plunging Boat against which no human foresight can guard them?

Peace, and "old school prejudices," probably had less to do with the end of the affair than the fact that the submarine boat, while an interesting novelty, was handicapped by the inefficiences of hand-power. *Nautilus* was simply too limited for the purpose for which it had been intended, and Fulton knew it.

5 1802–1815—The Mercenary as Savior

> *It is a most absurd visionary scheme that can be conceived to have originated in the brain of a man not actually out of his senses, & I am astonished that Congress should have suffered themselves to be so far imposed as even to notice it.*
>
> —*Commodore John Rogers to his wife, following a test of Fulton's torpedoes*

The British continued to follow Fulton's efforts with caution, if not with real interest. In May 1802, Earl Stanhope made a speech in the House of Lords in which he announced that submarine navigation had been perfected by a person in France, to "render the destruction of ships absolutely sure." A letter from the Admiralty warned senior officers:

> Mr. Fulton, an American resident at Paris, has constructed a vessel in which he has gone to the bottom of the water and has remained thereunder for the space of seven hours at one time—that he has navigated the said vessel, under water, at the rate of two miles and a half per hour; that the said submarine-vessel is uncommonly manageable, and that the whole plan to be effected by means thereof, may be easily

> executed, and without much risk. That the ships and vessels in the port of London are liable to be destroyed with ease, and that the channel of the River Thames may be ruined: and that it has been proved that only twenty-five pounds weight of gunpowder was sufficient to have dashed a vessel to pieces off Brest, tho' externally applied.

The Secretary of the Royal Society told Barlow, who had gone over to England in July for a short visit, that the submarine boat "would never be brought to use because no civilized nation would consent to use it." Slightly more than one hundred years later, Winston Churchill would say much the same, on the eve of World War I. That gets us ahead of our narrative, but not ahead of our theme.

Fulton turned his attention to other projects. Robert Livingston had been appointed the U.S. Minister to France, and Robert Livingston held a monopoly, granted by the state of New York, for the operation of steamboats on New York waters. Of course, at this point in history, there were no steamboats ready to operate in New York or anywhere else, but the concept had been validated fifteen years ealier by John Fitch, who operated a not-very-successful steamship line on the Delaware. Livingston—the businessman with the monopoly—and Fulton—the restless inventor who as a teenager had built a paddle-wheel-driven boat—were a natural team, and Fulton's greatest achievement, the commercially viable steamboat, was demonstrated August 9, 1803, for assembled officials of the French government.

Fulton wanted to sell the steamboat to the French for an invasion of England—it could tow a string of troop-laden barges across the English Channel. However, as Forfait, now promoted to Counselor of State, told Bonaparte, Fulton had attached "one condition without which he will not set to work, it is to have a very short conference with you. He has, he says, some political views of the very highest importance to share with you."

There is no record of such a meeting, nor, therefore, of the important political views. By October, Fulton was in secret negotiations with an agent in England, apparently on English initiative, to return to England. Many historians believe that the British lured him back during a brief period of "peace" to prevent further development of his weapons—at least, as they might be employed against the British. Other than to note that Fulton's inherent opportunitism would take him wherever he thought he might find reward, there is no reason to doubt that belief.

Fulton left Paris on April 29, 1804; by May 12th, England and France were once again at war. A war! Another opportunity! It did not matter to Fulton that he had so recently been on the other side. He drafted a thirty-one-page letter to British Prime Minister William Pitt, a manuscript copy of which, edited in Barlow's hand, is in the collected Barlow Papers at Yale University (although it bears no indication as to whether or not it actually was ever sent). The letter rehearsed a favorite Fulton theme: eliminate navies by giving each and all such a superior weapon that fighting would be futile. That weapon was the *Nautilus,* ready to put "an end to maritime wars with all the dreadful catalogue of crimes which they entrain." In Fulton's view, submarines could not fight each other, only ships of the line, so submarine-equipped navies would be at a virtual checkmate, subject to the "principle of equality among nations." So certain was he of a world free of war that he called upon the English to give up the aristocracy, with its "unnecessary titles and exclusive distinctions," and adopt the principle of democratic elections. In truth, he explained, only such a lofty goal could ever "have induced me to put my hand to a work which I should otherwise abhor." Robert Fulton, savior.

Operating under the assumed name of "Mr. Francis"—as if this would protect him from French reprisals—Fulton offered to help the British attack a French fleet then assembled at Boulogne in preparation for an invasion of England.

He laid down the plans for a larger, improved submarine—fourteen feet longer than the original, to thirty-five feet overall, ten feet wide, eight feet deep, with a double hull providing space for ballast in between. This boat would have a crew of six and be armed with thirty submarine bombs, each containing one hundred pounds of gunpowder. Instead of hydroplanes, Fulton planned to use a Bushnell-like haul-down screw at the bow. His proposed mode of operation: Wait in ambush until after dark, then surface to make the attack. The new boat, he wrote, "must be considered as a masked battery which can lie secure in the neighborhood of an Enemy, watch for an opportunity to deposit her cargo of Bombs, and retire unperceived." The "horn of the *Nautilus*" had disappeared from the plan, and the submarine would now operate as a clandestine minelayer.

Indeed, new schemes for setting minefields now took precedence in his thinking and in his proposals. He designed an anchored mine, adjusted to ride just below the surface. He suggested that a minefield could close off the English Channel, and tried to sell the idea to the British on the proposition that, otherwise, the idea might occur to the French:

> Ten lines of Instantanious Bombs or even a less number anchored in the British Channel would cut off the greater part of Commerce of London and of England . . . should France ever possess a means to cut off or interrupt such trade, England would be obliged to submit to any terms which Bonapart might think proper to dictate.

Fulton later noted that, "When Mr. Pitt saw the sketch of this engine of simple construction, easy application, and powerful effect, he observed that if introduced into practice it would lead to the annihilation of all military marines."

In what we may charitably assume was a concilatory gesture to permit the continuation of "military marines," Fulton offered to take his entire system—submarine boat and mines—off the market, to lie dormant for the value of one ship of the line, £100,000.

A government commission reviewed Fulton's proposals. They were interested in pursuing the Submarine Bomb but not, apparently, the submarine. While they accepted that underwater navigation was possible, they felt that it offered no advantage to England. The enlarged *Nautilus* was never built.

Fulton accepted this decision with apparent equanimity: A major power such as Britain, which already had control of the seas, could plant his anchored mines with ease and had no need of the submarine. It might be a different matter, he later wrote, for the "Sweeds, Danes, Dutch, Spaniards or Portuguise in a War against England," which would benefit from a method to secretly plant mines in the Thames River. The submarine—the ideal weapon of the weaker power against the stronger.

To address the immediate problem of the impending French invasion, Fulton offered the British yet another method of bomb delivery, a catamaran raft, which he described as made of "two pieces of timber about 9 feet long and 9 inches square placed parallel to one another at such a distance as to receive a man to sit between them on a bar which admitted of his sinking nearly flush with the water and occasionally immersing himself so as to prevent his being seen in the dark or by moonlight." The operator would be dressed all in black and with a black cap to pull down over his face, and equipped with a paddle. A different sort of canoe.

* * *

By July, after some discussion and much negotiation, the government had agreed to pay the inventor £200 a month and established a line of credit to cover the costs of development. If and when he had destroyed the first warship, he was to be paid £40,000. If for some reason the "plan" could not be executed, the scheme would go to arbitration. If the aribtrators determined that the plan could have worked, and that it "offered a more effectual mode of destroying the enemies fleet . . . than any now in practise and with less risk," Fulton might receive payment of £40,000 pounds "as a compensation for demonstrating his principles, and making over the entire possession of his submarine mode of attack."

For the subsequent destruction of other enemy ships, he would be paid one-half the supposed value of each "as long as he superintends the execution of his plan." Should either he or the government terminate his "employment," he was entitled to a royalty of one-fourth the value of each ship sunk by his methods for a period of fourteen years. He agreed not to pass his plan on to any other person for the same period.

A test was arranged: Attack the French fleet then at anchor in Boulogne Harbor. It seems to have made no difference to the inventor that the fleet commander was the same Admiral Bruix who, as the second Minister of the Marine with whom he had dealt almost six years earlier, had been the first to support the mechanical *Nautilus.* On October 2, 1804, a squadron of five catamarans approached the French unobserved, unloosed their weapons, and achieved the not very impressive destruction of a small boat and twenty-one French sailors.

Word of this Quixotic mission quickly reached England. The October 27, 1804, edition of the *Weekly Political Register* reproduced "The Catamarans, an excellent new ballad" which had just hit the streets, purporting to be the plaint of the Secretary for War:

See here my casks and coffers
    With trigger pulled by clocks!
But to the Frenchmen's rigging
    Who first will lash these blocks?

Catamarans are ready
    (Jack turns his quid and grins)
Where snugly you may paddle
    In water to your chins.

Then who my blocks will fasten,
    My casks and coffers lay?
My pendulums set ticking
    And bring the pins away?

Your project new? Jack mutters
    Avast! 'tis very stale:
'Tis catching birds, land-lubbers!
    By salt upon the tail.

In December, a smaller attack was mounted against the harbor at Calais. It was a wasted effort.

At this time, in addition to trying to fight a war, the British government was overwhelmed with internal problems. Fulton was more or less forgotten (although still on the payroll) until July, when he managed to get Pitt's attention and spark the planning for another October expedition against Boulogne. This had even less effect than the others: no sinkings, one unexploded bomb recovered the next day by the French (which, as in Bushnell's Battle of the Kegs, then exploded, killing four sailors). The French report of the incident described the bomb as "a lock like that of the fire machines which the English used last year with so much ridicule and so little success."

Fulton was desperate for a success—any success. He convinced the government to let him stage a dramatic public demonstration in Walmer Road, Deal Harbor, almost in front of the official residence of the Prime Minister, although Pitt was not present at the time. The government provided a brig captured from the Danish, the one-hundred-ton *Dorothea,* and, on October 15th, and in the presence of many ranking naval officers, government officials, and great crowds of people, Fulton blew her in two with a carefully positioned bomb.

Because there has been so much misunderstanding about this event, let Fulton speak for himself:

> Urgent business had called Mr. Pitt and Lord Melville to London [but] Admiral Holloway . . . and the major part of the officers of the fleet under command of Lord Keith were present. . . . Two boats, each with eight men, commanded by lieutenant Robinson, were put under my direction. . . .

Fulton put one submarine bomb in the stern of each boat, connected by a line eighty feet long; the boats, starting about a mile up-current from the brig, rowed down towards her, one to port and one to starboard, the line stretched out between them. When the line was snagged by *Dorothea*'s anchor buoy, the clockwork timer started and the bombs were pulled from the boats and carried under the ship's hull by the running tide. Eighteen minutes later, "the explosion seemed to raise her bodily about six feet; she separated in the middle, and the two ends went down; in twenty seconds nothing was to be seen of her except floating fragments. . . ."

However contrived the event, Fulton made his point: Ten minutes before the explosion, one of the Naval officers, affecting disinterest (or disbelief), had said that he would not be conerned just then to be sitting at dinner aboard *Dorothea.*

Out of this demonstration came one of the most frequently quoted jibes against the submarine—usually invoked to illustrate the supercilious incompetence of senior officers. Sometime after the event, Fulton gave a briefing to the Earl of St. Vincent, First Lord of the Admiralty, who likewise had missed the show. Fulton later reported that the earl had said that "Pitt was the greatest fool that ever existed to encourage a mode of warfare which those who commanded the seas did not want, and which, if successful, would deprive them of it."

Note well: While the comment is usually attributed to the earl, it's actually Fulton's phrasing and is a bit too close to Fulton's stock argument to be accepted as offered. Note also: The earl was talking about catamarans and submarine bombs, *not* about the submarine boats which had long since been dropped from the program.

Another minister, however, saw merit: "The success of Mr. Francis's experiment gives me great confidence in our means of annoying the enemy in their own ports with little comparative risk . . ." Thus encouraged, the British made one more attempt on Boulogne, on October 27th, with even smaller results than before. Soon thereafter, the government received news of Admiral Lord Nelson's crushing victory at Trafalgar on October 21st, which took the French Navy out of contention and nullified any threat of invasion from Boulogne.

After almost a year of sputtering inaction and some negotiation, the English settled accounts with Fulton. He had asked for a lump sum of £60,000 and £2,000 a month for life to keep his system secret from any other government; he settled for £1,653, 18s., 8d. Total remuneration for the two years was therefore £16,000, and he was released from the obligation of exclusivity and secrecy. He could sell his system to the next interested party. He could hardly wait.

Suddenly seized by allegiance to his homeland, and worried lest he fall victim to shipwreck en route, Fulton left a copy of his "latest" ideas for safekeeping with the American Consul in London. ". . . that the produce of my studies and experience may not be lost to my country," he wrote, "I have made out a complete set of drawings and descriptions of my whole system of submarine attack, and another set of drawings with description of the steamboat. These with my *will*, I shall put in a tin cylinder, sealed and leave them in care of General Lyman, not to be opened unless I am lost."

Lyman kept the plans so well that they remained hidden until 1870, when they were discovered sold at auction, then disappeared into a private collection for fifty years, eventually acquired by William Barclay Parsons and published in his book *Robert Fulton and the Submarine* (1922).

In November 1806, back "home" after an absence of nineteen years, Fulton was ready to demonstrate the submarine bomb to the next interested party. With the assistance of friend Barlow, also now back in the United States and living on his estate "Kalorama" in a suburb of Washington, Fulton met with Secretary of State James Madison and Secretary of the Navy William Jones and obtained permission to stage an American demonstration.

This trial was held on July 20, 1807, in New York harbor, and the result was not as conclusive as with *Dorothea*. Oh, the targeted ship was sunk, but this time it took three tries. On the first, the bombs turned upside down, dumping the priming powder out of the pan. On the second, the bombs missed the target. On the third, the ship was blown up.

Nonetheless, Fulton expressed his confidence in his system, and conveyed a Fultonian view of world peace in a letter to the governor of New York:

> Having now clearly demonstrated the great effect of explosion under water, it is very easy to conceive that by organization and practice the application of the torpedoes will, like every other art, progress in perfection. Little difficulties and errors will occur in the commencement, as has been the case with all new inventions: but where there is little expense, so little risk, and so much to be gained, it is worthy of consideration whether this system should not have a fair trail.

> Gunpowder within the last three hundred years has totally changed the art of war, and all my reflections have led me to believe that this application of it will in a few years put a stop to maritime wars, give that liberty on the seas which has been long and anxiously desired by every good man, and secure to America that liberty of commerce, tranquility, and independence which will enable her citizens to apply their mental and corporeal faculties to useful and humane pursuits, to the improvement of our country and the happiness of the whole people.

The New York demonstration came just before the launching of Fulton's American steamboat *Clermont,* which more or less occupied the inventor for the next two years.

Meanwhile, the British continued to monitor the activities of their sometime enemy/employee. Commodore E.W.C.R. Owen, RN, who had been a participant in the attempted attacks at Boulogne and whose skepticism was in large part responsible for the test in Walmer Road, admitted that his doubts had been overcome by the sinking of *Dorothea,* "which placed the power of the weapon beyond dispute." However, in a September 1807 letter widely circulated within the fleet, he suggested that, because there was no effective means of delivery, that "power" was inevitably tied to the action of the tides and could be ameliorated with some very basic defensive measures; for example, ". . . keeping the Lower Studding Sail Booms out at night in the same manner as for Guest Warps [i.e., as used for mooring a visiting ship's boats] and having a Rope thro' a Block at the end of each Boom bent to the Cable about 20 fathoms or more from the Bows which wou'd keep the Machines clear of the bows . . ."

Commodore Owen sent a copy of his Fleet letter to Admiral George Berkeley, Flag Officer Commanding the American Station, with the added caution ". . . should the Negociations between this Country and America take an unfavorable turn, [these weapons] may be used against our Naval Force upon that Station."

Admiral Berkeley in turn sent a note of his own to an American friend in London:

> The Author or rather projector of your Torpedos tried his hand upon John Bulls credulity, who possess full as much as his Transatlantik Children after a very expensive Trial The Scheme was scouted not

perhaps so much from its Failure, as for the Baseness & Cowardice of this species of Warfare. All strategems are however allowed in War, and there are certain Regulations attached to Ingenuity of this kind which I rather suspect Mr. Fulton is not acquainted with, at least in England he rather *Blinked* the question. Every Projector ought to be the Man, who first makes trial of his own Device, and then he is entitled to the Reward of his Merit. An officer who commands a Fire Ship, has a Gold Chain put round his Neck, if he is successful, But if he is taken, a Hempen One is the premium he is sure to receive, which I think Friend Fulton would rather be surprised at.

Of course, Admiral Berkeley had misjudged the target of his scorn; Fulton indeed had been willing not only to "make trial" of his own device but had himself made two attempts against the British at Brest, and had accompanied two of the expeditions against the French at Boulogne.

Back in England, an anonymous dispute erupted in the pages of the semiannual *Naval Chronicle* for 1809. One letter writer, signing himself "F.F.F.," was aghast that "*one* crazy murderous ruffian" could secretly blow up a crew. "Battles in future may be fought under water, our invincible ships of the line may give place to horrible and unknown structures, our frigates to catamarans, our pilots to divers, our hardy, dauntless tars, to submarine assassins, coffers, rockets, catamarans, infernals, water worms, and fire devils. . . . How honourable!"

A response in the next issue by "H." acknowledged that torpedoes "are understood to be employed in the darkness and silence of night, against a helpless and unsuspecting enemy." Were they clandestine? Yes. Reprehensible? Yes . . . unless "resorted to . . . for the accomplishment of some great and important purpose." The end, therefore, *could* justify the means.

In 1810, having diverted himself for a time from steamboats, Fulton made another play for government support. He published a pamphlet on *Torpedo War and Submarine Explosions,* addressed to the President and the Congress and bearing a title-page motto: "The Liberty of the seas will be the happiness of the earth." The submarine now has completely disappeared from view, although the variety of weapons has been expanded to include the spar torpedo—the petard gone to sea as a mine on the end of a long pole extending out in front of a boat. Delivery was simple. The boat would ram the target and, it was hoped, not herself be damaged by the encounter.

Fifty years later, during the American Civil War, that hope proved to be, well, hopeless.

Another pilgimage to Washington produced another independent commission and a $5,000 appropriation to fund a new test. Fulton's experience with the French and the British had taught him that independent commissions could not be depended upon to deliver the correct answer; they were, well, too independent. Fulton used all of his guile to rig this test. He recommended and ensured the appointment of five of the seven members of the commission, asked the Secretary of the Navy for advance notice of "the objections and plans of defence from the officers of the navy or persons connected with the Navy or Gentlemen opposed to the system," and even drafted—for the "consideration" of the committee, of course—the text of a proposed report.

In the event, the test was a failure. The Naval officer in charge, Commodore John Rodgers (1773–1838), followed the British plan (although he may have arrived independently at this solution) and arranged to have some booms and nets hung from the designated target ship, *Argus.* This simple tactic prevented Fulton's "torpedoists" from getting close enough to attach a mine. We presume that Fulton was not forewarned.

However, the failure did not prevent him from claiming victory; "an invention which will oblige every hostile vessel that enters our ports to guard herself by such means," he wrote, was clearly a defensive weapon of great importance.

The commodore's personal opinion has been preserved in a letter to his wife: "It is a most absurd visionary scheme that can be conceived to have originated in the brain of a man not actually out of his senses, & I am astonished that Congress should have suffered themselves to be so far imposed as even to notice it."

Undaunted as always, Fulton sent copies of the pamphlet to his former associates in both Britain and France—and, we may presume, to friend Barlow, now appointed to Jefferson's old post as Minister to France. Fulton's submission to the French included another clarion call, "to see commerce free, peace restored to the continent and the Genius of the Emperor relieved from the fatigues of war, directed to repair its losses, by pursuits as dear to his heart and interesting to his people by promoting the arts and converting France into a garden."

Thus, Citizen Fulton offered his "system" as a Citizen of the World. He would be beholden only to that nation which would pay him first, or top the last offer and pay him most.

Or, perhaps, just pay attention. Historian William Parsons, writing in 1922, suggested that Fulton's primary motive, in everything he did, was psychological, not pecuniary. He was desperate for public attention and approval.

Fulton again became fully absorbed—for a time—by the steamboats just then beginning service on the Hudson and Mississippi Rivers: Fourteen were in operation by 1813. But he returned to underwater warfare during the War of 1812. He briefly tried to develop a system whereby a mine could be planted on the bottom, to be detonated by an electrical charge (through wires running to the shore) when an enemy ship passed overhead. He could not, however, resolve two problems: how to protect the wires from environmental and human enemies, and how to know when an enemy ship was in position over the mine. He did provide mines themselves—his basic submarine bomb—to several warriors, notably Sailing Master Elijah Mix, USN, who with them made seven attempts to sink *Plantagent* moored near Norfolk. The first six were frustrated by wind, wave, and tide (Commodore Owen knew what he was talking about); on the seventh, the weapons were drifting nicely toward the target when they went off too soon.

There were a number of similar efforts, encouraged by passage of the Federal Torpedo Act of 1813. This promised a reward of one half the value of any British ship sunk by these means, and attacks spread over Long Island Sound, Chesapeake Bay, and the Great Lakes—so many, in fact, that by the summer of 1813 the *Niles Register* reported, "The much ridiculed torpedo is obtaining a high reputation."

A submarine—details lacking, but see page 72—attempted to fasten mines to *Ramilles*; the captain was so incensed that he brought American prisoners of war aboard, as a deterrent against any future attacks. He also warned that, in the event the Americans tried to use any form of exploding ship, he would retaliate by sinking every ship he could find, warship or merchant.

One mysterious submarine boat of the war, about which virtually nothing is known, was a *Turtle*-like craft sometimes associated, on no known evidence, with Bushnell. It should more likely be attributed to Fulton—as such it was described in a July 1814 letter from a British commander, taking credit for the destruction of "the Fulton turtle boat." This boat was designed to run awash while towing five torpedoes; she became stranded on a sand bar in Long Island Sound and was destroyed by her handlers before an approaching British patrol could take possession.

In 1813, Fulton came up with yet another device for destroying warships: cannons, mounted below the waterline of a ship and discharged underwater. Dubbed "Columbiads" in commemoration of a long epic poem of that title written by his friend Barlow, these were the subject of Fulton's last patent, to wit: ". . . for several improvements in the art of maritime warfare and means of injuring and destroying ships and vessels of war by igniting gunpowder under water or by igniting gunpowder below a line horizontal to the surface of the water, or so igniting gunpowder that the explosion which causes injury to the vessel attacked shall be under water."

They were not very effective—the projectile would travel only a few feet. Nonetheless, Fulton designed a special platform for the "Columbiad" which became the world's first steam warship, *Demologos* ("Voice of the People").

Fulton offered this as the ideal harbor defense ship—armed with a full battery of cannon as well as the Columbiads, able to move about at will, unlimited by vagaries of wind and tide. The recently formed New York City Coast and Harbour Defence Association, understandably nervous about the British and suitably impressed by Fulton's presentation, offered to put up the estimated construction cost of $320,000, provided that the Federal government would agree to reimburse them once it had proved successful. Congress relieved the Association of this burden even before the ship was built, with an appropriation of $500,000.

In the meantime, the good people of Philadelphia, concerned about the protection of *their* harbor—the next commercial port south of New York—wanted one for themselves. Fulton was willing but, as usual, condescending: "I must also remark that as this is a new Invention which *requires all my care* to render it as *complete and* useful as can reasonably be expected from my present experience, I *cannot trust the construction of the machinery or the fitting out of the* vessel to be directed by *anyone but myself* in which I will give every facility in my power to the Gentlemen of Phila. . . ."

Demologos was launched October 29, 1814. A week later, mindful, no doubt, of almost twenty years' frustration in having to deal with bureaucrats,

Fulton asked now-President James Madison to appoint him Secretary of the Navy in place of outgoing Secretary Jones. He was not, he wrote, "ambitious for office," but wanted to "have the power to organize, and carry my whole System into the most useful effect, in the least possible time: for which purpose, it is better to have the power to arrange and command."

He was not given that opportunity. The problems of both New York and Philadelphia were more easily solved: The war ended, settled with the Treaty of Ghent on December 24, 1814.

In February 1815, Fulton contracted pneumonia while visiting *Demologos* in the builder's yard. He died on the 23rd; he was fifty years old. There was a rumor that he was at the time working on a new submarine, called *Mute,* but the only apparent details which survive describe a craft like *Demologos.* We suspect that someone had tried to make a small joke at Fulton's expense, to our confusion.

Demologos went on sea trials in June 1815, but was never fitted out for service. She became a receiving ship at the Brooklyn Navy Yard—a way station for transient sailors—until destroyed by an explosion and fire in 1829.

6 1815–1860—Of Corporals, Shoemakers, Underwater Bands, and Samuel Colt

> *. . . on 1 February 1851, an event took place in Kiel harbour which filled the world with astonishment and admiration. It was the first—unfortunately unsuccessful—experiment of Wilhelm Bauer, a subaltern in the artillery, with the submersible fire-ship he invented. So important is this invention that it will not only compare with electrical telegraphy, but might even surpass it.*
>
> —Die unterseeische Schiffahrt *("Underwater Seafaring") by Ludwig Hauff, 1859*

Fulton had left his plans in France; some Frenchmen may have found them, and used them, in building *Nautilus*-like submarines. One, named *Nautile,* used oars for underwater propulsion (with a crew of nine), sails on the surface; Napoleon is reported to have inspected this craft. Later, after Napoleon had been defeated at Waterloo, deposed, and sent in exile, a group of his supporters may have commissioned Englishman Thomas

Johnstone to build a submarine by which Bonaparte might be rescued from the island of St. Helena.

Johnstone may—or may not—have had some involvement with Fulton's expeditions against the French at Brest; he may—or may not—have built a submarine in 1815 with tepid support from the government. He may—or may not—have been offered £40,000 for the effort on behalf of the Bonapartists; however, Napoleon died (1821) before the reported one-hundred-foot-long submarine was finished.

These are but two of several dozen submarine or semisubmersible boats of the period—some proposed, some actually built, most of which are today but names and dates on a list: French Navy Captain Montgery, *L'Invisible,* 1822—a *Demologos* clone, complete with Columbiads, but powered by a gunpowder engine; Shuldam, American, 1823—no information; Unidentified French Naval Officer, *Invincible,* 1825—eighty-six feet long, hand-cranked paddle wheels, never built; Cervo, Spanish, a spherical submarine, 1831—lost on trials; Dr. Petit, French, 1834—lost on trials.

We have a bit more information on submarines designed and built in mid-century by American shoemaker Lodner D. Phillips, although—surprise!—most accounts are not accurate.

Lieutenant Barber offered a description of Phillips's work in his 1875 lecture, noting initial success with two smaller boats and describing in detail two elaborate, unrealized schemes, one for a ship of war, one for peaceful commerce. Among other rather advanced features, Phillips proposed the use of steam propulsion, with a compressed air supply for submerged operation; of this, Barber was properly skeptical: ". . . it would seem that in practice the apparatus would heat the entire boat so much as to render the air within it unendurable to the operators."

Of the two boats Phillips actually built, Barber reported that one worked well enough for Phillips to take his wife and two children on an all-day underwater excursion in Lake Michigan. The other, he said, on a dive in Lake Erie, 1853, became Phillips's coffin.

Not quite: the Lake Erie experience was harrowing, but not fatal. Phillips shipped his "submarine propeller" by rail from his home in Michigan City, Indiana, to Buffalo, in an attempt to recover some valuables from a wreck resting on the bottom at 155 feet. At 100 feet, his own boat started leaking badly and he barely made it back to the surface.

By this time, Phillips had acquired considerable experience with submarines and seems to have swapped the trade of shoemaker for that of inventor. He had tested his first unmanned submarine boat in 1845, at the age of twenty. It collapsed at a depth of twenty feet. His next boat, 1851—that of the family excursion—was a forty-footer built with $241 of borrowed money. This was successful enough that Phillips offered to sell it to the Navy. In a letter dated April 7, 1852, he noted underwater speeds of "four or five miles per hour" and dives to 100 feet. The Navy's response of April 21, 1852, was remarkably timely but not encouraging. "No authority is known to this Bureau of purchase a submarine boat . . . the boats used by the Navy go *on* not *under* the water."

Undeterred, Phillips obtained a patent for "Steering Submarine Vessels," November 9, 1852:

> It often happens that when a submarine vessel is below the surface of the water, that sufficient headway cannot be obtained to steer her by, with a rudder of the usual construction, and the lives of the inmates of the vessels are from this cause often greatly imperiled. To remedy this difficulty is one of the chief objects of my invention which consists in

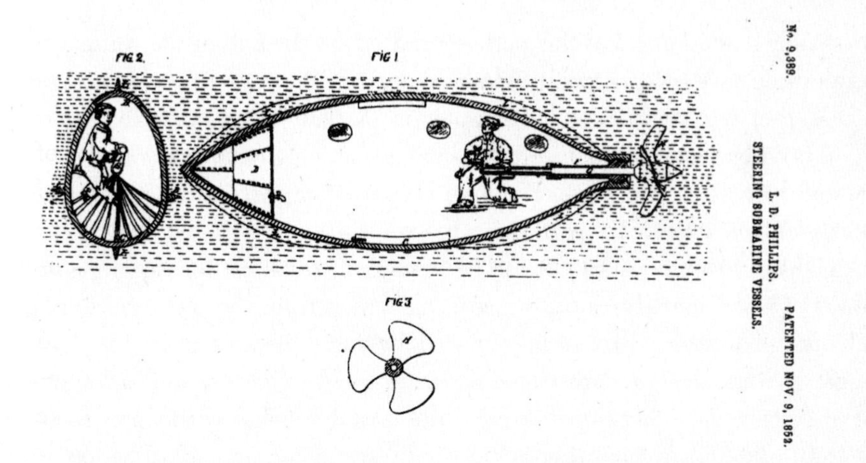

The drawing submitted with Lodner D. Phillips's 1852 patent application for "Steering Submarine Vessels." The rather sophisticated propeller is at one end of a combined tiller/crank shaft, the one-man propulsion system is at the other, and the shaft passes through a watertight ball-joint. Directional control is obtained by pushing the tiller left or right; depth control, by pushing it up or down, i.e., raise the cranking end, the stern will be driven upward, forcing the bow down and the boat to descend. The ribs along the deck (C) provide a foothold. The bow tank (D) holds compressed air; the spigot allows air to "be let out from time to time as may be required for the purpose of respiration, and to exclude the water." Other details, such as ballast tanks, were not included, "as every competent constructor of submarine vessels is capable without instruction to make such modifications." (U. S. Patent No. 9,389 of November 9, 1852)

> mounting the axis of the propeller on a universal joint so that it can be inclined in any direction for the purpose of applying the whole propelling power to the steering of the vessel . . . *[See illustration]*

Following the Lake Erie accident, Phillips moved to Chicago, and in 1856 obtained a patent for an armored diving suit—which he tested in the Chicago River. He may there have built another submarine boat, details unknown, which he may have sold, purchaser unknown. A submarine dredged from the bottom of the Chicago River in 1915, origin unknown, contained one human skeleton, identify not established. As advertised in the *Chicago Tribune,* February 23, 1916, the submarine—now dubbed the "Fool Killer"—was put on public display, ten cents a look. (It has long since disappeared.)

During the Civil War, Phillips again offered to design submarines for the Union Navy—presumably the source of the schemes presented in Barber's lecture—with apparently no more success than in 1852. His death—October 15, 1869—had nothing to do with submarines. The death certificate on file in his last place of residence, New York City, lists the cause as a variant of tuberculosis.

One major flaw in boats of this period—indeed, to the end of the century—was a simple and basic instability. Under most conditions of up or down angle or heel, the "free surface" effect of water in partially filled tanks (or in the bilges) is dangerously—fatally—destabilizing, suddenly rushing to the downside and increasing the angle. If you've ever tried to carry a flat pan full of water, you have experience with free surface.

In addition, submarines need to be "trimmed" to adjust for varying conditions of water density—not to mention changes from the consumption of fuel once manpower had been replaced by fuel-consuming engines and weapons. Some designers provided for trimming by the fore-and-aft shifting of ballast—but these boats were quite unstable, what with water rushing from one end to the other at the slightest encouragement. The key to stability, not fully understood until the beginning of the 20th century, was to keep the center of gravity in the center of the boat, completely filling the main ballast tanks and using small tanks for adjusting trim. Short-term adjustments were often made by the simple expedient of moving a few crew members forward or aft.

* * *

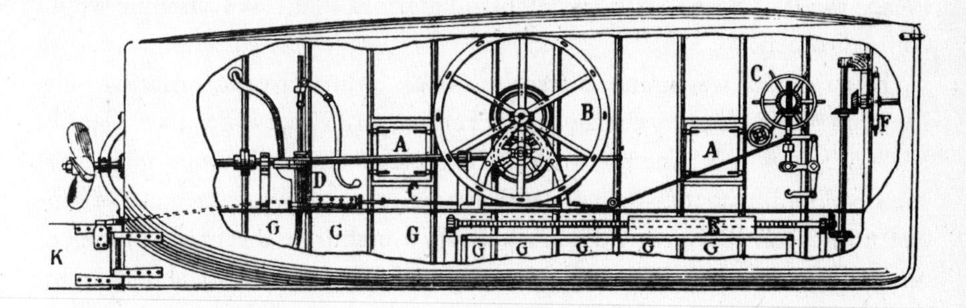

Corporal Wilhelm Bauer's 1851 *Brandtaucher.* The boat was powered by a treadmill (B); trim was adjusted by a weight sliding along a track (E), which—on a test dive—slid too far forward, taking the boat down to become stuck on the bottom. Bauer and his companions escaped by riding a bubble of air sixty feet to the surface. *Brandtaucher* was recovered in 1887 and is now on display in Dresden. Bauer was a persistent—and successful—designer and builder of submarines who never quite overcame his plebian origins. Ignored by his native Germany in his lifetime, Bauer was elevated to the Nazi Pantheon in World War II.

One inventor of this era was to spend some twenty-five years developing submarines on behalf of the governments of at least five nations, leaving behind plenty of details—and one submarine on display in a museum. In 1850, Prussian Army Corporal Wilhelm Bauer had an idea for combating a Danish naval blockade of the German port of Kiel, and persuaded a major shipbuilder to construct the little boat he called *Brandtaucher*—"Incendiary Diver." This was about the size and shape of a small sperm whale, with a riveted sheet-iron hull and ballast tanks. To adjust trim, Bauer installed a weight which slid fore and aft along a rod—like the sliding weight on a scale. The boat was driven by a propeller, but Bauer rejected the foot-treadle of Bushnell and the hand-crank of Fulton and provided a two-man-power treadmill.

On its first mission, *Brandtaucher* was sufficiently threatening to cause a blockading Danish force to move further out to sea. A short time later—February 1851—Bauer also moved out for a test dive into deeper water, and almost met disaster. His "sliding trim" feature slid forward, and the boat turned sharply down by the bow and hit the bottom at sixty feet.

The boat could not rise; the three-man crew could not open a hatch against the outside pressure; the hull was sprung and water was seeping in. In a remarkable display of iron nerve, Bauer persuaded his companions that their only chance for survival was to *wait*—for six and a half hours, as it turned out—until the incoming water had raised the internal pressure to match the external pressure to thus allow the hatch to be opened. The three men were swept aloft in the bubble of escaping air—the first, but by no means the last, time in the history of submarines that this technique would come into play.

Bauer's work was immortalized in a book on underwater seafaring published in 1859: "Eight years ago, on 1 February 1851, an event took place in Kiel harbour which filled the world with astonishment and admiration. It was the first—unfortunately unsuccessful—experiment of Wilhelm Bauer, a subaltern in the artillery, with the submersible fire-ship he invented. So important is this invention that it will not only compare with electrical telegraphy, but might even surpass it." The event did not lead to anything much in Germany, but then, Germany did not at that time have a navy.

Brandtaucher herself was rediscovered and salvaged in 1887; in 1973, she was put on public display at the Dresden (Germany) Military Museum.

Bauer was not to be deterred—although the Germans wanted nothing more to do with him. He approached the Austrian government, which was about to grant him £4,240 when objections by some in the government forced a cancellation of the appropriation. Bauer went to England, where he found tangible support in the amount of £7000, encouraged by Prince Albert and spurred by the proximate start of the Crimean War. His British submarine had one thing in common with the German—it also was propelled by men walking. However, in this case, the submarine was only a portable diving bell. The walking was along the bottom with the crew pulling the boat—their air supply—along with them. It didn't work too well; some men were killed. Bauer made (an unsuccessful) inquiry to the United States, and then went to Britain's Crimean War opponent, Russia.

As Fulton had earlier discovered, "clients" did not seem to care that the inventor had most recently been selling his services to the other side. For Russia, Bauer built the fifty-two-foot *Diable Marin* ("Sea Devil"), launched in 1855. Propulsion was once again by treadmill; the sliding-trim weight was there and the boat had diving planes. *Diable Marin* was a "real" submarine, not a figment or a prototype or a failed dream, and may have completed as many as 134 dives. One public performance was truly spectacular, unequaled to this day: In celebration of the coronation of the Tsar Alexander II in September, the boat took sixteen men underwater at the Kronstadt Naval Base—four of whom made up a brass band. Their rendition of the national anthem could clearly be heard by observers on the surface.

Bauer, however, did not enjoy his days in Russia. His autocratic style was unwelcome; his lowly former rank of corporal was *de classe.* His submarine had an accident, resulting in another "ride the bubble to the surface" escape and in Bauer's retreat—this time to France. Incredible, but true: back to the *other* side, as the French were allied with the British. The French were interested, but the war ended.

Possibly inspired by Fulton, the Russians used mines in the Crimean War, 1855–56, but not very well. The Fulton connection is tenuous but worth noting: He sent a set of his French plans to Catherine the Great in 1798, but she died before they could be presented. Those plans were squirreled away in Russia at least until 1855, when they were discovered in an archive and eventually given to the New York Historical Society.

The mines were barely useful. Those which were electrically detonated tended to go off prematurely. The contact mines—those few which did explode against the target—were too small to be of much effect.

Bauer went home to Bavaria. In 1864, his world had come full circle and Germany and Denmark were back at war. Bauer proposed a submarine to be powered by a visionary—and not yet practical—internal combustion engine which could run underwater; he next worked on a kerosene-fueled engine, which also was not to be perfected before his death in 1875. He has ever since been known as the most persistent—and *almost* the most successful—of the early submarine pioneers.

Fulton had tried to develop an electrically detonated, bottom-planted mine in 1811, but was frustrated by two problems: how to protect the wires, and how to determine when an enemy ship was positioned over the mine. Samuel Colt claimed to have solved both. And where Fulton was honestly amoral—openly willing to sell to anyone who would pay—Colt was devious and he, well, cheated.

Colt (1814–1862) built and tested his first underwater explosive as a teenager in 1829. Over the next twelve years, while at the same time perfecting the repeating pistol for which he was granted a patent in 1839, and upon which his fame chiefly rests, he continued the study of underwater weapons.

We're briefly interested in Colt for three reasons. While he did not directly work with a submarine vehicle, his work was on the continuum of submarine weapon development; his dealings with the government in general and the Navy Department in particular closely match the experience of inventors be-

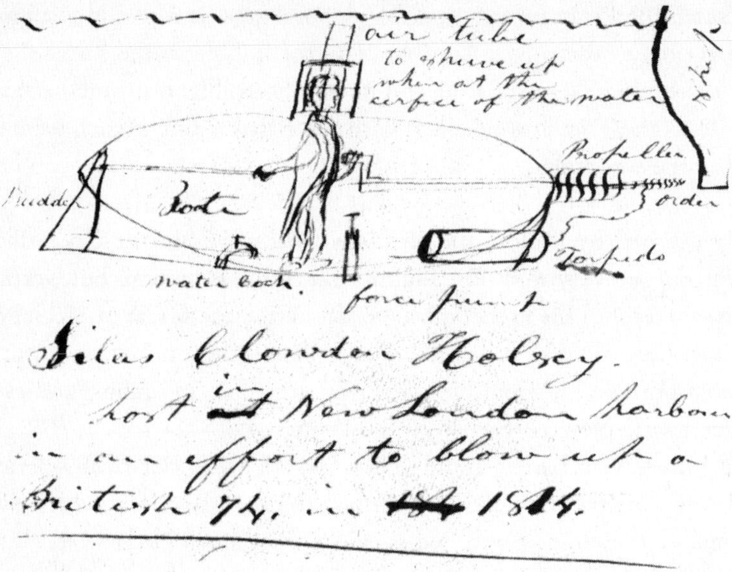

A mystery submarine of 1814—known only by this sketch, from the notebooks of Samuel Colt. Credit: Connecticut Historical Society.

fore and after; and he seems to have studied the record of every known submarine and in so doing, made the only known recorded comment on at least one.

Taking the last first: Colt's papers, held by the Connecticut Historical Society, contain fragmentary information about a submarine of "Silas Clowden Halsey. Lost in New London harbor in an effort to blow up a British 74 in 1814." Colt's sketch of the "boate" show *Turtle*-like attributes in a *Nautilus*-like hull: an air tube for use "when at the surface of the water," a "water lock, force pump," a rudder at one end and a hand-cranked propeller at the other. The weapon—labeled "torpedo"—is attached by line to an augur bit, pictured about to drill into the keel of a ship.

As for the other two reasons: By 1841, Colt had developed the philsophical if not the practical basis for a "submarine battery," a system of bottom-planted, electrically detonated mines. Colt—who by this time had already been bloodied trying to sell his revolvers to the U.S. Army—showed a bit more marketing savvy than predecessors Bushnell and Fulton. He lined up two key supporters before submitting his first proposal: Samuel L. Southard, president pro tempore of the Senate, and retired Major General William G. McNeil.

With their endorsement, Colt sent a proposal to President John Tyler. He started by noting that such luminaries as Jefferson and Pitt had been interested in underwater warfare, cited Fulton's unsuccessful efforts with mines in fixed position, and then asserted that he, Samuel Colt, had solved the problems which stymied earlier inventors. He was willing to make a bold offer: He would stage a demonstration and sink a ship, all at his own expense, provided that, if he was successful, the government would reimburse his costs and grant him an annuity.

Of course, this was much the same sort offer that Fulton had made—not once, but several times. And just as Fulton went through five French Ministers of Marine, Colt was put into the revolving door of American bureaucracy: President Tyler referred his proposal to the Secretary of the Navy—the first of six with whom he would have to deal over the next four years.

In the meantime, Colt's business dealings were about as clandestine as his proposed system: His most vocal supporter, Senator Southard, was a secret shareholder in the enterprise. With Southard's help and the connivance of the Chairman of the House Committee on Naval Affairs and the then-serving Secretary of the Navy, Colt obtained government support outside of his "at my own expense" offer and outside of the normal channel of appropriations.

He needed the money. His efforts at self-promotion had gotten a bit ahead of his efforts at invention, and for a time his soon-to-be-bankrupt revolver business, the Patent Arms Company of Paterson, New Jersey, absorbed most of Colt's energies and all of his spare cash. He also needed high-level support to fend off "the interference of any subordinate officers of Government" because, as he wrote to the next (and second) Secretary of the Navy in the string, William Upshur: ". . . the whims of old officers, checked by fears, prejudices, or motives equally adverse to improvments, too often smother inventions in the births, which if properly fostered would prove incalculably useful."

Colt stuck a deal with Secretary Upshur which continued the financial support and offered additional incentives; as he wrote to Southard a few months later;

> Mr. Upshur seemed to understand and appreciate my plans perfectly & it is the understanding between us that if I can prove to the President &

> other members of the cabinet that I can with positive certainty destroy a ship at pleasure by means of my submarine battery at a distance greater than the range of any known projectile he will recommend to Congress that immediate meens shall be appropriated to imploy my Engines for the general protection of our Coast & reward me for my secret.

Colt staged two public demonstrations in New York harbor in the summer of 1842: He exploded an electrically detonated mine and then used another to destroy a small gunboat. On August 20th, with President Tyler among the spectators, he sank a schooner in the Potomac.

John Quincy Adams, then serving in the House of Representatives, declared Colt's scheme to be "cowardly, and no fair or honest warfare," but was outvoted as the Congress rewarded Colt with an appropriation of $15,000. The money was to fund a more elaborate demonstration: Could a moving enemy ship be destroyed before coming within the range of her guns; could the mines be easily replaced; could the mines defend a harbor without hazard to friendly traffic? The $15,000 was to be taken from funds appropriated the year before—in other words, giving sanction to whatever monies Colt had already been advanced.

Now, where Fulton may have been overwhelmingly self-confident and "impudent," Colt was simply arrogant. With the appropriation in his pocket, so to speak, he sent a note to the chief clerk of the Navy Department suggesting that the Navy now stay out of his way:

> If I am permitted to conduct my experiments in my own way without being bothered by Navy Yard regulations, I will guarantee to accomplish all that is required of me by Congress for less expence & in less time than can otherwise be expected . . . The cost of *hack hire* (to say nothing of the personal annoyance & wast of time) was far greater than the advantages I derived from the Washington Navy Yard while making my late exhibition on the Potomac.

In the meantime, Upshur followed in the wake of Forfait and was promoted to Secretary of State. Four Secretaries of the Navy later, Colt was ready. On April 13, 1844, Congress adjourned and gathered on the bank of the Anacostia River to watch Colt sink a five-hundred-ton ship, under tow at five knots. The ship passed over a planted mine, the firing circuit was closed, the mine exploded, the ship sank.

Perhaps in reaction more to Colt's imperious attitude than to the merits of the case, several members of Congress joined Adams in trying to block any payments. They pushed through a resolution which implied that Colt was not entitled to compensation as he was offering nothing "new"—that his work was based on knowledge already commonly held—and challenged him to prove otherwise.

In a letter of April 22, 1844, Colt asked supporter McNeil to "come on here immediately & help me in my efforts . . ."

He continued;

> Movements are making to kill me of[f] without ceremony. A resolution was offered the other day in the House of Representatives I presume at the instance of some officer of the Army hostile to my new mode of fortification, calling on the Secretarys of War & of the Navy for information as to the plans of my invention the claims which I have in any to originality &c. &c. The Navy department I think will treat the subject fairly but the Sec'y of War has refered the resolution to the Ordnance Department & their Engineer beauroughs, people of all others the least calculated to give a just report in a matter directly hostile to their own profession. Col. Tolcott & infact nearly every officer of the ordnance department has been hostile to every invention I ever made & I can't hope for any other result in my present case.

The Congress had called on the secretaries "to communicate the fact, whether the combustible agent used by Mr. Colt was a secret before he made the same known at the seat of government, and whether the mode of application to harbor defense be new; and, if new, what objections are there to its adoption, if objections do exist."

Colt refused to give any details of his scheme to the War Department, on the apparent assumption that if he didn't provide any details, they could not judge his proposal. His confidence was bolstered by the fact that he had never claimed that the "combustible agent" (gunpowder) was exclusive to his scheme, or that he was the first person to suggest the use of underwater mines in harbor defense. He was offering a "system," which incorporated those elements with his own, secret methods.

However, in the absence of any information from the inventor, the War Department assumed that he *was* claiming originality on those two points, and in refutation, laid out (in admirable detail) the long public history of both. The Secretary of War, in his favorable endorsment of the report, asserted that:

> If the means assumed to be those employed by Mr. Colt (and if we are in error, he has his own caution to blame) are actually those which he uses, then I affirm any intelligent, scientific person, aided and encouraged by equal munificence and apprioriations, could, without invalidating any patent or exclusive right of other, and by merely applying means which have been gratuitously contributed to science by distinguished men of our country, accomplish all that Mr. Colt has achieved under the bounty and generous encouragment of his government, in his peaceful experiments against a defenceless and untenanted ship.

Colt protested. In a letter to the Chairman of the House Committee on Naval Affairs, he agreed that electrical detonation and the use of mines in harbor defense were not original; however, he affirmed, the key to his system was the secret method by which he could determine when an enemy ship had arrived over a mine. He refused to be just another engineer "hoist on his own petar" (in this case, his secrecy), and to prove that he indeed had claim to originality, accepted a recommendation from the House Committee on Naval Affairs that he submit an application for a patent.

In doing so, he described his whole system in great detail, but—clever Mr. Colt—left the application at the Patent Office only long enough for the issue of originality to be determined. He then withdrew the application before publication, while the Commissioner of Patents advised the Congress that the system was original and that a patent could be issued.

The "secret"—now it can be told—was a pair of hidden onshore observers using basic triangulation, or one observer using hidden mirrors, or both at the same time, to fix the moment when an enemy ship passed over the planted mines. The other problem which had thwarted Fulton—maintaining the integrity of the electrical circuit—had been solved by better insulation in the 1830s and first put to practical use by Samuel F.B. Morse in underwater telegraphy.

The Naval Affairs Committee chewed on the Patent Office findings for about six months, then decided that the Army's negative report had been based on insufficient evidence. They announced that the plan was "entitled to the favorable consideration of the government," but went no further than recommending the payment of all outstanding obligations and "compensation to Mr. Colt for the time he was engaged in his experiments."

In the meantime, Colt had barged ahead for three years, spending more of the government's money than the government had provided for him. He ignored a Navy request in May 1844 for itemization, and instead submitted a copy of Fulton's 1804 contract with the British, asserting that his inventions were superior to those of Fulton—as if this would excuse misuse of government funds. When the day of reckoning came, the $15,000 account was overdrawn by $9,914.28. The then-current Secretary of the Navy disclaimed responsibility—none of this had happened on his watch. Congress adjourned without taking any further action.

By 1847, Colt was back in the firearms business with an order from the U.S. government for one thousand revolvers. In 1848, he established Colt's Patent Fire-Arms Manufacturing Company, in Hartford, Connecticut.

Lacking any champion in the government—his original supporters having died or left government service—and lacking any proximate threats of invasion from abroad, Colt's system for harbor defense was never brought to Federal harvest, but simply left to wither on the vine. At his death in 1862, he was still trying to collect some government reward for his submarine battery. Ironically, whatever other seeds he may have planted were to generate into luxurious growth—in the Confederate States of America.

7 1861–1865—Of War, Civil and Uncivil

The whole international code is founded on reciprocity. The rules it prescribes are observed by one nation, in confidence that they will be so by others. Where, then, the established usages of war are violated by an enemy, and there are no other means of restraining his excess, retaliation may justly be resorted to by the suffering nation.

—*Henry Wheaton,* Elements of International Law *(1848)*

Throughout history, the development of submarine weapons has rarely moved forward unless "encouraged" by the advent of war—real or presumed imminent. With the exception of Colt's unproductive musings of the 1840s, Bauer's inconclusive efforts in the German-Danish conflict of 1850, and the ineffective employment of Russian mines in the Crimean War (1855–56), nothing much happened in underwater warfare between 1815 and 1861. Then, the American Civil War propelled arms development with an explosive force, and the submarine boat—and the weapons which it had been created to deliver—drew first blood.

For all intents and purposes, the Civil War began with the surrender of Fort Sumter, in Charleston harbor, on April 13, 1861, followed two days later by President Lincoln's call for the first 75,000 volunteers. Whatever the reasons for the conflict, and whatever the arguments for moral or cultural superiority on either side, there was an uncontested naval reality: The Federal government retained the maritime assets of the United States, along with the industrial capacity to maintain and renew them.

Thus, at an immediate disadvantage, the Confederate government resorted to a time-honored stratagem of weak navies against large—as practiced to great effect by the United States in the wars against Great Britain in 1776 and especially in 1812. On April 17, 1861, Confederate President Jefferson Davis invited ordinary citizens to become "privateers," operating privately armed warships under the authority of a government-issued Letter of Marque and Reprisal.

This form of auxiliary warfare had its origins in the Middle Ages, when rulers licensed bands of civilian warriors to cross the frontier—the "march," from Old English *mearc*—to deliver reprisals for an injury. The concept was extended to the high seas in the 14th century. A letter of marque gave a civilian crew aboard a privately armed ship the authorization to stop, search, and—if appropriate—seize merchant vessels and their cargoes. They did so in the expectation of a share of the profits from the subsequent sale of ships and cargo, to be held under the jurisdiction of a formal, legally constituted Prize Court.

In the most recent exercise of this concept, between 1812 and 1815, twenty-two U.S. Navy ships took 165 British prizes, while 526 privateers took 1,344 prizes. Pound for pound, the professionals were clearly more effective—there just weren't enough of them.

The applicant for a Confederate Letter of Marque had to prove ownership of a suitable ship, prove that a trained crew was ready for service, and put up a minimum bond of $5,000 as personal guarantee to conform, in letter and spirit, to law and custom. This latter requirement was the ingredient which always had set privateers apart from un-regulated bucanneers. Pirates answered to no one; privateers were, in theory, under surveillance by and answerable to their government, which in turn—in theory—was liable to the world community for the misdeeds of its agents.

Under the generally accepted international laws of 1861, a *Naval* vessel was authorized to stop, search, and seize merchant ships upon the high seas; to put a "prize crew" aboard the capture sufficient to ensure that she would reach a port under jurisdiction of the Prize Court; or, if necessary, to sink the capture and be prepared later to demonstrate, in court, the probable value of the ship. In such case, the captor had the moral and legal obligation to ensure that the crew, passengers, and ship's papers were put first put in safekeeping—taken aboard the captor, or put in lifeboats and provided with sufficient food, water, and navigational equipment to ensure safe passage to the nearest logical port.

In April 1861 there *should* have been a problem in allowing civilians to do the same: Privateering had been outlawed in 1856 by the Treaty of Paris. However, the Federal government had already resolved the issue in favor of the Confederates: The United States had pointedly refused to sign the treaty, reserving for itself the right to once again resort to the use of privateers, should conditions ever require. Thus, since neither the Federal government nor, or course, the Confederacy had been party to the agreement, it was not binding on either.

Response to the call for privateers was immediate, but not what the Confederate government expected. On April 19th, acting with remarkable speed but with an astonishing combination of hypocrisy and innocence, the Federal government condemned the use of privateers, threatened to hang as pirates any captured—and declared a blockade of Southern ports.

The hypocrisy is obvious; the innocence merits explanation. In international law, a blockade is an act of war which can only be directed against a belligerent *nation.* Incredibly—and perhaps unwittingly—the Union had given legal recognition to the Confederate States of America at the very time that Secretary of State William Seward was suggesting to the nations of the world that the Confederates represented only a "discontented faction."

In most wars—including the major conflicts of the 20th century yet to be examined—the historic impact of "blockade" is often lost in the temporal fascination with "battle." It can be argued—and not much argued against—that the Federal naval blockade, by severely reducing the importation of military and civilian supplies along the 3,500-mile coastline of the Confederacy, had more effect on the course of the Civil War than any battle, or all battles combined.

But as for the status of privateers in 1861: "I hereby proclaim and declare," said President Lincoln on April 19th, "that if any person under the pretended authority of the said States, or under any other pretence, shall molest a vessel of the United States, or the person or cargo on board of her, such person will be held amenable to the laws of the United States for the prevention and punishment of piracy." The referenced statute was the 9th Section of the Act of Congress of 1790; the punishment was death. Within two weeks, in an attempt to strengthen its hand, the Federal government tried to re-open negotiations on the Treaty of Paris. Secretary of State Seward offered unconditional acceptance of the provisions previously rejected. The offer was rejected.

As for would-be privateers, the Confederates received hundreds of inquiries—primarily from New Orleans and Charleston—and over the following year, issued Letters of Marque to fifty-four applicants. Of those, only about half actually put a ship into service. Of *those,* a bare handful saw any gain—and that marginal, at best—before the Union blockade effectively closed down the ports to all traffic. Where the privateers dreamed of intercepting Yankee gold from California, the few ships actually captured—about sixty, all told—typically had cargoes of lime, salt, and ice.

The Federal threat was put to the test in June 1861, when the crew of the captured privateer *Savannah* was taken to New York and put on trial for piracy. The matter was complicated by Confederate counterthreats of reprisal, to execute Union prisoners of war, selected by lot, one for one for any executed *Savannah* crew member. A standard text of the day, Henry Wheaton's *Elements of International Law* (1848), gave the pretext: "The whole international code is founded on reciprocity. The rules it prescribes are observed by one nation, in confidence that they will be so by others. Where, then, the established usages of war are violated by an enemy, and there are no other means of restraining his excess, retaliation may justly be resorted to by the suffering nation."

The threats were not idle; candidates for execution were selected and identified to the Federal government. However, as it happened, the *Savannah* trial jury could not arrive at a verdict; the crew was kept in jail for a time as prisoners of war, and was exchanged in February 1862.

On March 31, 1862, despite the blockade (or more to the point, perhaps because of the blockade) a consortium headed by Horace L. Hunley, a sugar broker from Mobile, Alabama, was granted a Letter of Marque—for a submarine.

Construction began in the fall of 1861 at a New Orleans machine shop owned by James McClintock—who was primarily responsible for the design—and was completed in a local shipyard. *Pioneer* was about twenty feet long, six feet wide, and built of iron plates cut from a boiler. There was to be a crew of three: one to steer and operate the diving planes mounted at the bow, and two to turn the hand-cranks which spun a small propeller at the stern.

Once in the water, the boat worked about as expected, although nothing could be done to induce the magnetic compass to operate inside the iron hull. The helmsman adopted a porpoising technique: Start out headed in the right direction, submerge, run ahead a short distance, come to the surface for some air and a quick look, submerge again. The method of attack: Dive under the target (taking aim on the shadow in the water), towing an explosive charge on a long tether. The weapon would detonate on contact with the enemy hull, while the submarine was safely clear on the far side.

In March 1862, *Pioneer* was ready for a full-scale demonstration, presumably to qualify for the Letter of Marque; at the same time, faced with the obvious fact that a submarine with a crew of three would be unable to operate under prize rules, the Confederate Congress authorized a bounty equal to one half the value of any Union warship sunk—an extension, if you will, of the Torpedo Act of 1813.

For the demonstration, *Pioneer* sank a barge moored in Lake Ponchartrain. That was to be the nearest that *Pioneer* would come to combat. New Orleans was captured by units of the Union Navy on April 26, 1862, and the consortium scuttled *Pioneer* to keep her from falling into enemy hands. A boat which may be *Pioneer* was discovered and raised in 1878, and is today on display at the Louisiana State Museum in New Orleans.

Hunley, McClintock, and friends, encouraged by their progress, shifted their base to Mobile and built another, slightly larger boiler-plate submarine. They made some improvements: more glass deadlights for light and visibility; a barometer for use as a depth gauge; two hatches instead of one, for better ventilation.

However, ventilation remained a critical shortcoming. After only a few minutes of submerged operation, the air became foul, with everything inside damp from exhaled breath and perspiration. The crew submerged with a burning candle. When the lack of oxygen had diminished the flame to a pinpoint, they knew it was time to come up for air. Somewhat ahead of his time—certainly ahead of the technology which would make the effort successful—McClintock invested time and money trying to develop a battery-powered electric power plant. He finally gave up, and returned to the hand-cranks.

The second *Pioneer* fell victim to rough weather in Mobile Bay; while it was being towed out toward the Union blockade, spray and chop began splashing through hatches left open for air. Despite frantic efforts at bailing, the boat quickly settled and soon sank. The crew was rescued. The consortium was undaunted, and in the summer of 1862 raised $15,000 to build yet another, larger submarine which came to be called *H.L. Hunley*—although that name probably was not used until more than a year later.

In the meantime, one Confederate naval officer was pushing another weapon of the weak navy against the strong: the mine. Commander Matthew Fontaine Maury had established an international reputation as the "pathfinder of the seas" during a thirty-six year-career in the United States Navy, when the circumstances of the spring of 1861 led him to take up arms with the Confederacy. There is little doubt that he was acquainted with the work of Colt—and perhaps with Colt himself—as a result of the activities at the Washington Navy Yard. Maury had direct experience with underwater explosives from his survey work for the laying of the first trans-Atlantic cable in the 1850s: He experimented with the use of underwater explosions in taking depth and topographic soundings of the ocean floor. Thus, he came to the Civil War with a background in an art unknown to most, and immediately set himself to developing an underwater-warfare program for the Rebels.

His first effort, on July 7, 1861, was with a pair of drifting mines attached by a line. It suffered the same fate as most of the similar efforts of Bushnell, Fulton, and Elijah Mix: The mines did not explode. Undeterred, Maury began a series of experiments, and by October 1862 had obtained sufficient funds and assistance to create a "Submarine Battery Service" within the Navy.

That he enjoyed considerable politicial success, when his philosophical predecessors in France, England, and the United States had run aground on traditional Naval thinking, can be attributed more to the weakness of the Confederate government than to any vision or powers of argument which might be credited to Maury. Like the Navy, the government itself was created out of nothing. A few experienced bureaucrats brought skills (and prejudices) over from the "old" government, but the supporting staffs were literally recruited from the streets. There was little institutional bias because there was no institution.

Maury's work was not fully embraced, however. Hunter Davidson, who followed Maury as head of the Submarine Battery Service, later noted that ". . . as late as the summer of 1863, some of the ablest men of the day did not regard torpedo warfare as worthy of consideration, and the very attempts of Fulton, and of Bushnell, and of the Russians, were used by those men in argument that my attempt would be useless."

There also was a strong moral undercurrent, in both the Confederacy and the Union, against any sort of clandestine warfare. This was an issue frequently raised, but not confronted, since the days of Gianibelli and especially during the various projects of Fulton and Colt. Professional warriors were cast in the chivalric mode, and the appearance (if not the substance) of war in the Middle Ages had carried down to the present day. Warriors went into battle fully identified, even down to the individual whose armor was emblazoned with his personal crest. Each knew who he was killing, or by whom he was being killed. One of the earliest authorities on international law, Alberico Gentili, suggested in 1598 that anything which "deprived men of the opportunity to show courage" was illegal.

Later authorities were not so narrow: Cornelius Bynkershock said in 1737: ". . . we may destroy an enemy though he be unarmed, and for this purpose we may employ poison, an assassin, or incendiary bombs, though he is not provided with such things: in short everything is legitimate against an enemy." To Christian Wolff, circa 1740, it made little difference *how* an enemy was killed, "since forsooth in either case he is removed from our midst." True sentiments, then and now, but rarely accepted—then and now.

Whatever opposition the Confederate government mustered against mines, whether on practical or moral grounds, was not strong enough to matter at the beginning, and dissipated like fog under a warm sun once success was demonstrated.

In the meantime, the Federal government had entered into an agreement for the construction and purchase of a submarine of its own—as a commissioned ship of the U.S. Navy. Virtually unknown today—because of a virtually undistinguished and short-lived career—*Alligator* was forty-six feet long, originally propelled by eight sets of self-feathering oars with hinged blades (open on the power stroke, closed on the return) and manned by a crew of sixteen oarsmen and a captain. The boat was equipped with a breathing pipe and two somewhat mysterious air-purification systems.

De Villeroi's 1861 submarine boat—operated, briefly, by the U.S. Navy as *Alligator*—as drawn by Lieutenant Barber.

This first submarine owned by the U.S. Navy was designed by a Frenchman and built under government contract by a private shipyard in Philadelphia. Designer Brutus de Villeroi remains something of an enigma: He may have demonstrated a submarine as early as 1832, in France, and apparently tried to sell a submarine to the French Navy during the Crimean War. Rebuffed, he moved to Philadelphia in 1859 and built a thirty-three-foot boat, financed by Stephen Girard, with the probable intention of salvaging the treasure from the British warship *De Braak,* lost near the mouth of the Delaware River in 1780. This submarine reportedly made at least one three-hour dive to twenty feet.

In May 1861, perhaps seeking encouragement to apply for a patent, perhaps seeking a government contract, de Villeroi demonstrated his submarine at the Philadelphia Navy Yard. He convinced the Navy that he could produce a submarine warship from which a diver could place an explosive charge under an enemy ship. He also offered a system for purifying and renewing the air within. The Navy negotiated a price for a larger boat—$14,000—as an engineering prototype. Construction began in November; *Alligator* was launched in April 1862.

About the same time, de Villeroi dissappeared—reason unknown—and with him went the "secret" of his uncompleted air-purification system. But there was the boat, more or less finished and floating in the Delaware River. It was accepted into service June 13, 1862, and immediately towed down to Hampton Roads.

Assistant Secretary of the Navy Gustavas Fox offered a promotion to Lieutenant Commander Thomas O. Selfridge if he would take command of *Alligator* and destroy the Confederate ironclad *Virginia II,* then in protected waters up the James River. Selfridge assembled a crew of eager volunteers, but during trials in August the air quickly grew foul, the crew panicked, and all tried to get out of the same hatch at the same time. Selfridge bowed out of the project.

He may have wished he'd stayed on with *Alligator.* The next ship under his command was the armored gunboat *Cairo,* which became the first major warship ever sunk by a mine—two mines, actually, floating in the Yazoo River and triggered from a concealed riverbank bunker on December 12, 1862.

The mine field was spotted the day before when one mine prematurely exploded in full view of an approaching Federal squadron. The next morning, the squadron commander sent two of the smaller ships on what may have been the first-ever American minesweeping mission—using sharpshooters to pick off the weapons and their handlers. The more valuable *Cairo* should have hung back, moving upriver only as the mine clearing proceeded, but Selfridge, responding to what he thought were sounds of an attack just around a bend in the river, rushed ahead. The "attack" turned out to be marksmen shooting at mines. *Cairo* sank within twelve minutes, without loss of life; rediscovered in 1956, she was raised and is now on display at the Vicksburg National Military Park.

Back in Hampton Roads, *Alligator* had proved to be too big to operate very far up the James, or any other local river. She needed six feet of water barely to be submerged, and another couple of feet if a diver was going to make a clandestine excursion. The propulsion system was not very effective either, so the submarine was towed to the Washington Navy Yard, where the oars were replaced by a three-foot hand-cranked propeller. President Lincoln watched a demonstration of the "improved" *Alligator* in the Potomac River. While under tow for operations off Port Royal on April 2, 1863, *Alligator* foundered in a storm and disappeared.

Construction of another Union submarine began in 1863 when a group of Northern speculators formed the American Submarine Company. Their motivation may have been similar to that of the *Pioneer* consortium; just about this time the U.S. Congress, adding hypocrisy to hypocrisy, authorized the President to issue Letters of Marque (he declined to do so).

The "Intelligent Whale" went through a period of disputed ownership, and was not completed until 1866; by that time, O.S. Halstead of Newark, New Jersey, was in control and offered to sell the boat to the U.S. Government. The Army Corps of Engineers conducted a trial in the Passaic (New Jersey) River which included submerging with a full crew of thirteen and using the boat as a diving bell. The government was interested; the asking price was $50,000. Three years later the Navy agreed to pay $25,000, with a 50-percent down-payment and the balance due on completion of an operational evaluation. Halstead also was required to provide full construction and operational details, including—shades of *Alligator*—a presumed air-purification system. Formal acceptance trials were finally held in September 1872, and the boat failed. Halstead was murdered, possibly by angry creditors, and the project was abandoned. The "Intelligent Whale" is today at the U.S. Navy Historical Display Center, Washington Navy Yard.

8 1862–1870—Of David and Goliath

The secrecy, rapidity of movement, control of direction, and precise explosion indicate, I think, the introduction of the torpedo element as a means of certain warfare. It can be ignored no longer.
—Admiral John Dahlgren, commanding the Union blockade of Charleston, S.C., to Assistant Secretary of the Navy Gustavas Fox

Meanwhile, back in the war . . . Charleston, South Carolina, became the focus of Confederate underwater warfare, because Charleston had become the focus, nay, the center of gravity of the Union naval blockade. The reason was partly symbolic—the war, after all, had started in Charleston Harbor—but largely because Charleston was the major seaport of the Confederacy.

In October 1862, a new weapon was brought to bear against the blockade—the steam-powered, low-freeboard warships known collectively as "Davids" (as in David vs. Goliath). Built to run awash, with just the deck, the funnel, and a small conning tower visible, these were designed to hit an enemy ship with a spar torpedo—the petard, gone to sea—or ram it directly at a speed of five knots.

The Davids were the brainchild of an Army officer, Captain Francis D. Lee. Captain Lee had trouble getting anyone to pay attention. The War Department was not interested: This was a matter for the Navy. The Navy was not interested: Lee was an Army officer. Lee finally got enthusiastic support from General P.T. Beauregard, commanding at Charleston and faced with the massive and continuing Federal blockade and bombardment.

A semisubmersible steamboat, able to approach an enemy ship unobserved until too late, seeemd to hold great promise. Beauregard wrote that the David concept "is destined ere long to change the system of naval warfare," and six months later acknowledged that he felt "as Columbus must have felt when he maintained that there was a New World in the West across the Atlantic, but could find no one to believe him or assist him in determining the fact."

The first David was cobbled together from a hulk, and lived up to no one's promise. The subsequent boats were built by the Southern Torpedo Boat Company in Charleston, assisted by Captain Lee, as a profit-making venture. They could look forward to the same government bounty which had been offered *Pioneer* for the sinking of enemy ships, but closer to home, John Fraser and Company of Charleston now offered $100,000 to anyone who could sink a mainstay of the blockade, such as *New Ironsides* or *Wabash,* and $50,000 for the sinking of any ironclad.

Back in Mobile, Hunley's group had finished their new submarine, which, not unmindful of the generous Fraser and Company prize, they now offered to General Beauregard. Beauregard accepted, not only with pleasure, but with some urgency. Calling it "Whitney's submarine boat," General Beauregard wrote to Mobile on August 7, 1863, asking that the movement of *Hunley* be expedited, adding, "It is much needed." *Hunley* arrived in Charleston a week later, aboard two covered flatcars, and a crew was assigned, commanded by Lieutenant John Payne, CSN. He was assured by the Chief of Staff that General Beauregard regarded *Hunley* as "the most formidable engine of war for the defense of Charleston now at his disposition & accordingly is anxious to have it ready for service . . ."

We know what *Hunley* looked like—there is a contemporary engraving made from a pencil sketch made at Charleston and dated December 6, 1863. The builders again started with a boiler, but rather than cut it up for raw materials, they cut it in half longitudinally and inserted a twelve-inch strip all

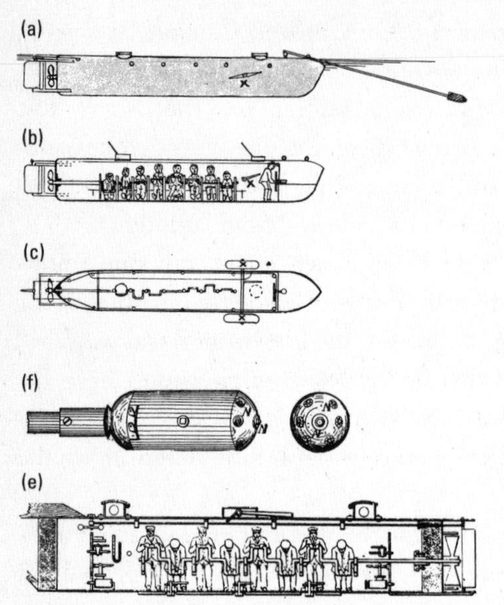

Post-war drawings of *Hunley,* based on sketches by co-designer Alexander. (a),(b),(c) and (d) are the more accurate; (e) is an "improved" version published in France that materially changed the cranking arrangement. The sketch of the warhead of the spar torpedo (f) does not show a barb or hook arrangement, which some authors speculate may have been included to allow *Hunley* to "plant" the weapon and then back off to a safe distance, using a lanyard to trigger detonation.

around to give in a more or less oval cross-section, then added tapered ends. There were ballast tanks fore and aft, filled by opening a sea cock and emptied by hand-pump. This operation required careful coordination between the men at each tank—and careful attention to fill-level, as the tanks apparently were open at the top and excess water would spill over into the machinery space. There was a heavy drop-keel which could be unbolted and released for emergency ascent. *Hunley* would carry a crew of eight (or nine—the records are inconsistent), of whom six or seven (or eight) manned the cranks which drove the propeller. On trials, *Hunley* managed a speed of four knots in calm water—sixty years had brought no improvement over Fulton. The boat had four-foot-long tubes to bring air in when running just below the surface. Underwater endurance—while cranking, and without the breathing tubes—was set at thirty minutes. As with the second *Pioneer, Hunley* sailors kept a candle burning in their dark, damp, and crowded quarters to give warning of dangerous air.

We are not so certain as to her size: various accounts put the length at twenty feet, twenty-five feet, thirty feet, thirty-five feet, and forty feet, with the longer dimensions being the more probable. Appearance aside, there are two things about *Hunley* that we know with certitude: She was the first submarine to actually sink an enemy warship in combat—and she was the first submarine to be lost in action with all hands.

Hunley's crew had planned to use the *"Pioneer"* attack—drag a ninety-pound floating torpedo astern on a two-hundred-foot line, take a course toward the target, submerge to pass underneath, and thereby pull the mine into the hull. This worked once during *Hunley* trials in Mobile Bay; however, on a second trial, the crew found that the torpedo was drifting forward faster than *Hunley* was moving, placing the submarine in jeopardy of being hit by her own weapon. They changed the tactic, and fitted *Hunley* with a spar torpedo. There is confusion in the records as to whether this had a contact detonator, or had a barb designed to catch on and hold the charge against the target while *Hunley* backed off to a safe distance, the explosion to be triggered by a lanyard.

Five days after *Hunley's* arrival, the first David went out to attack *New Ironsides.* This was not successful. The David was hard to maneuver, lost power, got caught in the tide, and ended up stuck alongside her intended victim. A Federal officer called down from the deck and struck up a conversation under the impression that the David was a new type of Union patrol boat. The Confederates finally got the engine started and escaped.

Hunley's crew had been getting ready, running drills, practicing their operations, when—only two weeks after arriving in Charleston—the boat sank at the pier.

It has been widely reported that *Hunley* sank *twice* at the pier, both times under command of Lieutenant Payne, both times by being swamped, and both times with the loss of a significant portion of her crew. The verified sinking was on August 29th, when *Hunley* was moored to the steamer *Etiwan,* which in turn was moored to the pier. *Etiwan* unexpectedly got under way, or surged in the wake of a passing steamer (another confusion), and moved away from the dock pulling *Hunley* onto her side. Water rushed into the open hatches, and the boat quickly sank. Lieutenant Payne and one other easily escaped; a third crew member was trapped when a hatch closed on his leg, and had to ride the boat to the bottom—forty-two feet—and wait until the boat was fully flooded. Then he was able to move the hatch and pull free. Five crew members drowned.

The second, un-verified, sinking at the pier has been attributed to "the wake of a passing steamer." Some sources note the sinking of a David, also under command of Lieutenant Payne and also by swamping, attributed to "the wake of a passing steamer." The *Etiwan* incident may have been caused by "the wake of a passing steamer." The facts are unknown, and probably unknowable.

Whatever the body count, Beauregard became disenchanted with this "formidable engine of war" and turned his attention back to the Davids. However, Horace Hunley was determined to prove the accident(s) to his submarine just that—accident(s). He wrote to Beauregard on September 19, 1863: "I propose that if you will place the boat in my hands to furnish a crew (in whole or in part) who are from Mobile who are well acquainted with its management & make the attempt to destroy a vessel of the enemy as early as practicable." The Chief of Staff replied for Beauregard, and ordered *Hunley* to be "cleaned and turned over" to Hunley.

On October 5th, an improved David made another attempt at *New Ironsides* with some success. One Federal officer on the deck was killed by a shotgun blast from the approaching boat, and the warship was damaged by the subsequent explosion of the spar torpedo. However, the David was almost swamped and took on enough water to put out the fires in the boiler. Half the crew abandoned ship, tried to swim for the shore, and was captured. One man who couldn't swim stayed with the David, and was rejoined by the engineer. Together, under rifle fire from *New Ironsides,* they managed to relight the fires and get under way for a safer berth in Charleston. The damage to *New Ironsides* was greater than first thought (or ever reported) and kept her out of action for a year.

Just as Bushnell's and Fulton's attempts on the British had prompted increased vigilance, so too with this attempt on the Union flagship. The attack also encouraged a threat from the Federal commander of the blockade, Admiral John Dahlgren, to hang the captured crewmen of the David "for using an engine of war not recognized by civilized nations." He added darkly, "It savors to me of murder."

But in writing to Union Secretary of the Navy Gideon Welles, Admiral Dahlgren was positively complimentary: "How far the enemy may seem encouraged, I do not know, but I think it will be well to be prepared against a considerable issue of these small craft. It is certainly the best form of the torpedo which has come to my notice. . . . The subject merits serious attention, for it will receive a greater development."

And to Assistant Secretary Fox, he added: "Among the many inventions with which I have been familiar, I have seen none which have acted so perfectly at first trial. . . . The secrecy, rapidity of movement, control of direction, and precise explosion indicate, I think, the introduction of the torpedo element as a means of certain warfare. It can be ignored no longer."

Horace Hunley arrived in Charleston with three of *Hunley*'s builders and several other experienced crew members. Two of the builders—Lieutenants George E. Dixon and W.A. Alexander—were members of the 21st Alabama Infantry Regiment, stationed in Mobile, and were more or less "on loan" to Beauregard.

Dixon was put in command, Alexander became the equivalent of the supply officer, and the new crew began to practice. On October 15th, Dixon was away on some temporary assignment, and Hunley took charge of the day's drill. An eyewitness reported: "The boat left the wharf at 9:25 a.m. and disappeared at 9:35. As soon as she sunk, air bubbles were seen to rise to the surface of the water, and from this fact it is supposed the hole in the top of the boat by which the men entered was not properly closed. It was impossible at the time to make any effort to rescue the unfortunate men, as the water was some 9 fathoms [54 feet] deep."

The boat was found three days later, bow down in the mud at a thirty-five-degree angle, and recovered after another week. The cause of this disaster was not from an open hatch: The bow ballast tank was full to overflowing, the sea cock open, the handle knocked loose and fallen into the bilges. No amount of pumping could empty the tank still open to the sea. The stern tank was empty—presumably pumped dry in a desperate attempt to float the boat free of the bottom. The bolts holding the keel weight had been loosened, but it did not drop away—possibly jammed in place by the angle of the hull. Bolts holding the hatches had also been undone, but the crew could not overcome the water pressure of about twenty-six pounds per square inch, and most certainly had never heard of Bauer's 1850 escape in an air bubble.

Beauregard canceled all further operations; Dixon and Alexander (who also had missed the fatal dive) harassed the general and his staff often enough that, after "many refusals and much discussion," the general agreed to one more attempt—but not as a submarine. *Hunley* was to attack, David-style, running awash. It was at this time that the boat was formally christened CSS *H.L. Hunley* in memorial to her deceased namesake.

Dixon and Alexander assembled a new crew of volunteers, possibly overcoming any reluctance by reminding them of the Fraser and Company offer, still valid. They held a series of successful drills—including one underwater endurance test which lasted two hours and thirty-five minutes.

They spent the next four months trying to get *Hunley* in position to attack. The Union fleet now kept station as much as seven miles away from *Hunley*'s base at Sullivan's Island, and the hand-powered boat barely managed to survive the varying conditions of tide, wind, and sea during various attempts. Alexander was recalled to Mobile before the final effort; Dixon stayed on as commanding officer.

Hunley finally made an attack on the steam sloop *Housatonic*, on the evening of February 17, 1864. The officer of the deck spotted *Hunley*, but did not immediately react to what he later said looked "like a plank moving along the water." By the time the alarm was given, it was too late to do much more than beat all hands to quarters. Some reports suggest that *Housatonic* was able to slip the anchor cable and back engines, to no avail; *Hunley* rammed her on the starboard side, just forward of the mizzenmast.

The blast of the spar torpedo may have triggered a larger explosion in a magazine, or merely punched a hole in her side; it mattered not which, as *Housatonic* settled to the shallow bottom in four minutes with the loss of five men. Survivors climbed into the rigging as the deck went under. However, *Hunley* followed her own sad tradition and disappeared.

A report of the attack in the *Charleston Daily Courier*, February 29th, suggested that the sinking of *Housatonic* had passed for some time unnoticed: "The explosion made no noise, and the affair was not known among the fleet until daybreak, when the crew were discovered and released from their uneasy positions in the rigging. They had remained there all night."

That much was not accurate; two boatloads of survivors had reached a nearby ship within a short time; flares had been fired, and ships that come over to assist. The balance of the account was correct: "Two officers and three men were reported missing and were supposed to be drowned. The loss of *Housatonic* caused great consternation in the fleet. All the wooden vessels are ordered to keep up steam and to go out to sea every night, not being allowed to anchor inside. The picket boats have been doubled and the force in each boat increased."

* * *

The fate of *Hunley* remained unknown, by either side; was she swamped by the force of the explosion, or trapped by the sinking *Housatonic,* or was she hiding? The Federals thought she might have escaped and returned home and was being kept out of sight; the Confederates, increasingly concerned as hours turned into days without word, hoped that the crew had been captured and was being held in some secret location.

A diver clearing wrecks some years after the war reported spotting what he thought was *Hunley* at some distance from *Housatonic,* but there was no verification. The issue was finally resolved during the summer of 1995, when *Hunley* was located by an expedition financed by Clive Cussler (author of, among other books, *Raise the Titanic*).

Cussler's fictional organization, "National Underwater Maritime Association" (NUMA), was given life and a charter to seek out and identify sunken historic ships beginning with an unsuccessful 1978 search off the English coast for the Revolutionary War *Bon Homme Richard.* After several attempts to find *Hunley*—the first in 1980—the submarine was found about a thousand yards southeast of the wreckage of *Housatonic.* Cussler speculates that *Hunley,* hatches open to revitalize an exhausted crew, was swamped by the wake of a passing steamer—in this case, the Federal ship *Canandaigua,* which was rushing past that spot to assist *Housatonic.*

The bare record would suggest that *Hunley* was a desperate failure, doomed to kill a crew on almost every voyage. As a Confederate engineer later noted, *Hunley* "would sink at a moment's notice and at times without it."

Hunley sank three times for certain—twice because of inexperienced or inattentive seamanship. If she indeed sank a second time at the pier in Charleston, it was for one of the same reasons. She may have sunk one more time—for a possible total of five—but here the record is even more ambiguous: A Confederate deserter later said that "the submarine boat" sank during trials in Mobile. He may have been referring to the second *Pioneer,* not *Hunley.*

A desperate failure? The full record shows a submarine which made repeated successful dives, and within the technical limitations of the day, had an impact on the disposition of an entire fleet—and in the precedent, on the entire maritime world.

* * *

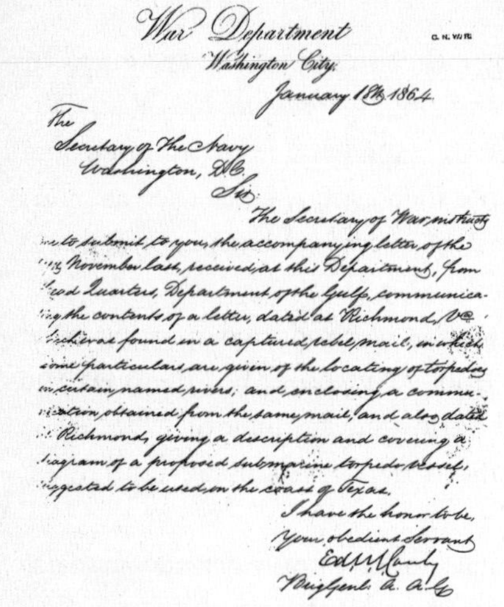

War Department
Washington City,
January 18th 1864.

The
Secretary of the Navy
Washington, D.C.
Sir:
The Secretary of War, instructs
me to submit, to you, the accompanying letter, of the
… November last, received, at this Department, from
… Quarters, Department of the Gulf, communica-
… the contents of a letter, dated at Richmond, Va.
… was found in a captured, rebel, mail, in which
some particulars, are given of the locating of torpedoes
in certain named rivers; and, enclosing, a commu-
… obtained from the same mail, and also dated
… Richmond, giving a description and covering a
… diagram of a proposed submarine torpedo vessel,
… suggested to be used, on the coast of Texas.
I have the honor to be,
Your obedient Servant
Ed. W. Canby
Brig. Genl. A. A. G.

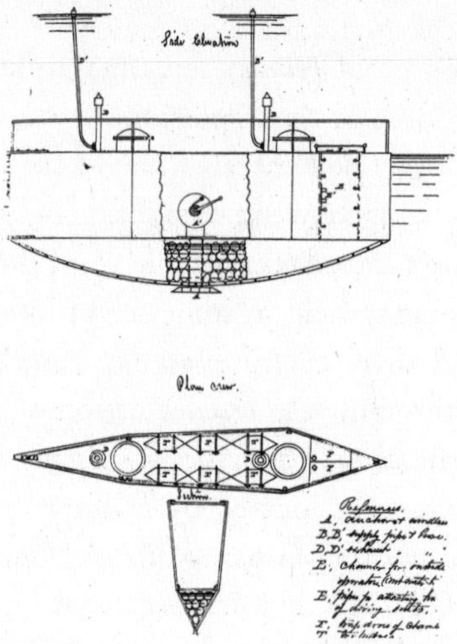

Confederate mail intercepted in 1863 included plans for a "proposed submarine torpedo vessel, suggested to be used on the coast of Texas." No other details are available.

The war moved on and brought other efforts, but no advancements in the art. Federal Lieutenant W.B. Cushing mounted a spar torpedo on a steam launch and rammed and sank the Confederate ram *Albemarle* at Plymouth, North Carolina, October 28, 1864. His boat was swamped by the rush of water, and most of his crew of thirteen were drowned or made prisoners, but he escaped, later to share in a prize award of almost $80,000.

A Confederate David-type boat, *St. Patrick,* attacked *Octorara* off Mobile Bay (January 26, 1865), but the spar torpedo failed to explode. A Union sailor tried get a rope around *St. Patrick*'s funnel and hold her captive, but he was beaten off by small-arms fire. Commanding Officer Lieutenant John T. Walker managed to get *St. Patrick* back to Mobile.

If nothing else, these incidents underscored the need for a weapon which could be delivered from a distance.

By war's end, mines and torpedoes in their various shapes and systems had sunk twenty-nine Union ships and seriously damaged another fourteen—the overwhelming proportion of the total destruction wrought by the entire Confederate Navy. Like seeds dislodged from a pod and carried by the winds—or like hornets scattering from a destroyed nest—veterans of the war, survivors of the war, students of the war took that message of the war—

the promise of underwater weapons—to parts near and far. Most of the "message" centered on mines and torpedoes because these were the destructive elements; the methods by which they were delivered to an enemy were—at that time—yet to be properly developed. Among other efforts: Matthew Fontaine Maury established a for-profit underwater warfare school in London, with students from France, Sweden, Norway, Russia, and Holland. Hunter Davidson became a "torpedo consultant" to Venezuela and Argentina. Major W.R. King, USA, published *Torpedoes: Their Invention and Use, from the First Application to the Art of War to the Present Time.* Problems with the weapon thus far could be attributed, he wrote, to the "slight attention which has been given to the subject by those who are competent to discover and correct them."

Attention now was being paid. The U.S. Army and Navy each established a center for underwater warfare in 1869.

Writing in 1866, Commander Henry A. Wise, USN, chief of the Bureau of Ordnance, addressed the moral issue raised at the beginning of the war: "There seems to be, no doubt, a want of fair play and manly courage in using a concealed and destructive weapon against an enemy, while you are secure from danger and quietly waiting for the castastrophe which launches hundreds of human beings into eternity without warning; but it is nevertheless perfectly legitimate warfare, and the danger must be met, and can only be overcome by caution, skill, and the free use of similar means." His position was affirmed in the postwar edition of Henry Wheaton's *Elements of International Law:* "Concealed modes of extensive destruction are allowed, as torpedoes to blow up ships, or strewed over the ground before an advancing foe, and mines; nor is the destructiveness of a weapon any objection to its use. . . ."

While so many were focusing attention on the torpedoes of the last war, an Austrian officer had begun creation of the torpedoes for the next and all future wars. In 1867, Captain Giovanni Luppis demonstrated a self-propelled bomb for the Imperial Austrian Navy—a small steamboat with an explosive charge in the bow. This bomb was steered by wires attached to a rudder bar and extending back to the launching site, rather like controls for a horizontal marionette. The warhead was ignited when a connecting rod from the nose pushed the trigger of a mounted pistol aimed at the explosive. The weapon was not very satisfactory, but the concept showed promise. Luppis teamed with English engineer Robert Whitehead, who helped turn the cumbersome boat into the cylindrical self-guided "automobile torpedo" with which his name has ever since been associated.

Whitehead first developed an engine powered by compressed air—this gave the torpedo a range of several hundred yards; he later shifted to a steam turbine fueled by alcohol. Also, since an explosion at the surface would largely be wasted (most of the force would take the course of least resistance, into the air), Whithead created a depth-control system which allowed the weapon to run at a depth of six or ten or twenty feet—however set—and thus use the water to hold much of the explosive force against the hull. He replaced Captain Luppis's pistol with a detonator designed for the purpose—but the name "pistol" stuck forevermore. He constructed a launching tube (to give directional stability).

The U.S. Navy began manufacturing Whitehead torpedos at Newport, Rhode Island, in 1869. This first U.S. model was a Mark II—14.5 feet in length, fourteen inches in diameter, with a dual range: one thousand yards at thirty knots, 1,500 yards at twenty-four knots.

The Whitehead torpedo soon was being installed on all classes of warship worldwide. By 1877 it had created its own class, the torpedo boat, which led, naturally enough, to the somewhat larger "torpedo boat destroyer" (1885) and eventually to an even larger ship, the "destroyer of torpedo boat destroyers," which became known simply as the "destroyer."

The first sinking credited to a torpedo was a Turkish customs boat in 1878, victim of an attack by two Russian torpedo boats. Chilean gunboats sank a renegade battleship in 1891. The Japanese had torpedo boats in action against the Chinese in 1895, and in the Russo-Japanese War of 1904–1905, Japanese torpedos damaged or sank five battleships and four heavy cruisers.

For the trivia file: The first submarine attack on a warship on the high seas took place on December 9, 1912, when Lieutenant Commander Paparigopoulos of the Royal Hellenic Navy, aboard a French-built submarine, fired a torpedo at the Turkish cruiser *Medjidieh.* He missed.

But to end this brief survey: The most influential of the Civil War-spawned progeny was a work of fiction, Jules Verne's 1870 novel *Twenty Thousand Leagues Under the Sea.*

The tale opens in 1866, with the USS *Abraham Lincoln* searching for a mysterious sea monster. The monster turns out to be a submarine boat, not of fantasy but one well crafted by Verne (1828–1905). He had read of Bourne, Drebbel, Bushnell; studied the work of Fulton; read Maury's *Physical Geography of the Sea;* followed newspaper accounts of the war.

Verne's *Nautilus*—so named in homage to Fulton—was a thoroughly practical submarine. The mad Captain Nemo described his boat to the prisoner-narrator of the story: "It's in the shape of a very elongated cylinder with conical ends—rather like a cigar." The hull was 230 feet long, twenty-six feet wide, and had a surface displacement of 1,492.13 tons; the weight-to-volume ratio was calculated so that "when in a state of equilibrium in the water" 10 percent of the hull would remain above the surface. The boat sank and rose by filling and emptying "supplementary reservoirs" equipped with pumps strong enough to overcome outside pressure. It had a carefully calculated reserve buoyancy (upon which Verne—through Captain Nemo—elaborated in exact and accurate scientific detail, even down to the factor for determining the slight compressiblity of seawater: .0000436 for each thirty-two feet of depth). *Nautilus* had self-perpetuating electrical propulsion (the necessary chemical elements were extracted from seawater). Hydroplanes, set at the center of gravity on each side of the hull, allowed for vertical maneuvering. As long as the propeller was driving the boat forward, it would climb or dive "according to the degree of this angle and the speed at which it is travelling." To rise quickly, the captain told his prisoner, "I merely disengage the propeller" and *Nautilus* will "rise vertically like a hydrogen balloon going straight up into the air."

With the exception of the "self-perpetuating electrical power," *Twenty Thousand Leagues* can be read as an instruction manual on submarine design. It can also be read as a political document, a pleading, under color of Captain Nemo's bold assertion:

> The sea does not belong to tyrants. On its surface, they can still exercise their iniquitous rights, fighting, destroying one another and indulging in their other earthly horrors. But thirty feet below its surface their power ceases, their influence dies out and their domination disappears!

Of course, this is but an echo of Lord Byron:

> Roll on, thou deep and dark blue ocean—roll!
> Ten thousand fleets sweep over thee in vain;

Man marks the earth with ruin—his control
Stops with the shore;

But within Verne's lifetime, saner men than Captain Nemo would turn the fantasy into practical reality; soon, other megalomaniacs would turn Captain Nemo's pleading upside down. The poet and the novelist were both wrong. Tyrants indeed could exercise power—great power—beneath the waves.

9 1870–1886—Of Irish-American Schoolteachers and Success

I decided to blow out the ballast and come up. . . . The green blur on the ports in the conning tower grew lighter as I gazed through them until suddenly the light of full day burst through, almost dazzling me. . . . a cheer burst from the crowd of observers on the dock, among whom opinion was equally divided as to whether we would emerge alive from our boat or not.

—*John P. Holland, first submergence of* Fenian Ram, *1881*

The man who, more than any other, created the modern submarine was John Phillip Holland, born in Ireland, probably in 1841. He entered the religious order of the Irish Christian Brothers in 1858, remaining, largely as a teacher, until 1873, when he left the order (not having taken his final vows) and emigrated to the United States to join other members of his family. He became a lay teacher for the order in 1874 in Paterson, New Jersey.

Influenced by reports from the American Civil War, Holland may have been interested in submarines as early as 1863, and by 1870—still in Ireland and perhaps having come across Jules Verne's just-published novel—he had

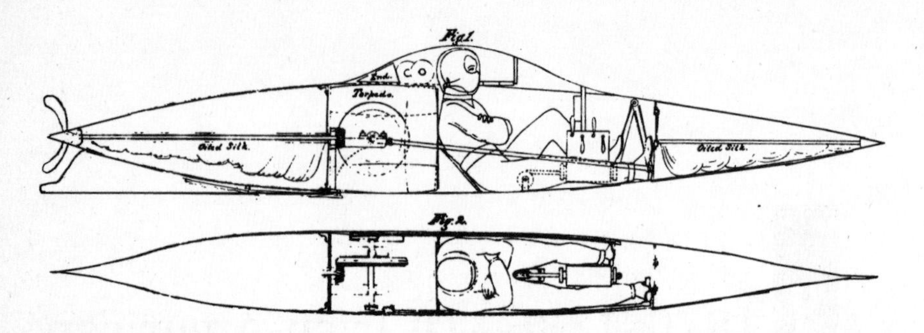

Lieutenant Barber reproduced—without the inventor's knowledge or permission—the first design Holland submitted to the Secretary of the Navy. A foot-treadle operated not only the propeller, but also a pump to empty both the one-cubic-foot ballast tank and to discharge "used" air. This latter feature was part of an elaborate scheme incorporating a helmeted operator and fore-and-aft air chambers (subdivided by the "oiled silk" noted on the drawing).

sketched out his first ideas. Soon after he had settled down in Paterson, he offered his services to Secretary of the Navy George M. Robeson, along with a "design" for a 15.5-foot one-man, man-powered submarine boat.

The Secretary of the Navy did what most Secretaries of the Navy do with enthusiastic entrepreneurs; he handed him off to the staff. In this instance, Holland's proposal was sent to Captain Edward Simpson, the commanding officer of the Naval Torpedo Station at Newport. He rejected Holland's concept out of hand: There was no way for the operator to see where he was going once underwater. Holland responded with a bit of laconic wit (perhaps unwittingly repeating Bushnell's response to similar criticism). He would use his compass to navigate underwater—an activity in which he doubted that Captain Simpson had any practical experience. (We doubt that Holland, likewise, had any practical experience with a magnetic compass in an iron hull, and he probably had not heard of the compass problems in the pioneering *Pioneer.*)

Captain Simpson advised Holland to drop the subject. No one would willingly go underwater in such a craft, he suggested, and even if the idea had merit, "to put anything through Washington was uphill work."

Soon, Holland was startled to discover that his complete scheme was included in a lecture and subsequent pamphlet on "Submarine Boats" delivered at the Torpedo Station by Lieutenant Francis Morgan Barber. The lecture covered all known examples of submarine development, from Drebbel on down through the Civil War—and on to Holland's latest addition. The known consequences of this unauthorized disclosure were two: Holland ever after became wary of revealing anything, to anyone, until he was ready, and the officer who delivered the lecture became a lifelong friend, supporter, and eventually associate of Holland.

Be that as it may, in July 1876, Holland met some men who had a better appreciation of his ideas than Captain Simpson. Through a connection arranged by his brother, Holland was passed on to the leadership of a group of anti-British Irish exiles. "I introduce you to a Mr. Holland of Patterson [sic] New Jersey," the letter read. "He visited me in connection with a torpedo he has invented, and I wish you would have a few minutes conversation with him."

The "conversation" proved fruitful. The exiles, who were the American wing of the Irish Revolutionary Brotherhood, styled themselves as the "Fenians." Their leaders were impressed by this fellow expatriate: "He was well informed of Irish affairs," one was to write many years later, "and was anti-English and with clear and definite ideas of the proper method of fighting England. He was cool, good-tempered, and talked to us as a schoolmaster would to his children."

The Fenians had established a "skirmishing fund," encouraging a minimum contribution of ten cents a week from every member nationwide to underwrite actions against the British. Receipts were averaging $2,000 a month, to a total of some $90,000 by 1883. Holland's "torpedo"—his submarine boat—could give them a powerful weapon against Great Britain's most vaunted military force: the Royal Navy.

While they were revolutionaries, they were not wild-eyed fools and wanted some proof that this amateur engineer could deliver. Holland built a thirty-inch working model, powered by a wind-up motor, which served the purpose. The Fenians agreed to fund a prototype, but this was not to be a crash project; the months turned into years.

Some of the brotherhood grew tired of waiting—they wanted to take immediate action and blow something, anything, up; others counseled waiting for a war to break out between England and someone, anyone—Russia was a candidate at the time—and then intervening on the other side. One key leader, frustrated with the arguments and exasperated by continual leaks to the newspapers, resigned in February 1878. "I do not propose to fritter away my life in endless squabbles," he wrote, "nor do I think it safe to go into serious revolutionary work with men who cannot keep a secret."

The 14.5-foot-long Holland *No. 1,* known as the "wrecking boat" to shipyard workers and newspaper reporters, finally was launched May 22, 1878.

It sank.

More properly, it *appeared* to sink; one local newspaper made great sport of the success of the submarine boat, which "went immediately to the bottom; and this without even the assistance of the captain."

Holland may have computed the displacement for salt water, rather than the fresh water of the Upper Passaic River, or some error in construction or preparation may have tipped the scales, but no matter. *No.1* was so close to neutral buoyancy that she was easily brought back to the surface by two men pulling on the tending line, to the reported amazement of onlookers.

Two days later, *No.1* seemed ready for dock trials.

Except for the leaks, which took another five days to locate and repair.

Next, taking no chances, Holland had the submarine slung between two large rowboats while balance, trim, and buoyancy were all evaluated and adjusted.

Then he had to deal with the problem of the engine. It would not work.

Originally meant to be man-powered, *No.1* had been converted to use one of the world's first internal-combustion gasoline-fueled engines, recently patented by George Brayton. However, Holland could not induce it to run, and realizing that any expanding gas would do, had it connected, with a rubber hose passing through a watertight joint, to the steam boiler on an accompanying launch. It worked just fine.

On June 6th, he was ready—*No.1* was ready. Holland squeezed into the operator's compartment—not more than three feet wide, not quite four feet long, and just over two feet high. His head stuck up into a "conning tower" turret, ringed with glass ports through which he could see whatever there was to see. The engine was between his knees.

After moving her under her own power into a clear stretch of the river—the steam-providing launch keeping pace—Holland flooded the two ballast tanks (unlike those in *Hunley,* these were controlled by a central valve) and the boat sank—as he had calculated it would—to the level of the port lights. Then, with the diving planes set at a down angle, forward momentum drove the boat to a depth of perhaps twelve feet; Holland reversed the angle and rode back to the surface. *No.1* had worked, just as he had planned.

No. 1 remained only a prototype—and even though the existence of the submarine was well known, the details of design and operation were hidden from public scrutiny. Therefore, once the basic operation had been proven in another underwater excursion lasting about an hour, and to keep the secrets secret, Holland removed the engine and some other machinery and scuttled the boat in the Passaic River.

Shortly thereafter someone tried to salvage the boat for the value of the scrap metal, using the structure of a nearby bridge as anchor for a winch. The would-be salvors only managed to rip the turret from the hull before city authorities, afraid the bridge would be pulled down to join the two-ton boat in the water, called a halt. Thus, Holland *No. 1*—minus the turret—was preserved for another salvage effort, almost half a century later, by a group of students. They donated their find to the Paterson Museum.

The Fenians—most of the Fenians—were happy with the progress, and gave Holland an order for a larger, operational submarine. There was talk of an eventual fleet of the boats, perhaps even prepositioned in the British Isles to "lie idle until the opportunity should present itself." The Delamater Iron Works in New York City promised to handle the job with discretion and for less than $20,000, and Holland quit his teaching job—with some regret. As he was later to tell his daughter: "I loved the children because they were the only ones who didn't think I was crazy."

The new boat was more than twice as long as *No. 1,* at thirty-one feet, but geometrically larger at nineteen tons. The size, someone later estimated, had been calculated so that the boat might fit in a standard railroad box car, to facilitate shipment to a Fenian war zone.

An improved Brayton engine was installed, which Holland thought could be operated underwater to a depth of about forty feet. Compressed air would provide oxygen for both combustion and the crew; the engine exhaust would be expelled into the water. Holland predicted submerged operations, with the engine running and three-man crew healthy, of seventy-two hours. It was fortunate (for the crew) that this proposition was never tested.

If Holland thought that this boat could be built in secrecy, he soon was disabused. Delegations from Italy, Germany, Russia, Sweden, and Turkey passed through Delamater's yard—along with, anonymously, British Naval Attache Captain William Arthur. The British Foreign Office was very much aware of the Fenian activities in the U.S., and in fact had hired private detectives to keep them (and Holland) under observation. However, when Captain Arthur described Holland's strange little boat to his superiors in London, they were neither threatened—nor impressed. "There seems to be no reason to anticipate that this boat can ever be a real danger to British ships," wrote the Director of Naval Construction. "We should not recommend the spending of any money in order to obtain information."

Construction of the boat took more than two years, and by then the Fenians were running out of money—almost two-thirds of their fund had gone to support Holland's projects. Some members of the group were also running out of patience: When were they going to hit the British? The boat finally was launched on May 1, 1881. A few months later, a reporter for the *New York Sun,* broadly hinting at the ownership and probable employment, dubbed the boat the "Fenian Ram," and *Fenian Ram* it became.

Fenian Ram submerged for the first time in June 1881; as described by Holland:

> . . . the engineer and myself entered the boat and closed the hatch. This shut us off from the air, and our breathing now depended entirely on the compressed air reserve. After waiting a few moments and finding no ill-effects from the compressed air, I decided to submerge. I drew back the little iron levers on either side of my head. . . . Almost immediately the boat began to settle giving us the suggestion of slowly descending in an elevator. I now looked through the ports in the superstructure and observed that the bow had entirely disappeared and the water was within a few inches of the glass. A second or two later everything grew dark and we were entirely submerged and nothing could be seen through the ports except a dark green blur.
>
> Our next suggestion was a slight jar when the vessel struck bottom. It might also be mentioned here that we had no light except the glow that came through the conning tower. This just about sufficed to read the gauges, but was too poor to be of interest to the engineer. . . .
>
> . . . I decided to blow out the ballast and come up. Accordingly I opened the valve admitting air to the ballast tank, and at once heard a hiss that told me that the air was driving out the water. The green blur on the ports in the conning tower grew lighter as I gazed through them until suddenly the light of full day burst through, almost dazzling me. . . . I now opened the hatch and stood on the seat, thus causing my head and shoulders to protrude from the tower. As soon as I was observed doing this, a cheer burst from the crowd of observers on the

dock, among whom opinion was equally divided as to whether we would emerge alive from our boat or not.

Since he was building a weapon of war, Holland included a ram bow and a novel 9-inch air-propelled gun. The boat was rigged for use as a diving bell—shades of *Alligator* and "The Intelligent Whale"—and that capability was tested at least once. The air gun was Holland's own invention, and one he continued to design into his boats for many years. The gun tube ran through the center of *Fenian Ram* and out at the bow. In addition to the operator and the engineer, the boat carried a third crew member: the gunner.

To fire the weapon, the gunner would ensure that the tube was empty and the bow cap was closed, thus sealing the tube; open the breech, load the projectile; close the breech, open the bow cap. Then, the operator would come near to the surface, point toward the target, and elevate the bow to an angle (predetermined by the estimated range to the target). The gunner would open a valve to admit 400 psi of air into the tube, the projectile would be projected, water would flow in behind and fill the tube, and the gunner would close the bow cap and drain the tube into a trim tank, thus replacing the displacement lost with the ejected projectile. The gun was tested, using dummy warheads, at least twice. One dummy, launched from underwater, traveled about ten feet underwater and then burst forth for another three hundred yards through the air. It was cumbersome, hardly accurate—but fitted on later Holland designs, even after he was including a Whitehead torpedo.

Tests of the *Ram* continued for two years, up and down the rivers, in and out of the upper harbor, including dives to sixty feet and as long as one hour. Holland estimated her surface speed at nine knots, and thought the submerged speed was not much different although it was never measured.

Some years later, Holland was to offer what amounted to a long-delayed apology for his flip comment to Captain Simpson: "The writer had no suspicion that his boat could not be steered perfectly [by compass] until he had tried it. . . ." All attempts at compensating the compass (by the careful placement of small magnets near the compass) failed, because compensation adjusted for a boat on a static, even keel. Diving, or rising, or heeling, altered the magnetic influence so much that a 90-degree error was not uncommon. The compromise solution, which covered the next twenty years, was to mount the compass outside the hull, to be viewed from within by a telescope or mirrors.

It was not until the invention of the gryo-stabilized compass in 1915 that this problem was resolved.

The newspapers followed every move of *Fenian Ram* with great interest; once, when the boat had dropped out of sight, a newspaper speculated darkly that she had been sent off on a mission against the British. She merely had been shifted to a shipyard to be painted. In the meantime, Holland was gathering data, refining his scheme, and building improvments into a sixteen-foot working model.

During this period, the engineer had taken the *Ram* out for an unauthorized joy ride; running with the hatch open, she swamped in the wake of a passing tugboat, sank, and cost the Fenians salvage fees of $3,000. The Fenian squabbles intensified; one member took the Brotherhood to court to enjoin any further expenditures from the fund. Fearing that the court would attach the boat, a group of Fenians crept into the yard under cover of darkness, stole their own boat, and towed it out to New Haven on the Connecticut shore of Long Island Sound. They had started out also with the sixteen-foot model in tow, but apparently the hatch was not properly sealed and the model sank en route in 110 feet of East River water, where she remains, somewhere near today's Whitestone Bridge.

Holland was later to write: "I received no notice of the contemplated move then, nor was I notified after. . . . As a result, I never bothered again with my backers nor they with me." The two submarines and the sixteen-foot model had cost the Fenians about $60,000.

A brief footnote to this part of history: Legal actions clouded the future of *Fenian Ram*, and for thirty-five years she sat hidden in a shed on New Haven's Mill River. The engine was removed and put to work running a forge in a foundry owned by one of the Brotherhood. Eventually—in 1916—what remained was brought back to public view. *Ram* was put on a truck and carried to New York City's Madison Square Garden, where she was exhibited as part of a fund-raiser for the victims of the British suppression of the Irish Easter Uprising. Later, *Ram* was donated to the city of Paterson and placed in West Side Park.

Meanwhile, back in 1883, Holland, then forty-two, had made friends outside the Fenian Brotherhood. He had come to public attention as America's leading submarine inventor. He had become an American citizen. He had begun collecting patents on his work, but now, to support himself took a job as a draftsman in Roland's New York City Iron Works. Among other assignments, he worked with an aging George Brayton on improvements for the engine.

Just at this time, Holland received an invitation to dinner at the Brooklyn Navy Yard from Lieutenant William W. Kimball, USN. Kimball had read the published version of Barber's Newport lecture and had seen the newspapers accounts of the *Ram,* and wanted to learn more. Over dinner, Kimball was impressed with Holland's practical approach and his vision for the future, and offered to get him a job as a draftsman with the Navy Department.

Unfortunately, the Navy Department was not in a hiring mode. Kimball had also arranged a meeting between Holland and Lieutenant Edward Zalinski, U.S. Army, who was involved with the Pneumatic Gun Company (apparently a separate and distinct business, not connected with Holland's own air gun). After waiting for several months without word from the Navy, Holland went to work for Zalinski. He took with him fresh ideas for yet another submarine, and had hopes of making a sale to the French (just then at war in Indochina). Zalinski, Holland, and some investors created the Nautilus Submarine Boat Company.

Zalinski thought Holland's boat well suited to his air-powered weapon: The submarine became, in effect, a disappearing gun carriage. Zalinski bragged to anyone who would listen that his dynamite gun could send an explosive charge of two hundred pounds—and more—for almost half a mile—or more. In theory, the explosion would sink or disable an enemy ship from a direct hit or from the underwater concussion of a near miss.

The Sino-French War ended too soon. The company prototype, dubbed the *Zalinski Boat,* was launched September 4, 1885, but she was too heavy for the launching ways; they collapsed and the boat bounded down the slope and smashed into some pilings. Almost thirty years later, Holland misremembered the event, and wrote that "the cost of repairs would exceed the amount of money still on hand in the company's treasury. Accordingly the wrecked boat was broken up where she lay, the engine and fittings removed and sold, and the proceeds used to partly reimburse the stockholders for the money they had invested." The project actually stumbled along for almost a year, the patched-up boat making a few token runs up the river before being broken up and sold for the parts.

The Nautilus Submarine Boat Company went bankrupt. The separate Pneumatic Gun Company continued in business, and the dynamite gun found its eventual platform in the (circa 1888) "dynamite cruiser" *Vesuvius.*

Holland's ill-fated *Zalinski Boat,* 1885, built for a war that ended too soon, heavily damaged on launching, soon enough sold for scrap. Credit: *Scientific American*

Equipped with three pneumatic guns, each intended to fire a 250-pound TNT projectile for 1.5 miles, *Vesuvius* saw scant and useless service as a dispatch boat.

Meanwhile, Holland had not been the only submarine designer pushing the state of that art. The French had been working on submarines through most of the century—picking up, to some degree, where Fulton left off. In 1863, a French team had launched a submarine of exceptional size: *Le Plongeur* ("The Diver") was about 140 feet long, twenty feet wide, ten feet high, and had a displacement of more than four hundred tons. She was powered by engines run on compressed air, stored at 180 psi in some twenty-three tanks. Ballast tanks were filled to reduce buoyancy to just above the neutral state; final adjusments were made by cylinders which could be run in or out of the hull to vary the volume—Bourne's prescription by different means.

While it may have seemed like a useful combination, the realities of free surface effect, wave action, marked differences in the salinity of water in the coastal region (often varying from one depth to another), even the movement of a crew member, often sent the boat into radically unstable operation. One report stated that while operating in thirty feet of water, *Le Plongeur* would make progress "by alternately striking the bottom and then rebound to the surface like an elastic india-rubber ball."

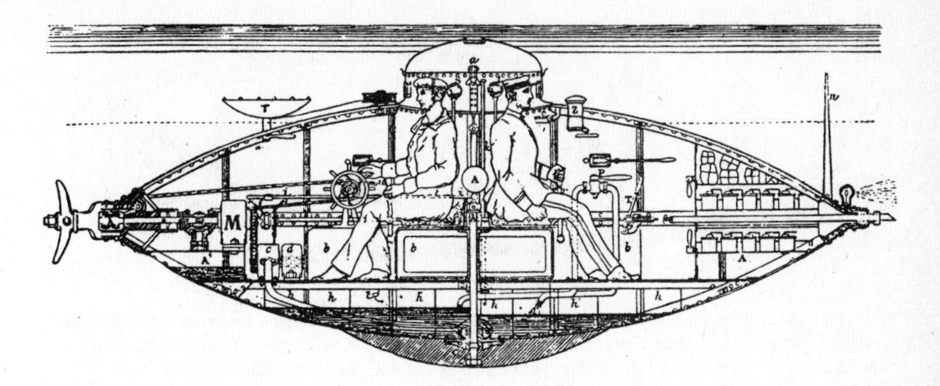

The battery-powered French submarine *Goubet,* circa 1885. Awkward, unstable, unused.

More recently, about the same time that *Ram* was being launched, Holland had seen reports of a French boat which ran on batteries. There had been rumors—and rumors only—that many of these *Goubets* were built and sold to Russia, but in reality the small two-man boat was too unstable for safe operation. In 1889, designer Claude Goubet launched an improved *Goubet II,* also small, electric, and not effective. When the French Navy declined to purchase *Goubet II,* Goubet sold the boat to a British venture, which turned it into a tourist attraction.

Holland had not been impressed, and would later write: "The designs for these boats, I am sure, were based on certain fundamental points of my *Fenian Ram* design. . . . there were a number of foreign officers present at Delamater's Yard from 1879 to 1881, while the boat was in course of construction, and it is hardly to be expected that they failed to take notes. However, the knowledge they secured did them very little good, because, while they secured a lot of valuable data, their inexperience caused them to disregard the most vital points, with the result that their boats never attained any degree of success."

Closer to home, another submarine had been seen operating in the Hudson River while the *Zalinski Boat* was being completed. This was *Peacemaker,* and was actually the second boat built by Josiah H.L. Tuck. His first, a thirty-foot model built in 1883 and powered by electricity, proved impractical. His new boat was powered by a chemical (fireless) boiler. This avoided the problem of

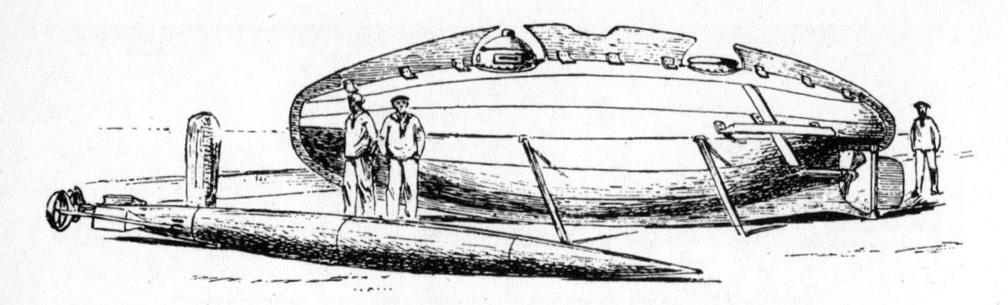

Professor Tuck's second submarine torpedo boat, circa 1885, was powered by a chemical "fireless boiler," in which 1,500 pounds of caustic soda provided endurance of five hours. The first, smaller boat, did not have an enclosed helmsman's station: the operator stood in a well, wearing a full diving suit. Tuck's inventing days ended when relatives—noting that he had squandered most of a significant fortune—had him committed to an asylum for the insane.

providing oxygen for combustion—but 1,500 pounds of caustic soda would only drive the boat for five hours. Nonetheless, this was a working submarine.

But most significantly, also in 1885, a new boat built by the Nordenfeldt Company in Sweden caught the public eye—not so much because it was a superior submarine, but rather because the Nordenfeldt Company knew how to catch the public eye.

The Nordenfeldt boat was awkward: It used three separate engines, one for propulsion and one each for horizontal port and starboard haul-down screws. It was powered by steam on the surface, and used steam accumulated in pressurized hot-water tanks to run the engine while submerged.

This strange scheme had been invented by Anglican Reverend George W. Garrett, and then tested in his own 1879 boat. *Resurgam* passed initial trials but, just as many other not-quite-seaworthy boats, sank while under tow. (*Archaeology* magazine reports that the wreck was discovered in 1996, and may be salvaged for a museum display.) Undeterred, Garrett followed the path of many underfunded inventors and in 1881 took his ideas to a wealthy Swedish arms manufacturer, Thorsten Nordenfeldt.

The Garrett-Nordenfeldt system, however, had a serious flaw: There was too much free surface in the hot-water and ballast tanks. According to a contemporaneous U.S. Navy intelligence report, any sudden change in speed or direction triggered "dangerous and eccentric movements." Also, the boat was not very effective as a submarine: It took as long as twelve hours to generate enough steam for submerged operation—and once ready, it took about thirty minutes to dive.

But Thorsten Nordenfeldt applied marketing savvy to push this barely useful submarine onto the world stage. The first Nordenfeldt was publicly demonstrated in September 1885, before an audience of royalty and naval observers from as far away as Japan and Brazil. One observer was the former British Naval Attaché to the United States (and now an admiral) William Arthur, who had "visited" Holland's *Fenian Ram* a few years earlier. Admiral Arthur was impressed enough to suggest that all future torpedo boats should be able to submerge for some brief period of time.

Also impressed: the Greek Navy, which became the first to buy a Nordenfeldt. They took delivery of *Nordenfeldt I* in 1886 (sixty four feet, sixty tons, armed with one external torpedo tube), whereupon their bitter rivals, the Turks, ordered two of a larger model, *Nordenfeldt II* (one hundred-feet, 160 tons, armed with two tubes). The Greek boat went to thirty feet, and once stayed submerged for six hours, but appears to have done nothing else worthy of note; the Turkish boat was rated for fifty feet, but proved so unstable that it was never put into service. Among other problems: When a torpedo was fired forward, the boat tipped backwards and sank, stern-first, to the bottom. The second Turkish boat was left unfinished.

Nordenfeldt was by then connected with the Barrow Shipbuilding Company of England, which built his next boat: the 123-foot, 245-ton *Nordenfeldt III.* Launched in March 1887, she was rated to one hundred feet with an advertised surface speed of fourteen knots, and was the first submarine to have

Nordenfeldt III, 1887—powered by steam on the surface, and from reservoirs of compressed steam while submerged. Too much free surface in the tanks contributed to a dangerous instability, but good public relations overcame bad design, and, for a time, the Nordenfeldt boats were regarded as the "world standard." Credit: *Scientific American*

internal torpedo tubes. This boat was sold to Russia after another impressive public demonstration (more crowned heads and government officials), but it ran aground en route and the Russians refused to accept delivery. The boat was scrapped.

Nordenfeldt's public relations emphasized the positive—high speed, demonstrated depth capability, modern weapons—and overpowered the reality of a dangerously erratic platform. Never really proven, the Nordenfeldts nonetheless had now set an unobtainable standard by which many nations—including the United States—came to evaluate submarines.

For it was just at this time that the United States Navy began to give serious consideration to submarines. While the Fenians were taking *Ram* into historical obscurity, the U.S. Navy had begun a long climb out of a post-Civil War nadir. The Civil War-spawned government interest in submarines had not died, but without the impetus of the war it largely went into hibernation—along with the rest of the U.S. Navy. From a wartime high of seven hundred ships, the Navy began a slide which by 1880 had reduced the fleet to twenty six effective warships worldwide, only four of which were so modern as to have iron hulls.

By comparison, the British had almost four hundred ships in service, fifty-six of them armored. Even Japan, which had not started a navy until 1866 with the purchase of a ship from the United States, had nineteen ships, five of which were armored.

In an 1878 speech in the House of Representatives, Congressman (later President) James A. Garfield had defined his philosophy of national preparedness: "Our fathers said: 'Though we will use the taxing power to maintain a small Army and Navy sufficient to keep alive the knowledge of war, yet the main reliance for our defense shall be the intelligence, culture, and skill of our people.'"

However, in meeting its obligation to "keep alive the knowledge of war," the U.S. Navy had rejected most of the advances of the Civil War, advances which had been invented by, introduced by, or combat-tested by the U.S. and Confederate Navies—armor on ships, steamships without the encumbrance of sails, guns with rifled barrels, guns mounted in revolving turrets. The submarine.

The nation was almost bankrupt, and in the name of economy full sail rigs were ordered to be put on steamships; efficient screw propellers were replaced with low-drag models that were hardly useful under steam but would minimize interference under sail. Armored ships were put in reserve. Economy was a reason, but professional intransigence may have been of equal importance. Tools of naval warfare which were being changed almost daily in other navies of the world were branded "costly experiments" by the U.S. Navy and the Congress.

Among those "experiments" were the elements for the creation of a truly effective submarine warship: rechargeable electric batteries, the automobile torpedo, and the internal combustion engine. The threads of technology began to converge around Civil War experience; *Hunley* had demonstrated a real, albeit flawed, potential for warfare.

Finally, after more debate than anyone today has time to read—much of it centered on Congressional concerns about creating a blue-water navy with the range and capacity to attack the European shores from whence so many voting constituents had so recently come—Congress passed the Navy Revitalization Act of March 3, 1883, adding three modern cruisers and a gunboat. Two years later, in 1885, Congress authorized two more cruisers and two gunboats. Two battleships, *Texas* and *Maine,* were authorized in 1886, and the battleship *New York* in 1888.

About this time, Holland wrote an article, "Can New York be Bombarded?"—"with the intention," he was later to say, "of bringing before the public the pitiable condition of our fleet and coast defences, and showing how a few submarines would place us in a position to ward off an enemy's attack from mostly any point on our coast as effectively as if we had an adequate shore defence and a fleet equal to Great Britain's."

This touches on what was to become Holland's major marketing theme: Congress, not the Navy, was the real target.

10 1887–1900—Of Entrepreneurs, Politicians and Businessmen

The Holland *boats are interesting novelties which appeal to the non-professional mind, which is apt to invest them with remarkable properties they do not possess. . . . in my opinion, they are ingenious contrived craft of an eccentric character which mark a step in the development of an interesting science but nothing more.*

—*Rear Admiral Charles O'Neil, chief of the Navy's Bureau of Ordnance, Congressional testimony, 1900*

Into this time of revival and growth, the news from abroad and the work of Holland and Tuck were sufficient to encourage the Navy Department to announce the first-ever open competition for a submarine torpedo boat. A $2 million appropriation was the incentive.

The November 1887 announcement was drafted by Holland's friend, Lieutenant Kimball. While the basic document outlined an operating system almost the duplicate of Holland's, Kimball's superiors had been sufficiently taken with the Nordenfeldt claims that they insisted on requiring presumed Nordenfeldt-level performance: surface speed fifteen knots, submerged speed of eight knots with two hours endurance; depth capability to 150 feet. The size was left to the designer, although the Navy expected a power plant of one thousand horsepower. The only way then to achieve one thousand horsepower was with steam.

The Navy received at least four proposals, including submissions from Holland, Tuck, George C. Baker—of whom more in a moment—and Nordenfeldt, offering his *Mark III.* Holland's design won, but when the Navy's designated shipbuilder could not offer performance guarantees, the award was withdrawn.

Nordenfeldt, by this time having delivered only two boats in eight years, lost interest in submarines. Soon thereafter, the Nordenfeldt Company was taken over by another gun maker, Hiram Maxim (itself soon to become part of the Vickers Company which, in its Barrow Shipyard, would later build the first submarines for the British, designed by Holland).

The competition was re-opened the next year, the submissions already in hand were reviewed, and Holland was again the winner. However, before a contract was issued, a new Administration and a new Secretary of the Navy took office (shades of Fulton and Colt!). Secretary Benjamin Tracy created a special Policy Board, the Board called for a massive reinvigoration of the fleet, and the $2 million appropriation was diverted to surface ships.

About this time, another competitor dropped out—literally. Professor Tuck had invested so much money in his submarines that his relatives had him judged insane and committed to an institution. Holland went to work as a draftsman for the Morris and Cummings Dredging Company in New York, earning $4 a day.

Four years later, when Grover Cleveland returned to a second term in the Presidency, Congress quickly appropriated $200,000 for an experimental submarine and the Navy announced a new competition. There may have been two reasons for this speedy turnaround. For one, Chicagoan George Baker was well connected with the Democrats, was not shy about exercising his connections, and had just completed a submarine which he was demonstrating on Lake Michigan. The Baker boat had some interesting features: For example, he had a clutch between the steam engine and the electric motor so the motor could be used as a dynamo to recharge the batteries used for submerged propulsion. The Baker boat also had some troubling features: For example, both forward and vertical motion was provided by a single pair of amidships-mounted propellers which swiveled up or forward, going through a difficult and clumsy transition in the process.

The other possible reason: The U.S. had just avoided a war with Chile and was at odds with Great Britain over Venezuela. Rumors of war were circulating. As the Police Commissioner of New York City remarked to a friend in the Senate, "Let the fight come if it must; I don't care if our seacoast cities are bombarded or not; we would take Canada. The clamor of the peace faction has convinced me that this country needs a war." Thus wrote Theodore Roosevelt to Henry Cabot Lodge. With politicians thus engaged, rumors of war tended to influence defense planning.

Whatever the impetus, the new competition called for a $7,500 deposit and a performance bond of $90,000 from each entrant. To raise the necessary capital, longtime Holland friend and supporter E.B. Frost joined with Holland to establish the John P. Holland Torpedo Boat Company. Holland assigned the rights to his patents to the company, and was given a large block of stock and appointed manager, with a salary of $50 a month. (This was a substantial cut from his previous $4 a day, although by the end of the year, his salary had been raised to $150.) Lawyer Frost was secretary-treasurer and handled investor/stockholder relations. By way of comparison: The monthly pay for a Navy ensign on sea duty was $100; a rear admiral earned $500.

This competition had been announced on April 1, 1893. Submissions were opened June 30th. This time, there were three entries: Holland, Baker, and a new competitor, Simon Lake—of whom more in the next chapter.

Before long, the *New York Times* had reported that Holland once again was the winner. Baker, however, had not been sitting around waiting for a decision. His agents had been pushing the fact that *he* had a *boat* ready for inspection; Holland's was only a plan. Baker persuaded the Secretary of the Navy to have the Board run an evaluation of his in-the-water-ready-to-go boat. In fairness, therefore, the Board asked Holland to provide a boat of his own for a sail-off.

Holland answered the challenge with logic—legally, he pointed out, the competition was for a design, not for a finished boat; he *had* a boat (*Fenian Ram*) but it was in such disrepair that refurbishment would be costly and his small company's resources had already been stretched to the limits by the administrative delays. He had noted, he wrote, that the *New York Times* had declared he was the winner. If so, was the Board now changing its mind? Finally, with more of his Irish wit: "If the newspaper description of Mr. Baker's boat is anywhere near accurate," he wrote, "I entreat you to examine the structure carefully before you submerge its center below 20 to 25 feet. My motive for this request is—I admit a very selfish one—of objecting to the risk of having to wait for a decision until a new Board can be appointed."

The Board was not charmed, nor did it appreciate the *New York Times* being viewed as an official organ of the Navy Department. The Board asserted that, since Baker did indeed have a boat, it was proper and prudent that they give it a trial.

So the Board went to Chicago and evaluated Baker's boat; it did not do well. The swiveling propellers were not a good idea. The Board's "final" decision reached Secretary Hilary A. Herbert in September. As anticipated, Holland had won again. Baker was furious; he lodged a protest and tried to have the award canceled. Time passed. Rumors of war evaporated. The Secretary was in no hurry—possibly out of "courtesy" to Baker. The Secretary expressed grave concern over the probable safety of the crew of any submarine when exposed to an underwater explosion.

The Navy set up a test, putting a cat, rooster, rabbit, and dove in a watertight tank exposed to blasts as near as one hundred feet. The cat and rooster survived; the more delicate rabbit and dove did not. The Secretary's concerns had been answered, but 1893 passed into 1894 and nothing happened. Holland took a pay cut, to $100 a month. Nothing happened in 1894 either as far as the Navy Department was concerned.

But now the John P. Holland Torpedo Boat Company took an initiative sure to attract the attention of Congress. Over the years, Holland had been approached by foreign governments, but had never entertained any offers. Now, the company—really being guided by Frost—ensured foreign patent protection and actively began offering the boat for sale to any and all bidders. Zalinski, Holland's former partner and fellow dynamite-gunner, was signed on as a traveling company representative, and over the course of a year set up agents in Argentina, Chile, Ecuador, Peru, and Italy, then moving on to the Far East to meet with Commander Francis Morgan Barber—the man who had included Holland's first design in the Newport lecture and was now "unofficial" Holland agent for Japan.

The ploy worked: On March 3, 1895, the John P. Holland Torpedo Boat Company received a government contract for $200,000 to build a submarine to be called *Plunger*. This was twenty years after Holland's first approach to the Navy, seven years after the competition had opened, and two years after the once and final winner had been selected.

* * *

The Congress was pleased, and not waiting for any signals from the Navy—nor even for the first submarine to be tested—the Senate Committee on Naval Affairs held hearings intended to support future authorizations. The witnesses were firmly in the submarine camp, and the theme was firmly "harbor defense." Lieutenant Kimball was there: "Give me six Holland submarine boats," he said, "the officers and crew to be selected by me, and I will pledge my life to stand off the entire British squadron ten miles off Sandy Hook without any aid from our fleet." Of course, he might be considered biased. But what about Rear Admiral James E. Jouett, a retired veteran of the Civil War Battle of Mobile Bay? "If I commanded a squadron that was blockading a port," he said, "and the enemy had half a dozen of these Holland submarine boats, I would be compelled to abandon the blockade and put to sea to avoid destruction of my ships from an invisible source, from which I could not defend myself."

Perspective was maintained by the testimony of Captain Alfred T. Mahan, who "wrote the book" on grand naval strategy (*The Influence of Seapower on History 1660–1783,* published in 1892). The captain glumly observed, "In our present unprotected condition, the risk of losing the money by reason of the boat's being a failure is more than counterbalanced by the great protection the boat would be if a substantial success."

The hearings resulted in the Act of 10 June 1896, which authorized two more submarines of the *Plunger* type for $175,000 each, just as soon as the first had been accepted for service.

But therein lay a major problem. Holland's contract was not for his winning design—it was for a design upon which the Navy imposed a number of changes. Holland had proposed a relatively simple, single-screw boat which would make changes in depth with hydroplanes. He had included a steam plant, as this was the only way to meet the required surface speed of fifteen knots. But the Navy insisted on three screws for forward propulsion, added two haul-down screws in the manner of *Nordenfeldt* for vertical propulsion, and required three steam engines to drive them all.

Launched in August 1897, *Plunger* failed before ever leaving a dock. The temperature in the fireroom reached 137 degrees before the plant had passed two-thirds of its rated output. A few years later, during another series of Congressional hearings, a Holland employee testified, "They forced us to put steam in the *Plunger* against Mr. Holland's advice. When we built

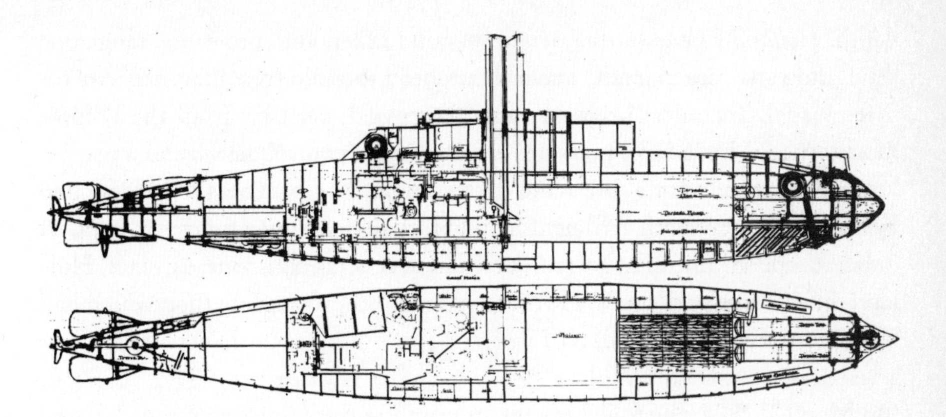

Holland's plan drawings for the never-finished, steam-powered *Plunger.* "When we tried to submerge, it was so hot no one could live in her." (See also, photograph 8)

the *Plunger* and launched her and put the steam on, we found it was so hot we could not live in her." Lieutenant Barber's prediction of 1875, in commenting on the Phillips plan for steam propulsion, was cleary ignored. Or more likely, forgotten.

In what must be an unwitting irony, the first U.S. Navy submarine to be designed with built-in air-conditioning was the 1935 SS-179, *Plunger.*

Early on, Holland suspected that *Plunger* would never work, and he became increasingly frustrated with almost daily interference from the Navy Project Officer. The Navy, it seemed, knew what it wanted but didn't know what it was doing—and before Holland could make any changes of his own to *Plunger,* he first had to get Navy approval.

Encouraged by the tone and tenor of the 1896 hearings—there was indeed support for a submarine fleet in the Congress—Holland suggested to his Board of Directors that they should build a new submarine of their own design, on their own, with no government contract, no Navy "help." The board agreed.

The new design cut the unrealistic speeds of *Plunger* in half, eliminated any hint of steam, and put in an Otto gasoline engine. Holland had abandoned the unsatisfactory Brayton engine, with personal reluctance but out of professional necessity.

While *Plunger* was still on the ways in a Baltimore shipyard, *Holland VI* was built and launched in New Jersey. The boat was smaller than *Plunger*—fifty four feet, compared with eighty five feet—and was rated for only half the depth. *Holland VI* had both a dynamite gun (222-pound projectile, range one thousand yards, seven loads) and a Whitehead torpedo (one tube, capacity for three loads). Technical improvements (borrowed, perhaps, from the Whitehead torpedo) included a pressure-sensing depth-control device and a pendulum rigged to prevent a too-acute diving angle. The boat was to be operated by a six-man crew, with sufficient habitability—including a toilet—to support onboard operations up to forty hours. There was no periscope—in fact, Holland did not like periscopes. He preferred the "porpoising" method which had been pioneered, if you will, on *Pioneer.*

On May 17, 1897, the following bit appeared in the *New York Times:* ". . . the *Holland,* the little cigar-shaped vessel owned by her inventor, which may or may not play an important part in the navies of the world in the years to come, was launched from Nixon's shipyard this morning." Newspapers had great sport guessing at the new submarine's ultimate owners: Cuban refugees, wanting to take the fight to the Spanish colonial government? The Spanish colonial government, faced with insurrection in Cuba? The French?

Holland was invited to describe his work in the British *Cassier's Magazine.* Unlike, say, Fulton, who claimed all credit for himself, Holland freely acknowledged those inventors who had helped lay the tracks along which he could run: Drebbel, Bushnell, Fulton, Lodner Phillips, Charles Burn and Simeon Bourgeois (*Plongeur*), James McClintock and Baxter Watson (*Pioneer* and *Hunley*), O.S. Halstead ("Intelligent Whale"), Claude Goubet, and Thorsten Nordenfeldt.

And then, the curse which seems to have hovered over Holland submarines hit *Holland VI.* On October 13, 1897, a yard worker left a valve open, and overnight the boat slowly sank, unnoticed, at the pier. When she was raised some eighteen hours later, it was apparent that all electrical equipment, especially the motor, was damaged, perhaps ruined, by the salt-water immersion. Holland had two choices, one of which was simple but problematic—clean up everything, dry out the windings, if possible, and pray. The other choice was akin to major surgery and might be too costly—cut open the hull and replace the old motors with new.

The struggling company wasn't ready to support major surgery, so they brought heaters in—lots of heaters—and tried to dry the motor. For a month. It didn't work; the motor wouldn't run. They asked the manufacturer of the motor for help; the Electro-Dynamo Company of Philadelphia sent an electrician who easily solved the problem. He reversed the flow of current through the system and created heat in the windings themselves, thus drying them from the inside out. His name was Frank T. Cable, and in the years to come he would not only take a job with the Holland Torpedo Boat Company, but would become *the* test captain for all Holland boats and one of the world's leading authorities on submarine operations. For now, he went back to Philadelphia. *Holland VI* was ready for trials toward the end of February.

But in the middle of the month—February 15, 1898—the new battleship *Maine* was blown up, or blew up, in the harbor of Havana, Cuba; 260 American sailors were killed. Questions surrounding the cause of the explosion have never been resolved (it may well have been from spontaneous combustion in a coal bunker next to a magazine) but at the time, blame was laid squarely on the Spanish colonial government. Rumors of war began circulating; more than that, demands for war began dominating the front pages of most every newspaper in the nation.

Holland VI got under way for the first time on February 25th. Newspapers noticed; speculation abounded. Holland *really* was going to sell his boat to the Cuban insurrectionists. Holland was going to torpedo the Spanish cruiser *Vizcaya,* then on an unfortunately timed "goodwill" visit to New York. Of course, *Holland VI* had not yet even started sea trials, but rumors of war have a way of getting in the way of logic.

On March 11th, Holland made a first trial submergence with *Holland VI* at the dock. When next he tried to submerge out in the bay, he found that the ballast was not heavy enough to compensate for salty water. He refigured—but did not repeat Day's fatal error of 1773. Lead bars, molded to fit, were gradually added to the bilges until the displacement was correct.

Finally, on March 17th, St. Patrick's Day—thought by some observers to have been a setup in recognition of Holland's Irish birth—*Holland VI* made her first real dive and an elated Holland invited the Navy to send up an observer. Lieutenant Nathan Sargent—coincidentally, one of Fulton's crew members in France had been Nathaniel Sargent—watched a full demonstration on March 27th as *Holland VI* submerged and fired a dummy dynamite charge three hundred yards. Lieutenant Sargent offered a relatively unsophisticated report. *Holland VI,* he wrote, "fully proved her ability to propel herself, to dive, to come up, to admit water to her ballast tanks, and to eject it again without difficulty." But he added, "I report my belief that the *Holland* is a successful and veritable submarine torpedo-boat, capable of making an attack on an enemy unseen and undetectable, and that, therefore, she is an Engine of Warfare of terrible potency which the Government must necessarily adopt to its service."

The Holland Torpedo Boat Company courted the press, freely giving interviews and encouraging sensational stories. The publicity prompted an immediate offer from France, but at $100,000, it was below the manufacturer's cost.

The former Police Commissioner of New York—now an energetic Assistant Secretary of Navy—also was paying attention; he sent a note to his boss:

10 April 1898

My dear Mr. Secretary:

I think that the Holland submarine boat should be purchased. Evidently she has great possibilities in her for harbor defense. Sometimes she doesn't work perfectly, but often she does, and I don't think in the present emergency we can afford to let her slip. I recommend that you authorize me to enter into negotiations for her, or that you authorize the Bureau of Construction to do so, which would be just as well.

Very sincerely yours,
T. Roosevelt

A special Board of Inspection was in New York the next day (we know that T. Roosevelt had a lot of clout, but because of the dates, we presume that the visit had already been scheduled). *Holland VI* performed well, and the inventor made a major point: He remained submerged almost an hour, much longer than called for in the schedule, while worried observers on the surface were trying, without success, to follow his movements. The Board, however, was not entirely friendly. Why, after all, were they asked to evaluate a boat built on "speculation" when a government-financed submarine (designed by the same firm) was not yet finished? In the event, the Board quibbled over details and the report was inconclusive.

It was about this time that Holland offered the following definition of a Board: "something that is long, narrow and wooden."

The day of the trials, President McKinley asked Congress for wartime powers; four days later, Spain declared war. Thenceforth the attention of the Navy Department was diverted; submarines were not on the daily schedule. The Assistant Secretary of the Navy soon diverted himself to an assignment as the executive officer of a regiment of Volunteer Mounted Cavalry Reserves which the newspapers called, variously, "Teddy's Terrors" and "The Rocky Mountain Rustlers" until settling on "The Rough Riders."

The commanding officer of the unit was a surgeon, Colonel Leonard Wood. He was later to be appointed Chief of Staff of the U.S. Army. A strange transition? Perhaps. His executive officer, Lieutenant Colonel Roosevelt, USVMCR, former New York City Police Commissioner and former Assistant Secretary of the Navy, soon after the war become Governor of New York, and in 1900 was elected Vice President of the United States—a post he held briefly until President McKinley was assassinated.

Roosevelt was a strong advocate of a strong Navy: in his oft-quoted remark "Speak softly—and carry a big stick," the "big stick" was the U.S. Navy. In 1905, Roosevelt would become the first Chief Executive to take a dive in a submarine. After having done so, he instituted "submarine pay" for all crew members . . . but that gets us ahead, again, of our story.

The Navy Department did not have the staff or resources to argue with the Spaniards and the Holland Torpedo Boat Company at the same time; Holland understood, and offered to take *Holland VI* to Cuba for an attack on the Spanish fleet then bottled up in Santiago Harbor. "If the government will transport the boat . . . to some point near the entrance to the harbor of Santiago, and a crew can be secured to man the boat, Mr. Holland will undertake the job of sinking the Spanish fleet . . . commanding the boat in person. If his offer be accepted, and he is successful in his undertaking, he will expect the government to buy his boat." Thus reported the *New York Sun,* May 27, 1898.

The public, as reflected in the newspaper accounts, was thrilled. The Navy, we may assume, was horrified. A private citizen taking a private warship into a foreign harbor to sink enemy ships? A lot of things had changed since 1812.

In the first few weeks of the war, every community along the Eastern seaboard of the United States expected at any moment to be attacked by the Spanish fleet, and demanded immediate "protection" from their elected representatives. Congress ordered the Navy to pull some obsolete ships out of reserve and station them, piecemeal, along the coast. As weapons, they were totally ineffective, but this was an exercise in public relations, not combat readiness. The public was mollified; the Congress basked in a reinforced sense that harbor defense was more important than blue-water endurance. Harbor defense could be obtained with ships which seemed warlike but were cheap enough to be stationed within a few miles of every seaboard voter. This philosophy would have a major, negative impact on submarine design and deployment for the next fifty years.

The war ended in August. While awaiting further word from the Navy, Holland continued to run trials—and host a constant stream of visitors, some two dozen of whom rode the boat underwater. They included Holland's friends, of course, like Commander Kimball (just back from duty commanding a torpedo-boat squadron in Cuba), a *New York World* reporter, and several Japanese and Norwegian naval officers.

This had been a period of great activity in France. Unfortunately, it was a period of unfocused activity. The French put a lot of submarines into service, but the boats ranged in size from tiny to eight hundred tons, with almost every imaginable power plant in one or another or in combination: electricity, steam, diesel, gasoline.

They began with designer Gustav Zede's *Gymnote* in 1888. She was a sixty-foot, thirty-ton electric-powered boat capable of eight knots on the surface, but lacked any provision for recharging her batteries while running. Some writers, perhaps unfamiliar with the ambiguous career of the Civil War *Alligator,* claim that *Gymnote* was the first submarine ever accepted into the naval service of any nation. Whatever, her actual "service" was limited to experimentation.

Not long after, the French had built the 148-foot, 266-ton *Gustav Zede*—named after the recently deceased designer. Despite problems with stability, she managed on several occasions to demonstrate the probable utility of this addition to the naval arsenal. In 1898, while on maneuvers, *Gustav Zede* "torpedoed" the battleship *Magenta. Gustav Zede* was a marginal design, at best, and the battleship was at anchor, but the event prompted outrage, consternation, and pride within the French Navy.

Soon after, the Minister of Marine held an international competition for a submarine with a surface range of one hundred miles and a submerged range of ten miles. There were twenty-nine entries; the winner was *Narval,* submitted by former chief engineer of the French Navy, Maxime Laubeuf.

Narval was somewhat smaller than *Gustave Zede*—188 feet, 136 tons—and more manageable. She began life with steam power, which soon enough demonstrated the problems of steam and was replaced with a diesel engine. But the next class of French boats went back to steam. It was, indeed, a very mixed bag.

However, submarines were such a popular item in France that the newspaper *Le Matin* launched a public subscription to raise money to build submarines for the Navy. With the 300,000 francs thus contributed, two submarines were built—the 118-foot-long, 143-ton *Français,* launched in January 1901, and her twin, *Algérien,* launched in February 1902.

We might note that in 1901 *Gustav Zede* demonstrated her capabilities for French President Emil Loubet and an entourage which included the Minister of Marine. If we accept that Alexander the Great's excursion was a myth and that James I did not go for a ride in Drebbel's submarine boat, M. Loubet became the first chief executive to have the nerve to go underwater. He did so, in full formal dress, frock coat and all; the French do have a certain sense of style. On maneuvers three months later—in the Bay of Ajacico, three hundred-plus miles from her base at Touluon—the submarine put a practice torpedo into the side of the moving battleship *Charles Martel* to the reported "general stupefication" of those on *Martel.*

One other foreign competitor deserving of note but not discussion was Spaniard Isaac Peral's electric *Peral,* launched 1889. She successfully fired three Whitehead torpedoes while on trials, but internal politics kept the Spanish Navy from pursuing the project.

The official Navy trial of *Holland VI* was finally set for November 12, 1898. Lieutenant Sargent returned as recorder of the Board, carefully noting every dive (there were nineteen in all), the surface speed (six knots), and the successful firing of a Whitehead torpedo to hit a target three hundred yards distant. But the Board's report wasn't a clear "pass." More work would be needed.

For the first time, Holland was not aboard at the controls of his boat. The insurance underwriters had declined to cover the inventor—whose continued good health was, at this time, the key to everything—if he was to go out on operations. His absence was certainly one reason that the Board's report noted: "The boat turns quickly, but the steering and diving gear did not work satisfactorily owing, we believe, to the inexperience of the crew." Before another trial, the Board recommended that the crew be "exercised by actual practice as to be able to make required submerged runs and steer a straight course."

But the trial had uncovered some problems in the boat, not just with the handlers. Steering and depth control *were* erratic, and after some argument with his own staff, Holland finally was convinced to make some changes: shorten the stern, move the rudders, remove the after dynamite gun the better to accommodate the gasoline engine and exhaust.

Now Holland was desperate. His struggling company did not have the money to continue and now, after twenty-five years of continual development and with the technical problems almost solved, success remained just out of reach. Just then, he found an angel—a man whose motives Holland and some in Congress would later call into question, but for now, he was an angel indeed. Among other business holdings, Isaac Rice was president of the company which made the submarine's batteries.

Rice offered to underwrite the cost of the improvements, and to guarantee Holland five years' employment, under an arrangement whereby a new company would be created and given the rights to all patents then held by the Holland Torpedo Boat Company. By February, the Electric Boat Company had been incorporated, with the Holland Torpedo Boat Company as a wholly owned subsidiary. Holland stockholders were given equivalent shares in the new company.

But there was another problem. The Navy believed that, under the provisions of the Act of 10 June 1896 which authorized purchase of two more *Plunger*-type boats, it could not buy *any* new boat from Holland until *Plunger*

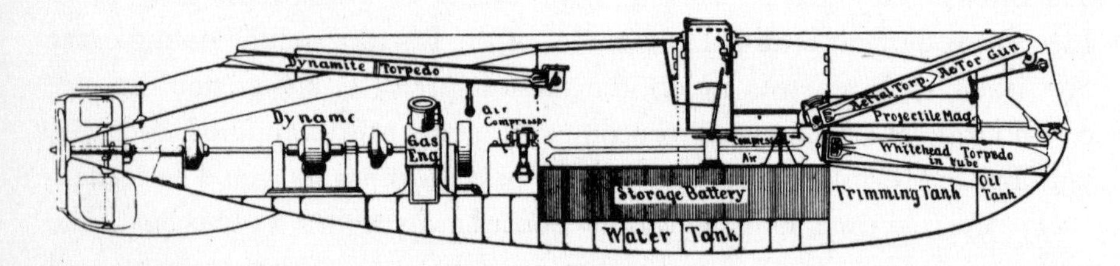

Holland VI, 1898, "which," as reported in *Scientific American,* "has just undergone a series of official tests in New York Bay and has given good satisfaction." Actually, the boat failed, demonstrating erratic control. Holland's associates believed this was because the aft control surfaces were *forward* of the propeller—as shown here—rather than the traditional position. The boat was modified, and passed the next trial. An historic irony: Holland's original scheme was proven on the U.S. Navy's 1953 experimental high-speed submarine *Albacore.* Credit: *Scientific American*

had passed all trials. The company was willing to complete *Plunger,* but since it would never pass anything as a steamboat, they offered to convert the power plant to diesel. That would take eight months and, in the end, the Navy would gain nothing but a useless compromise. Finally, a friendly senator from the great maritime state of Nevada pushed through an amendment authorizing the substitution of *Holland VI* for *Plunger.*

Trials of the modified boat were held for a five-man Board of Inspection and Survey, at the eastern end of Long Island, on November 6, 1899. The boat passed.

There was one near-disaster just before the official Navy trials. The crew was overcome by carbon monoxide, and as a result a cage of white mice became standard equipment aboard gasoline-powered submarines. Like the canaries taken into coal mines to signal deadly methane gas, a mouse would pass out—and die—before a human. British gasoline-powered boats carried on the practice of carrying mice, although the pet-loving sailors seem typically to have so overfed the mice that they spent much of the time in a food-induced stupor.

Two weeks after the trials, Frost made a formal offer to sell *Holland VI* to the Navy. Frost included specifications for an "improved" version (slightly larger, and with greatly increased power) for the second boat authorized under the revised Act. Then, perhaps borrowing a leaf from the playbooks of Fulton, Colt, and Nordenfeldt, the company arranged to demonstrate *Holland VI* to the American equivalent of crowned heads. The boat was moved from New York, through New Jersey's Raritan River and Canal (buoyed up with pontoons on either side to accommodate the shallow waters), to the Delaware River, through the Delaware Canal to Chesapeake Bay, then up the Potomac River to Washington, D.C., and a berth at the Navy Yard. The boat arrived just at Christmastime. The month-long trip was more like a royal progress than a trip; publicity was heavy, and large crowds turned out to gawk at each stop.

A trial course was set between Fort Washington in Maryland and George Washington's estate, Mount Vernon, on the Virginia shore. The channel was about four hundred yards wide, and depths ranged from thirty to thirty-six feet. Ballast was adjusted to compensate for the fresh-water environment. The morning of March 14, 1900, four boats left the Navy Yard and headed for the test range. A Navy tug had *Holland VI* in tow, and two other boats carried the official observing party. The senior officer in the Navy, Admiral George Dewey, members of his staff, the Assistant Secretary of the Navy, members of the House and Senate Committees on Naval Affairs, and naval observers from abroad (including Japan) all watched the boat dance in the water like a trained porpoise. Admiral Dewey's aide, Lieutenant Harry Caldwell, was aboard the submarine. Caldwell would become the first commanding officer of USS *Holland.*

There were four more demonstrations, attended with ever-increasing hype. The "Nordenfeldt touch" was at work: The company prepared and distributed printed programs for the trials, adorned with advertising slogans—"The Monster War Fish," "Uncle Sam's Devil of the Deep," "The Naval Hell Diver."

On April 11, 1900, the Navy bought *Holland VI* for $150,000—the original contract price of *Plunger.* The new boat had cost the company $236,615, but in marketing terms, *Holland VI* was a loss leader. The boat was rechristened USS *Holland.*

The Navy had a submarine; what next? Ah, thought some in Congress, the submarine! Warlike, inexpensive, ideal for harbor defense. Congress held yet another round of hearings. A star witness was Admiral George Dewey, who commented on his favorable impressions during the Potomac River demonstration. "I said it then, and I have said it since, that if they had two of those things at Manila, I could never have held it with the squadron I had. . . . With two of these in Galveston, all the navies in the world could not blockade that place." He suggested one benefit that the other witnesses overlooked: "The moral effect, to my mind, is infinitely superior to mines or torpedoes or anything of the kind. Those craft moving underwater would wear people out."

Rear Admiral Charles O'Neil, chief of the Navy's Bureau of Ordnance, was not so impressed. "The *Holland* boats," he testified, "are interesting novelties which appeal to the non-professional mind, which is apt to invest them with remarkable properties they do not possess. . . . in my opinion, they are ingenious contrived craft of an eccentric character which mark a step in the development of an interesting science but nothing more."

Congress ignored the minority opinion, and by August, six improved "Hollands" were on order. The Navy had not asked for them; in this case, Congress knew best. Electric Boat (which did not itself build any submarines until after World War I) contracted with building yards in New Jersey and

San Francisco. To clean up any lingering issues with the Congress or the Navy, the company refunded the $93,000 which had been paid on account for *Plunger* in exchange for another contract for one more new Holland, and the *Plunger* contract was formally canceled. The name was assigned to a newer boat (later designated A-1) which became the submarine in which President T. Roosevelt made his historic plunge. The original *Plunger*—whatever remained—would eventually serve as a target for diver training in World War I.

During war games off Newport in September, Lieutenant Caldwell and his fledgling crew made a bold effort to put the Navy on notice. *Holland,* not to be given the status of a commissioned ship of the U.S. Navy for another month, made a quiet attack on the battleship *Kearsarge* just after dark, seven miles offshore, and claimed a kill. The umpire discounted the score, noting that *Kearsarge* had already been sunk by a pair of torpedo boats; since the battleship was out of action, he reasoned, the lookouts would hardly have been operating at "full combat alertness."

However, the *Kearsarge* commanding officer, Captain William Folger, put it all in proper context. "It is clear," he said, "that the *Holland* type will play a very serious part in future naval warfare. There is no doubt whatever that the vessel at Newport can approach a turret ship unseen, either by night or day." He added, with an unwitting clairvoyance, "Her only danger is she may be run over herself by picket or larger vessel."

As for John P. Holland: there was good news and bad news. Good news: His salary was increased substantially, beginning in June 1900, to $10,000 a year—$2000 more than that of the Secretary of the Navy. Bad news: He was given his promised five-year contract, to date from April 1, 1899, in which he was "demoted" from manager to chief engineer. The businessmen were taking over the business.

11 1900–1904—Of Marketing, Foreign and Domestic

While my book, "Twenty Thousand Leagues Under the Sea," is entirely a work of the imagination, my conviction is that all I said in it will come to pass. . . . The conspicuous success of submarine navigation in the United States will push on underwater navigation all over the world. . . . The next war may be largely a contest between submarine boats.

—Jules Verne, September 1898

The Electric Boat/Holland Torpedo Boat Company had a contract—and a problem. What had only recently been an operation of one or two men, then a bare handful, had now become an industry. Plans had to be refined and duplicated; improvements had to be pursued and tested. Books had to be kept. Coordination had to be maintained with the government contracting office, and contracts had to be administered. John P. Holland himself had been eased out of control—and some of his associates believed that, as he approached sixty, age was not only catching up with him, but was about to pass him by. He had become somewhat forgetful; once, he was found wandering off in the Long Island countryside, apparently disoriented and certainly lost. He had been replaced as trial captain by Frank

Cable—because of the "insurance" issue already noted, but also because more and more he seemed too easily distracted from the job of running the boat by any and all minor problems within the boat.

Within four years, Holland would be out of the company altogether, his contract ended, striking out once again on his own with a visionary zeal but without a practical appreciation of the naval needs of the day. Two years after that, the lawyers—his and theirs—would be arguing over what part of his life's work he owned, or didn't own, and the Congress would be brought in to ponder antitrust issues.

In the fall of 1900, the company's prospects were bright; the seven boats under contract were like money in the bank. Almost. However, that business was predicated on the successful delivery and operation of an entire class of submarines which at this point was nothing more than a set of plans. The company decided that it should build, on its own account, a prototype of the "improved" design.

Thus, the problem of cash flow—there wasn't enough cash on hand. The solution was to sell or license the Holland design overseas. Zalinski's earlier efforts had planted the seeds, and Holland himself had made overtures to the British the year before. When Cable later wrote his own account of his adventures with Electric Boat, he noted the irony of a man formerly in the pay of Irish terrorists, now willing to sell his secrets to their enemy. "About this time [April 1899] he was in England in consultation with naval and shipbuilding officials. The humor of such a contact could not have been unrecognized by either party. The man who had set his heart on devising an instrument to cripple Great Britain's sea power had seen the way open to introducing his device into her navy, not as a destroying foe, but as a protecting auxiliary."

Financier Rice—now firmly in charge—began to bring in the harvest. Great Britain, Japan, Russia, Italy, Norway, Austria, Chile—all were to buy Holland boats, or boats licensed by Electric Boat, over the next dozen years.

Great Britain was to be the first international customer for a Holland. In Electric Boat's darkest hour, the shipbuilding company of Vickers, Sons and Maxim came through with cash transfusions in exchange for stock. By the end of 1903, Vickers held 30 percent of Electric Boat, an ownership position second only to that of Rice. Vickers was the licensee which would build the boats for Britain.

As noted earlier, official English interest in *Fenian Ram* was less than casual, and while the actions of Nordenfeldt may have influenced American naval planners, they had no apparent impact on the English even though *Nordenfeldt III* was built in the Vickers shipyard. However, when *Gustav Zede* "torpedoed" the battleship *Magenta* in 1898, the event got the startled attention of France's traditional rivals just twenty-one miles across the English Channel.

The English Ambassador in Paris sounded a warning: ". . . belief in the success of the invention is very likely to encourage Frenchmen to regard their naval inferiority to England as by no means so great as it is considered to be [by us]." His naval attaché underscored the point: "These submersible vessels have now reached a practical stage in modern warfare and will have to be reckoned with, and met, in future European war." He noted that it was now possible for a French submarine to launch an attack in English waters and return to port "unaided." He added, "This fact is carefully hid from the public by the authorities, though considered the greatest triumph of this new vessel."

The Royal Navy's generic disinterest notwithstanding, British politicians were cast in much the same mold as their American cousins and were very much aware of developments in France and Holland's recent progress. When Isaac Rice came calling, they were ready for the conversation. First Lord of the Admiralty George Goschen, in an address to the House of Commons 1900, first made a play to his Naval constituency: "The submarine boat, even if practical difficulties attending its use can be overcome, would seem . . . to be eventually a weapon for Maritime powers on the defensive." Then he entered into the private negotiations with Rice out of which came a contract, in October 1900, for five Holland-type boats, at a price to the government of $165,000 each. Vickers was given a ten-year sole-source position to build the boats.

The Royal Navy was, at first, an even more reluctant partner than its American cousin. The ultra-conservative Navy still practiced sail drill on armored steam warships; a proposal to modify the practice, this same year, was met with this comment; "You have got an established system and a time-honoured one, so why alter it?"

But Navy leadership finally agreed to give submarines a trial. First Sea Lord Admiral Kerr conceded, ". . . we have not only adopted the best course that was open to us, but also have done all that we can prudently do. . . . we are bound to follow up the development of the submarine boats, and thus have at our disposal whatever advantages they may possess. . . ." The modest approach was best, he added: " . . . it is not desirable to plunge too heavily as it must first be in the dark, not until experience points us in the direction in which we should work."

Third Sea Lord and Controller of the Navy Rear Admiral A.K. Wilson echoed the sentiment: "Our primary objective is to test the value of the submarine boat as a weapon in the hands of our enemies." Wilson, however, was not quite ready to go quietly into compromise. It was Wilson's personal view that the submarine was "underhand, unfair, and damned unEnglish" and that the Government should "treat all submarines as pirates in wartime and . . . *hang all crews.*" He tried to squeeze the following caveat into the contract announcement: "HM Government considers it would be to the advantage of all Maritime nations of the world if the use of the submarine boat for attack could be prohibited. . . ." The statement was not used.

In the unfortunate tradition of Holland submarines, the first British boat, *No. 1,* sank upon launching, in October 1901. Well, once again, "sank" may be a bit strong; the boat only capsized and remained floating, but it was not an auspicious beginning. There was a flaw in the plans (in which the boat was identified as *Holland No. 8*) which changed the relationship between the center of gravity and the center of flotation—in technical terms, the "metacentric height" of the boat. If the center of gravity ever is moved to a point *above* the center of flotation, in any vessel, capsize is the unavoidable result. Ships are never purposely built with negative metacentric height—the condition usually comes from putting too much cargo topside, or taking too much fuel from below without adding compensating ballast, or taking topside ice in a winter storm.

Now, the origin of that "flaw" has long been a matter of dispute. The popular version—a re-awakening of Holland's Irish sympathies—was advanced in an obituary written on the occasion of Holland's death (in 1914) by one of his former Fenian associates. John Devoy claimed that Holland had been opposed to the sale to England and, when overruled by the company, had managed to sabotage the plans. It looks plausible in the script, but doesn't play well on the stage.

Holland did indeed draft the plans for the "improved" boats, but the first of these to be laid down, several months before construction began on British *No.1,* was the company-funded prototype, *Fulton.* There were problems with the calculation of weights and volumes of *Fulton,* but these were discovered and corrected before she was launched in June 1901. It is probable that the changes never reached England. Either Frank Cable—in England for pretrial assistance and training—or the British officer in charge of submarines, Captain Reginald H.S. Bacon, spotted the errors in *No.1* (after, that is, she capsized) and made the necessary changes. *No.2* was right behind, and by 1903, the Royal Navy had all five "Hollands" in service and had started building follow-on designs of their own.

Following retirement in 1913, *No. 1* sank while under tow en route to the breaker's yard; found and recovered in 1981, she was put on public display at the Gosport Submarine Museum, near Portsmouth. However, by 1993 the combination of entrapped salt and the weather had ruined the unprotected boat almost beyond the point of no return. *No. 1* was put in a large enclosed tank through which a solvent was constantly circulating, to remove the salt and other chemical debris. Once the deterioration has been stabilized, display will be resumed.

Fixes initiated on *Fulton* were rushed to the U.S. boats then already under construction, but this "just in time" process ultimately delayed the delivery of some boats as much as two years; although six had been laid down in 1900, and *Plunger* in 1901, none were in commission until 1903.

These first U.S. submarines were at first given names, which was later changed to an alpha-numeric scheme (and still later to a combination of both). The first boats after *Holland*—in the order in which commissioned, all from January through September 1903—were: *Adder* (A-2), *Moccasin* (A-4), *Grampus* (A-3), *Pike* (A-5), *Plunger* (A-1), *Porpoise* (A-6), and *Shark* (A-7). (In the 1930s, nomenclature became even more confused. Submarines then had names, numbers, and altogether different identifying numbers painted on the sail and bow. *Squalus,* SS-192, was also S11.)

There was another issue with *Fulton,* but it involved Holland and his relationship with the company rather than the boat herself. While *Fulton* was a company-funded "prototype," she had to conform to a great degree with the boats for which the Navy had already contracted. Therefore, the Navy assigned a supervisor of construction to "guide" the project. Lawrence Y. Spear was a thorough, conservative engineer and managed to thoroughly alienate Holland; it was the *Plunger* experience all over again. Then in 1902, to Holland's amazement and chagrin, the company induced Spear to leave the Navy

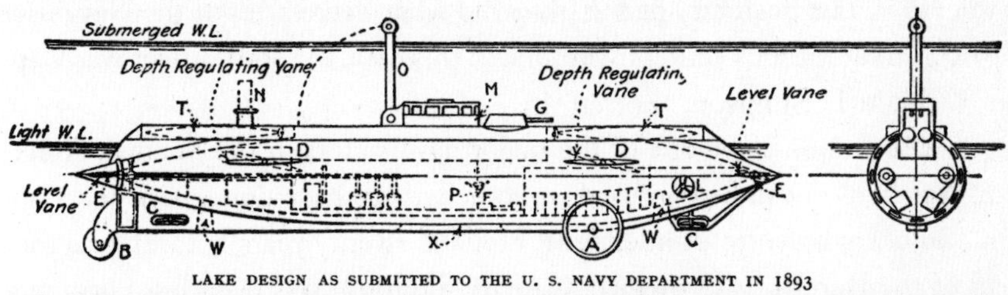

LAKE DESIGN AS SUBMITTED TO THE U. S. NAVY DEPARTMENT IN 1893

Novel features consisted in: (A) wheels for running on the bottom; (B) rudder forming also a steering wheel when navigating on the bottom; (C–C) propellers for holding vessel to depth when not under way; (D–D) depth regulating vanes or hydroplanes for causing vessel to change depth while under way and to accomplish the changes of depth on an even keel; (E–E) horizontal rudders or "leveling vanes" designed to automatically hold the vessel on a level keel when under way; (F) a weight automatically controlled by a pendulum; (P) mechanism to correct trim; (G) gun arranged in watertight revolving turret for defense purposes or attack on unarmored surface craft; (L) propeller in tube for swinging vessel at rest to facilitate "pointing" her torpedoes; (M) conning tower; (N) telescoping smoke-stack; (O) observing instrument arranged to turn down on deck when under way; (T–T) torpedo tubes, two firing forward and two aft; (W–W) anchoring weights to hold the vessel at rest at any desired depth between the surface and bottom; (X) an "emergency keel" which would be automatically released if the vessel reached an unsafe depth. She was a double-hull vessel, water being admitted to the space between the inner and outer hulls and in trim tanks forward and aft to effect submergence. A diving compartment was also provided to enable the crew to leave or enter the vessel while submerged.

Simon Lake's 1918 presentation of his 1893 proposal, which bears, indeed, some "novel features." In addition to wheels, Lake included a periscope, bow thrusters (for adjusting aim of the torpedoes) and a revolving gun turret. Innovation was Lake's strong suit; his problem was execution. Credit: Simon Lake

and take a position as vice president and naval architect. Spear became, in effect, Holland's boss.

Electric Boat worked hard to protect its monopoly position as the only supplier of submarines to the United States Navy, but there were competitors who were unwilling to abandon the field. Prime among them—and the only one to survive long enough to count—was Simon Lake.

Lake was a spiritual offspring of Jules Verne: ". . . when I was a mere schoolboy," he would later write, "I had become interested in the submarine by reading Jules Verne's 'Twenty Thousand Leagues Under the Sea'" Lake claimed to have designed his first submarine at the age of fourteen, but his father dissuaded him from pursuing a line of endeavor "that great engineers had given a lot of attention to."

In 1893, Lake's was one of the three submissions in the submarine competition which had been in part engineered by George Baker. Lake's scheme included a set of wheels under the boat by which it could run along the bottom. Some years later, he was told by one of the members of the selection committee that the majority favored the Lake design but that the senior member objected out of concern that such a boat "running on the bottom on wheels . . . might run off from a precipice and go down head first . . ."

Soon afterward, in trying to accommodate to what he had observed were the ways of the government, Lake tried to get an audience with the Secretary of the Navy. The Secretary of the Navy did what Secretaries of the Navy usually do with ambitious young inventors; in this case, the staff officer was Captain William T. Sampson.

Sampson, soon to be one of the major naval actors in the Spanish American War (and not to be confused, as some writers have done, with the Captain *Simpson* who corresponded with Holland twenty years before), was then head of the Bureau of Ordnance. Captain Sampson was even less encouraging to Lake than Simpson had been to Holland; Simpson, at least, had been cordial.

Lake struck out to find support on his own—and struck out. Finally, his aunt and uncle put up the money to finance his first, very crude, homemade underwater "test vehicle" in 1894. *Argonaut Jr.* was a self-propelled diving bell—a weighted, waterproofed wooden box on wheels which was moved along the bottom by a hand-crank. Public demonstrations—the thing actually worked—brought in enough money to finance the building of *Argonaut I* in the same Baltimore yard—and at the same time—where the abortive *Plunger* was taking shape. *Argonaut I* used a gasoline engine for both surfaced and submerged propulsion (drawing air from the surface through breathing tubes). By adjusting buoyancy to suit bottom conditions, the boat could run over almost any type of ocean bed. At this point Lake envisioned his boats more for use in salvage and exploration than as warships, although that was not ruled out.

In September 1898, the thirty-six-foot-long *Argonaut I* made an open-ocean passage from Norfolk to Sandy Hook, prompting a cable from Jules Verne:

> While my book, 'Twenty Thousand Leagues Under the Sea," is entirely a work of the imagination, my conviction is that all I said in it will come to pass. A thousand-mile voyage in the Baltimore submarine boat is evidence of this. The conspicuous success of submarine navigation in the United States will push on under-water navigation all over the world. . . . The next war may be largely a contest between submarine boats.

Lake had some problems with gasoline fumes in *Argonaut I,* so he shifted the fuel tanks outside the pressure hull. Soon after, Laubeuf adopted a similar arrangement—similar enough that Lake would later claim patent infringement, but without resolution. *Argonaut I* was enlarged and improved in 1899, and redesignated *Argonaut II.*

Lake began his first truly sophisticated submarine—*Protector*—in 1902. At sixty five feet, and 130 tons, *Protector* was a substantial effort. She had two pairs of hydroplanes, a diver's airlock, wheels, and Lake's patented "omniscope"—a form of periscope. Holland still resisted using periscopes; he thought they were too limiting, preferring to use the "porpoising" method for sighting on a target.

The pre-World War I submarine periscopes were indeed limiting—and because they were of a fixed height, the submarine still had to porpoise to bring the head above the surface. There were several issues with the length of the periscope tube. Too short, and the boat might be almost awash and perhaps revealed as waves passed over. Too long, and the image was dim and, as experience soon proved, movement of the shaft through the water set up vibrations which rendered the periscope unusable. Vibrations could be prevented if the submarine was not moving, but most submarines could not maintain depth without some forward motion. "Housing" periscopes, which could partially extend from and retract into the hull, were an improvement—although the instant change in submarine volume raised some depth control problems, and it was soon discovered that a periscope tube of any useful length would start vibrating (thereby creating a useless image) if the submarine was moving. The depth control problem more or less went away as submarines became ever larger. Vibration was cured by having a thicker tube, with just the last few feet narrow to minimize visible wake. A narrow head, of course, limits the optics which can be installed, but over the years, increasingly sophisticated optics allowed the head on a World War II attack scope to be as narrow as 1.5 inches.

The early periscopes also were mounted in a fixed ahead position; if the operator wanted to view some other direction, the sub had to change course. The first rotating periscopes brought an optical dilemma: If the operator's viewing position was fixed while the periscope head turned, the image would rotate vertically as well as horizontally. At 90 degrees (the head pointed on the beam) the horizon would be vertical; at 180 degrees (the head pointed astern) the image would be upside down. If the head and the viewing position were locked together to create a "walk-around" periscope, there had to be space in the control room for the operator to walk around, which required a larger control room, which added both volume and weight to the submarine.

These problems were resolved by the invention of the "self-erecting" periscope, which interposed a series of geared prisms in the optical path and by the steady growth in submarine size.

The U.S. Navy first tested a periscope in 1902–03, comparing British units with the Lake "omniscope." The Lake unit was superior; the Navy asked for pricing, but Lake wanted to sell submarines, not bits and pieces.

However, Lake first had to overcome a law which, as written, gave Electric Boat a de facto monopoly. His successful challenge resulted in the Act of 3 March 1903, under which the next submarine procurement would be an open competition.

It soon enough developed that this would be a sail-off between *Protector* and an upgraded *Fulton.* Then, when it appeared that Electric Boat might not have its boat ready in time and wanted a delay, Lake cried foul and forced a change in the rules: There would be a "competition" even if only one boat was entered.

Lake was to be hoist on this, his own, petard. By the time the Navy was ready for the competition, winter weather had closed in forcing a delay. Lake, who had his own cash-flow problems, took the bird in the hand and accepted an offer to sell *Protector* to the Russians, just then at war with the Japanese. He went along to Russia to demonstrate the operation of his boat.

Protector was renamed *Oestr,* put on two railroad cars, and taken six thousand miles from the Baltic to the Pacific naval base at Vladisvostok. The Russians also ordered five more Lake boats, to be built at the Newport News Shipbuilding and Drydock Company in Newport News, Virginia. Lake used the revenue to finance his next designs, and sold other boats to Russia and to Austria-Hungary, and as early as 1905 conversed with (and later claimed to have influenced the submarines of) Germany, France, and Italy.

However, Lake was the inveterate tinkerer, unable to keep his hands off a design even when the boat was almost finished. There was always one more improvement to make. As a result, his boats never met delivery schedules. The first submarine he managed to sell to the U.S. Navy—*Seal,* G-1, laid down in February 1909—was delivered two years, five months, and fifteen days late. G-1 was virtually obsolete by the time she entered service in 1912, although she set a depth record, 256 feet, in 1914. The Lake Torpedo Boat Company went out of business in 1924.

12 1904–1912—Of Entrepreneurs vs. Businessmen

These suits have had the effect of frightening off the capital that I had enlisted . . . to prevent me from building a boat and going into competition before the Navy Department with the submarine boats now being built by the Electric Boat Company under my old patents.

—*John P. Holland, letter to the Naval Affairs Committee, U.S. House of Representatives, February 8, 1906*

The U.S. Navy submarine sail-off began on June 1, 1904, with only one entrant; thus *Fulton* was an easy winner. The trials board recommended continued purchases from Electric Boat, which led to the next two classes of U.S. submarine, "B" and "C." However, by the end of the month *Fulton* herself was aboard an English steamer en route to Russia. In the impartial (one might say immoral) mode of the man for whom their prototype had been named, the company played both ends against the middle.

Japan already had proven to be a ready customer for Holland submarines, perhaps because it had a weak navy, perhaps because it had an infant navy, lacking tradition and therefore lacking professional bias. Several very senior officers had climbed all over *Holland VI* in July 1897, leaving propriety behind and stripping to the waist in the broiling summer heat. Another took a submerged ride in October. While Commander Francis Morgan Barber was stationed in Japan, he had "represented" company interests, buttressed by Zalinski's visit in 1894. A junior officer, Lieutenant Kenji Ide, was Japanese Naval Attaché at Washington during *Holland VI's* public demonstrations.

Thus, it was no surprise when, shortly after the start of the war with Russia, Japan signed a contract for five "improved" Hollands, to be built at Massachusetts's Fore River Shipyard and shipped in sections to be finished near Tokyo. Now, on June 25th, Electric Boat shipped *Fulton* to Russia just a short time behind *Protector.*

There was a legal issue to be confronted—or avoided—by both companies. With a war going on and with the United States as a neutral bystander, the law prohibited U.S. corporations from selling warships to belligerents. Lake skirted the issue by removing the batteries from *Protector* (thereby, in theory, removing her ability to operate as a submarine) until she made rendezvous with a Russian-chartered ship, to be hoisted aboard and taken to a port on the Baltic. Electric Boat skirted the issue with the submarines for Japan by shipping them as components. In the case of *Fulton,* their illegality was more straightforward. The company simply loaded *Fulton* on a barge and towed it out past the three-mile limit to a waiting freighter in international waters. Electric Boat's trial captain, Frank Cable, took his crew to the Baltic to train the Russians, missing Lake by only a few weeks, and then was off to Japan to supervise assembly and training for the other side.

As it worked out, none of the submarines made it into the Russo-Japanese War, which effectively ended with the overwhelming Japanese naval victory at the Battle of the Tsushima Straits in July 1905. Mines, however, were a major factor, sinking eighteen warships on both sides. The Japanese tried using torpedo boats, but these did not prove to be very effective. They fired 370 torpedos and scored only seventeen hits. This poor showing led naval planners—especially in England—to denigrate the value of the Whitehead torpedo, by whatever means it was carried into battle, for at least the next ten years.

A longer-range outcome of the war: Japan turned its interest east and began studying the nation which it viewed as its natural enemy in the Pacific.

Just a year after Japan's submarines were launched, the other major aggressor of the 20th century began using submarines. Germany had taken no serious interest in a handful of homegrown and inconsequential submarines over a period of fifty years, and was now interested only because of the efforts of her neighbors.

The limited German experience began with Bauer's *Brandtaucher* in 1850, followed by a boat built by Frederich Otto Vogel—which sank on trials in 1870. Two *Nordenfeldts* were built under license in 1890, to no great purpose, and the Howaldt yard at Kiel built a boat in 1891—one of the classic disappearing submarines, as nothing further is known. In 1897, Howaldt built another: This boat, designed by a torpedo expert, looked like a torpedo with a diver's helmet attached as a sort of conning port. The boat was scrapped in 1902.

Spanish designer Raimondo Lorenzo d'Equevilley had worked in France with Maxime Laubeuf, but when the French submarine industry became confused, he went to Germany in search of work on his own account. He was, at first, given a most tepid reception. As Admiral Alfred von Tirpitz recently had said (correctly, at the time), "The submarine is, at present, of no great value in war at sea." But von Tirpitz did not share the British Navy's willingness to at least give the submarine a trial. "We have no money to waste on experimental vessels," he said. "We must leave such luxuries to wealthier states like France and England."

Rebuffed by the Navy, d'Equevilley sold his plans to the Krupp Germania shipyard, which on speculation built *Forelle* ("Trout") in 1902—forty feet long, 15.5 tons. Powered only by electricity, like contemporary French boats, *Forelle* was not a practical warship. The only way the batteries could be charged was when the boat was in port. But Kaiser Wilhelm II was impressed, and his brother, an admiral, went out for a ride.

D'Equevilley knew something about marketing. He published a book in Germany (and in German) in which he traced the history of submarines and offered an echo of Fulton. "As exaggerated as it may sound," he wrote, "who knows whether the appearance of undersea boats may put an end to naval battles."

Krupp engineers worked with d'Equevilley on an improved design, which became the Karp-class: gasoline-powered on the surface, with an onboard battery-recharging system. They also adopted a double-hull design, which allowed fuel to be carried outside the pressure hull, which reduced fuel hazards while leaving more room for supplies and crew. Krupp was later to be sued on this feature by Simon Lake, claiming patent infringement, but by then Germany was at war and the suit was rather meaningless. Russia ordered three of these 205-ton boats—which, like all others Russia purchased at this time, arrived too late to be of any value in the war with Japan.

The German Navy finally ordered its first submarine—a slightly modified Karp, burning kerosene rather than gasoline. A description of the first U-boat, then under construction, was carried in the February 1905 issue of the German Navy League's magazine, *Die Flotte.* The author was not very enthusiastic:

> Even today, the utility is quite limited. Darkness and sea state make it almost completely blind, and its short range doesn't permit it to go far from its support base. The modern submarine can therefore only be regarded as an imperfect weapon which only has prospects of success under specific conditions and for limited purposes like harassment of a blockading fleet.

U-1, launched in April 1906, was 139 feet long and had a surface displacement of 239 tons, a surface speed of eleven knots, nine knots submerged, with a range of two thousand miles. The German Navy did not take much encouragement from sea trials: "Her small displacement renders this boat unfit for operations at any distance from the coast, as observations taken showed that her employment in the high seas is attended with danger." In 1908, the virtually identical *U-2* was added to the fleet without much notice; however, faced with a French force of sixty submarines and almost as many British submarines, Germany was persuaded to began a gradual development of the U-boat.

Soon enough, the Navy magazine, *Nauticus* welcomed the new additions to the fleet: ". . . the nimbus of secrecy which had earlier surrounded the submarine politics of the leading nations like a thick veil has to a considerable degree now been lifted. . . . the undersea boat has gained access to all the navies of the world—not as an object of experimentation, but as a fully fledged weapon of war."

As his five-year contract came to an end, with an effective date of March 31, 1904, John P. Holland resigned from the Electric Boat/Holland Torpedo Boat Company. One of his last contributions was to design a depth-safety regulator, which automatically would blow out a 1200-pound reservoir of water when a pressure sensor was triggered by a preset depth. It was installed in the *Plunger* (A-1) in which President Roosevelt made his underwater excursion in 1905. (In later years, the device came to be viewed with horror, and was removed by wartime commanders who *did not want* arbitrarily to pop to the surface while under attack, no matter how much beyond the rated limit the boat might have gone.)

Holland probably could have continued at Electric Boat, but on their terms, not his. He was not happy with the changes in management, not happy that his ideas were increasingly set aside—or ignored. Holland, having solved the basic problems, was ready to push the state of the art—underwater speeds easily above twenty knots. Holland the inventor wanted underwater sprinters, but Spear the naval constructor better understood what the Navy needed; he wanted marathon runners. They would submerge only when necessary.

The real problem, which Spear understood but Holland did not address, was that there was no propulsion system then in existence or contemplated which would allow a boat to operate submerged at full speed for more than an hour or two. A submarine on patrol would spend infinitely more time slogging through the waves than dashing underwater to take a firing position.

Spear began a shift away from the short, fat, streamlined Holland boat toward the long, skinny hull of a surface ship, with a buoyant, flared bow to ride up over the waves. Holland's experience with surface ships was limited to three trans-Atlantic crossings as a passenger, and he seems not to have appreciated the difficulties his porpoise-shaped hulls would have on open-ocean voyages. In truth, submarines which retained the Holland-type hull had problems in a seaway; the British A-boats would nose into a large wave—with a tendency to start heading down. Several of them kept on going, with fatal results.

Holland took his ideas along with his independence, and quickly gathered a few former associates into a new, as-yet-unorganized venture. His first post-employment contract (which may have been based on work he did sub rosa while still on the payroll at Electric Boat) was with the Japanese. They bought a set of plans for two enlarged, high-speed submarines, to be built in Kobe, Japan, under the supervision of a Holland man. These boats were seventy five and eighty-seven feet long, and according to one source could make sixteen knots *submerged.* Japan's five "improved" Hollands—the boats ordered from Electric Boat in 1904, built in sections in Massachusetts, and assembled in

Yokosuka in 1905—had a maximum submerged speed of seven or eight knots.

Holland's next hull form was tested as a scale model in the experimental tank at the Washington Navy Yard; the results pointed to calm-sea surface speeds as high as twenty-two knots, and Holland predicted that a submerged speed of twenty-two knots was within reach. He submitted a proposal to the Navy Board, which gave him a hearing but offered the opinion that such a fast boat would be too hard to maneuver and therefore dangerous—to herself—when submerged; the board argued that a submarine should never go faster than six knots underwater.

Nonetheless, Holland was satisfied that the world needed an innovative, forward-thinking, clearly focused submarine company. However, he did not launch himself into that abyss before checking with friends, colleagues, and former associates who had left Electric Boat out of their own frustrations. He found sufficient support to incorporate "John P. Holland's Submarine Boat Company" on May 18, 1905.

He solicited invitations from shipbuilders in England, Holland, and Germany. However, the Electric Boat Company made certain that each shipbuilder knew that Electric Boat, not Holland, controlled the patents then key to submarine construction. Each shipbuilder sent Holland a letter of demurral.

Holland was stunned to learn that he no longer held international rights to his patents. He had registered them years before in Britain, Germany, Sweden, and Belgium, but had neglected (or was unable, the result was the same) to keep them current. In 1898, Frost, acting in behalf of the company, had paid five years' worth of back taxes and at the least acquired a unassailable lien. No European shipbuilder was interested in challenging the claim.

Did the Electric Boat Company play fair in all dealings, especially those involving John P. Holland? That question is well outside the scope of this book, but three points might be offered. One, favoring Electric Boat: Holland had assigned his patents to the Holland Torpedo Boat Company, long before Rice and Electric Boat entered the picture, to give the fledgling business some tangible assets. Prudent investors want to own a piece of *something* beyond the promised energy and skill of one man. Patents endure for a finite period; the man may not. In return, Holland had been given financial security which increased to a respectable level well above average.

Two, not favoring Electric Boat: There is a private record—a letter—dated April 7, 1899, and preserved in the Library at the Submarine Force Museum at Groton, CT, from a former president of the Holland Torpedo Boat Company, in which he complains that he was not given an equitable share of Electric Boat stock in the acquistion of the former by the latter. Unless properly compensated, he threatened to expose the names of Naval officers for whom the company was holding stock in trust. This matter presumably was settled, as it never reached public attention.

Three, neutral: There is a public record, laid down in *The Hearings Beginning 9 March 1908,* wherein Holland and Rice each had the opportunity to make his own case. Like the transcript of any courtroom drama, the published version makes interesting reading, but like a courtroom drama which ends in a hung jury, the unresolved issues must remain, well, unresolved.

This drama began when Electric Boat went to court, seeking an injunction to block Holland's new business. Holland laid it all out in a letter to the Chairman of the House Committee on Naval Affairs, February 8, 1906:

> I am the inventor of the Holland submarine boat, now in use in the United States Navy and in Europe. My old patents, to the number of about twenty, are owned by the Electric Boat Company. . . . Since the expiration of my contract with the Electric Boat Company I have devoted myself to remedying the defects of my old inventions, and perfecting designs by which the low speed of the present Holland boats can be increased three or four times. Having perfected these inventions until I was sure I could obtain about 25 knots per hour submerged, and after making numerous other alterations, greatly improving the efficiency over my submarine boats now in use in the Navy, I procured the organization of a company, "John P. Holland's Submarine Boat Company" May 18, 1905, with sufficient capital to build a boat under my new plans and inventions, and was about to start work, when the Electric Boat Company filed a suit against me in the Court of Chancery of New Jersey, applying for an injuction, and claiming substantially that I had agreed to assign to them all my inventions and patents during the term of my natural life. Two other suits have been started, one against my new company in the United States Circuit Court to enjoin the use of the name "Holland"; the other against me personally, alleging a verbal contract never to compete with the Electric Boat Company. . . .

> My contract with the Electric Boat to act as their engineer, and to give them my patents and inventions, was for the five years during which I acted as engineer, and no longer. . . .
>
> These suits have had the effect of frightening off the capital that I had enlisted, and I have not been able to get the capital to build my new boat, by reason of these suits. The only object of these suits was to prevent me from building a boat and going into competition before the Navy Department with the submarine boats now being built by the Electric Boat Company under my old patents.

Holland asked for an opportunity "to present a proposition to the Secretary of the Navy to cause my plans and new inventions to be thoroughly examined by a board of experts, and if favorably reported on, that the government may build the same in its yards under my supervision, and pay me a reasonable royalty." In September, he made a similar request directly to Secretary of the Navy Charles Bonaparte (another American submarine inventor pleading with another Bonaparte!) to no avail. In time, the lawsuits were all dismissed, but by that time, it was too late. His business never recovered.

In an interview with the *New York World* circa 1906, Holland made a comparison—unfavorable—between the submarines then being delivered to the U.S. Navy by his former company and those just going into service in Japan.

> They are building boats designed and fitted to accompany a fleet in any kind of weather for any distance and at any speed. . . . Our boats cannot travel with a fleet, and they cannot venture far from port. Japan's boats work; they don't do stunts. Our submarines, I am sorry to say, are now a joke. My patterns have been subjected to the treatment of young, inexperienced engineers who professed to know more about problems I had battled for years, and ruined.
>
> It is amazing how the United States can spend millions for submarines and then get really nothing compared to what skillful Japanese engineers are building for their country.

Holland was being disingenuous—or downright dishonest. The only submarines "skillful Japanese engineers" were then building were the two he had designed, or perhaps the first of a series of boats Japan subsequently purchased from Great Britain, France, and Italy. Later, at the end of World War I, Japan acquired seven German U-boats as war reparations and hired as many as eight hundred German specialists to help them develop—and build—submarines in Japan. Then, and only then, did "skillful Japanese engineers" move out on their own.

In April 1910, one of Holland's Kobe-built submarines—Japanese *No.6*—sank in sixty feet of water, without hope of rescue. *No.6* had been running submerged using the gasoline engine, taking suction through a primitive version of the "schnorkel," when waves washed over the open end, the float valve did not close, and water flowed into the boat. Commanding Officer Lieutenant Sakuma kept a running log of the almost three hours that he and his fellow crew members awaited the inevitable. "Words of apology fail me for having sunk His Majesty's Submarine No. 6," he wrote. "My subordinates are killed by my fault, but it is with pride that I inform you that the crew to a man have discharged their duties as sailors should with the utmost coolness until their dying moments."

He asked that this accident not be held against Japan's adoption of submarines, urged the Emperor to continue the search for the ideal boat, said farewell to friends and relatives, and closed the log: "My breathing is so difficult and painful. It is now 12:40 pm."

In January 1909, Ensign Chester Nimitz reported for duty for training in submarines. He had not volunteered for this duty, nor was he interested in serving on something which was, in his words, "a cross between a Jules Verne fantasy and a humpbacked whale."

Nimitz had just left command of the destroyer *Decatur*—an assignment marred by a grounding in ill-charted waters. Submarines were not a punishment, although at first he may have thought so. Nonetheless, he soon had commanded three submarines in a row: *Plunger, Snapper,* and *Narwhal,* and almost singlehandedly began—and won—a campaign to replace the noxious and dangerous gasoline engine with the newly available diesel.

Well, sort of newly available. In 1899, Holland had approached the U.S. agent for inventor Rudolf Diesel, Adolphus Busch of St. Louis, but for reasons unknown, nothing happened. In any event, the fourth submarine Nimitz commanded, *Skipjack,* became the first U.S. diesel boat in 1911, and Nimitz was then perhaps the leading U.S. Navy authority on diesels.

In June 1912, the twenty-seven-year-old lieutenant addressed the Naval War College on "Defensive and Offensive Tactics of Sumbarines"; an unclassified version was published as an article in the December issue of the *Proceedings* of the U.S. Naval Institute. His concepts were routine for the day—harbor defense, fleet support. He offered one innovative method for forcing enemy surface ships to pass through a submarine zone: "drop numerous poles, properly weighted to float upright in the water, and painted to look like a submarine's periscope."

Following the successful introduction of diesels into submarines, Nimitz was assigned to supervise a trial diesel installation in a surface ship, the oiler *Maumee*—a job which lasted until October 1916. Nimitz was then assigned as executive officer, and was a key player in the development of underway refueling techniques which continue in use to the present day.

13 1904–1913—Of Young Whippersnappers and Professional Intransigence

It is astounding to me, perfectly astounding, how the very best amongst us fail to realize the vast impending revolution in Naval warfare and Naval strategy that the submarine will accomplish!

—*Admiral John A. Fisher, Commander in Chief Portsmouth, April 20, 1904*

For more than one hundred years, the development of submarine weapons—boats and torpedos—had been largely centered in the United States. However, once the Holland boats began to enter service around the world, the focus shifted from invention to implementation and western Europe became the proving ground.

Of course, while individual officers and politicians were pushing into the future, the corporate navies of western Europe were no more enamored of the submarine boat than were their contemporaries in America. Some were even more handicapped by the past than others, having inherited styles and attitudes based on four-hundred-year-old traditions and a firm sense of what was right and proper, probably sent from God and therefore proof against professional heresy. Many leaders of the British Navy—and the French and German and Russian as well—did not see, did not *want* to see, the signs warning of change.

It is not as if the signs were hidden: On their first fleet maneuvers in March 1904, the five British Hollands were assigned to defend Portsmouth and managed to "torpedo" four warships—including the battleship on which Admiral A.K. "Damned UnEnglish Weapon" Wilson now flew his flag. The admiral, it was intimated, was not amused. He also, apparently, was not enlightened.

On a more somber note, those maneuvers brought the first submarine disaster—the accidental sinking of *A.1.* This was a brand-new boat, first of a British-designed class of enlarged Hollands. On March 18, 1904, while maneuvering to take position for a shot at a warship, *A.1* became the fulfillment of Captain Folger's prophecy of 1900 and was run over by an unwitting passenger liner, *Berwick Castle.* The master—who knew that maneuvers were being conducted—thought he had struck a practice torpedo, and sent a radio message to that effect as he continued on his way to Hamburg. A few hours later, that message was handed to the submarine flotilla commander just as he was becoming concerned about the whereabouts of *A.1.* The ship sent out to search the area found a patch of white water, bubbles rising. There was no method by which any survivors could be rescued.

A.1 was salvaged a month later and refitted for service, and went on to establish a memorable record in World War I.

We have from time to time pointed a critical finger at many naval leaders; but "many" is not "all." It certainly does not include the single most influential British naval officer of the day—or the decade, Admiral John Arbuthnot (Baron) Fisher, called "Jacky" in a profession which cherished nicknames almost as much as tradition.

Admiralty House
Portsmouth
April 20th, 1904

My Dear Friend,

I will begin with the last thing in your letter, which is far the most important, and that is our paucity of submarines. I consider it the most serious thing at present affecting the British Empire. That sounds big, but it is true. Had either the Russians or the Japanese had submarines the whole face of their war would have been changed for both sides. It really makes me laugh to read of Admiral Togo's eighth attack on Port Arthur! Why! Had he possessed submarines it would have been all over with the Russian Fleet caught like rats in a trap. Similarly, the Japanese Admiral Togo outside would never had dared to let his transports, full of troops, pursue the even tenor of their way to Chemulpo and elsewhere!

It is astounding to me, perfectly astounding, how the very best amongst us fail to realize the vast impending revolution in Naval warfare and Naval strategy that the submarine will accomplish! (I have written a paper on this but it is so violent that I am keeping it!). Here, at Portsmouth, just to take a single instance, is the battleship *Empress of India* engaged on maneuvers and knowing of the proximity of submarines, the Flagship of the Second Admiral of the Home Fleet, nine miles beyond the Nab Light (out in the open) so self-confident of safety and so oblivious to the possiblities of modern warfare that the Admiral is smoking his cigarette, the Captain is calmly seeing defaulters on the half-deck, no one caring an iota for what is going on, and suddenly they see a Whitehead torpedo miss their stern by a few feet! And how fired? From a submarine of the 'pre-Adamite' period, small, slow, badly fitted, with *no periscope at all,* and yet this submarine followed that battleship for a solid two hours under water, coming up gingerly about a mile off every now and then like a beaver! Just to take a fresh compass bearing on her prey and then down again.

Remember that this is done (and I especially want to emphasize the point) with a Lieutenant in command of the boat out in her for the first time in his life on his own account, and half the crew never out before either. Why, it is wonderful! And so what results may we expect when bigger and faster boats and periscopes more powerful than the naked eye (such as the latest pattern one I saw the other day) and with experienced Officers and crews and with nests of the submarines acting together?

I have not disguised my opinon in season and out of season as to the essential, imperative, immediate, vital, pressing, urgent (I cannot think of any more adjectives) necessity for more submarines at once—at the very least twenty-five in addition to those now ordered and building and 100 more as soon as practicable, or we shall be caught with our breeches down just as the Russians have been!

And then, my dear friend, you have the astounding audacity to say to me: 'I presume you only think they (the submarines) can act on the defensive'!! Why, my dear fellow, not take the offensive? Good Lord! If our Admiral is worth his salt he will tow his submarines at eighteen knot speed and put them in the hostile port (like ferrets after the rabbits) before war is officially declared. Just as the Japanese acted before the Russian Naval Officers knew that war was declared.

In all seriousness I don't think it is even faintly realized the immense, impending revolution which the submarine will effect as offensive weapons of war.

When you calmly sit down and work out what will happen in the narrow waters of the Channel and the Mediterranean—how totally the submarines will alter the effect of Gibraltar, Port Said, Lemno and Malta, it makes one's hair stand on end.

Yours,
J. A. Fisher

Admiral Fisher was then Commander in Chief at Portsmouth; soon, he had been appointed First Sea Lord and was putting his ideas of a formidable fleet—surface and submarine—into reality. Fisher is best known for the so-called "all-big-gun" battleship, wherein secondary batteries, which he came to realize had little utility in a major ship duel, were eliminated. All firepower was concentrated in the main battery; more large-caliber guns on deck, more large-caliber ammunition in the magazines. Any smaller guns were intended only for defense against torpedo boats. The first in this class, and the ninth British ship to bear the name, was *Dreadnought*.

But Fisher had perhaps an even greater influence on British submarine development. From 1906 to 1914, he put 5 percent of the total shipbuilding budget into submarines, and by 1914, Britain had the largest submarine fleet in the world—seventy-four boats in service.

Having submarines and *using* submarines were two quite different matters. All navies in all nations were caught up in the notion of "harbor defense."

Even if the submarine were to be given great range, to what purpose? Writing toward the end of the last century, the American Naval officer Alfred Thayer Mahan had set the theme:

> It is not the taking of individual ships or convoys, be they few or many, that strikes down the money power of a nation: it is the possession of that overbearing power on the sea which drives the enemy's flag from it, or allows it to appear only as the fugitive: and by controlling the great common, closes the highways by which commerce moves to and from the enemy's shores. This overbearing power can only be exercised by great navies.

This was the theme that drove shipbuilding budgets, defined strategy, underscored training. "Control of the seas" by overbearing power; the fleet-in-being. Mahan agreed that the *guerre de course*—raiders sent out to disrupt commerce—was useful, but only to a point. Raiders could annoy, discomfort, cause damage to an enemy, but never be decisive. This was the tactic of the "weaker nation," which had no recourse. The stronger nation—here, and always, Great Britain—would immediately take the attack to the enemy raiders, and render their efforts ineffective; for the short time it would take for this to be effective, the merchant ships would be directed to sail on diverse and unusual routes, and thus complicate the enemy's search problem.

Mahan is the embodiment of the prophet without honor at home; his writings were largely ignored by the U.S. Navy. A superior officer complained that Mahan spent too much time in nonproductive pursuit, and noted in his annual report of fitness, "It is not the business of naval officers to write books." And yet, the British, the Germans, the Japanese devoured his writings, scouring every page for wisdom.

Thus, Mahan was influential—but he wasn't always right. The turn-of-the-century naval historian and journalist Fred T. Jane punctured one balloon. No nation, he wrote, had ever launched a *guerre de course* against a nation as vulnerable as modern-day England, which imported nearly two-thirds of its food supply and did not, as a rule, have more than a month's or six weeks' supply of food on hand at any one time.

Well, if merchant commerce was at risk, how best could it be protected? There weren't many choices. Most effective, of course, would be to assign a military escort to each individual ship—effective, but impossible.

In 1905, a British conference studied three methods for protecting merchant ships against raiders, including submarines: patrolling trade routes, establishing fixed mid-ocean stations, and convoy.

For the first, naval forces would patrol the most heavily traveled areas and merchant ships could be left each to seek her own course. They would be scattered all over the ocean in such disarray that an enemy could only pick them off one at a time, and not frequently.

For the second, merchant ships would be routed through designated stations, each station patrolled by naval forces on alert, ready to respond to a radio call for help.

For the third, merchant ships would be required to sail in convoy, grouped together under the protection of a few naval escorts. Convoy had always offered the maximum protection for the merchant fleet. The British had employed convoy since at least the year 1215, and the Spanish had guarded their New World treasure fleets from British marauders by convoy. So helpful was convoy in time of war, that by the end of the 18th century the insurance industry was assessing significantly higher premiums for ships electing to sail independently rather than in convoy.

But there was something not very, well, *naval* about convoy. Mahan was invoked; the capital ship was the capital of naval warfare. The conference offered so many arguments against convoy that convoy was last choice—or no choice:

The concept was out of date, a leftover from the days of the sailing ship.

Sailing ships of yore were at the mercy of the wind, were slow, and needed constant protection when sailing through danger zones. Modern ships moved so quickly that they would pass through danger zones untouched.

Convoy offered too a rich and tempting a prize for the attacker, easily discovered by spies on land (reporting the assembly of the convoy) and smoke trails at sea.

Because of time lost in assembling the convoy and in conforming to the speed of the slowest member, followed by delays in unloading because of port congestion, convoy posed an unacceptable financial burden on ship owners.

Convoy kept escorts tied down, rather than being free to track the enemy.

And yet, in one of his less frequently read books, *The Influence of Sea Power upon the French Revolution and Empire 1793–1812,* even Mahan admitted that convoy

> . . . when properly systematized and applied . . . will have more success as a defensive measure than hunting for individual marauders—a process which, even when thoroughly planned, still resembles looking for a needle in a haystack.

But note well: He wrote *as a defensive measure.* Naval professionals did not wait for the enemy to come to them, they went out after the enemy. The 1905 conference chose the second alternative: Confine the merchant traffic to clearly defined routes, passing through focal points guarded by cruisers on station. Through the use of radio, the cruisers would be altered to enemy activity.

Wilson, now Commander in Chief of the Channel Fleet, had his own objection to convoy: British ocean trade was simply "too gigantic" to be handled in regimented fashion. But he also rejected the conference plan, although not for what would seem the logical reason—that it might give the enemy a road map for success. He chose instead a modification of the first alternative: Let the merchant ships disperse, and let the naval forces take aggressive, active, immediate attack to the enemy, to pick off the raiders as they put to sea. It was the tactic of a bold tactician. It did not, however, make allowance for any initiative on the part of the enemy.

The concept was tested in the 1906 war games: Merchant ships were moved across the northeast basin of the Atlantic, while the Fleet went hunting. In the first two weeks of the "war," fifty-two of the ninety-four merchants ships were "sunk," but no one seemed bothered by this loss rate because, as the Chief Umpire predicted, "It is practically certain that the commencement of the third week of the war would have seen all commerce-destroying ships either captured or blocked in their defended ports. . . ."

The Admiralty agreed: ". . . although a temporary commercial crisis might possibly be caused in London by this form of attack, the complete defeat of the aggressor could not long be delayed, with the result that public confidence would be quickly re-established and the security of British trade assured."

Admiral Wilson was also in agreement. "Attacks on commerce," he wrote, "might cause loss and annoyance, but with ordinary care they are not likely to do us any vital injury, and the enemy's vessels engaged in them will be sure to be gradually sunk, captured, or interned in neutral ports." But the same time, he acknowledged that it would be foolish to expose the fleet to torpedo attack—the great ships should be held back in protected harbors while the destroyers went forth to take care of "the small craft menace."

The submarine, of course, had barely been counted in any of these deliberations. The submarine was limited to harbor operations. Of course. But whose harbor? In the 1910 maneuvers, one submarine (the first of the D class), operating more than five hundred miles from base, "torpedoed" two cruisers as they left port.

The Admiralty set up an Anti-Submarine Warfare Committee in March 1910. They conducted a range of experiments—firing machine guns at periscopes, trying to hit a submerged boat with torpedoes launched from destroyers, testing the ability of an aircraft pilot to spot a submarine. No useful recommendations seem to have emerged.

A few years later, a former member of the committee, now a vice admiral, complained that the Navy appeared to have no means to hand of preventing the destruction of "surface vessels" by submarines. It was not the fault of the committee, he implied, as he somewhat bombastically added, "It is high time we put the fear of God into these young gentlemen who lie about the North Sea attacking all and sundry without let or hindrance."

In other words, the problem was young whippersnappers who didn't know their place. When World War I began, the British Navy had not one method for detecting the presence or position of a submarine underwater, nor any effective weapon against it. The accepted defense was to surround major fleet units with a screen of protecting destroyers, and to keep everyone moving as fast as possible and changing course as often as feasible.

In the 1912 Fleet Maneuvers British submarines slipped into a fleet anchorage—twice—and "torpedoed" three ships; a staff evaluation warned that in a war with England, enemy submarines might prove a serious menace to the fleet in the North Sea. The Navy Board scoffed. Next, an Admiralty Report calling for war risk insurance for merchant shipping included an ominous warning: "There is every probability that prices in the United Kingdom may rise to prohibitive levels, that panic may occur, and that great pressure may be exercised upon the Government . . . to submit to peace on unfavourable terms." Because of the Navy's confidence in its ability to deal with surface raiders, the warning was passed over.

However, some politicians—and a few Naval officers—were becoming, in the 1913 phrase of former Prime Minister and future (1916) First Lord of the Admiralty Arthur Balfour, "more than ever convinced that the days of the dreadnought are numbered." Some politicians, and a very few Naval officers.

14 1907–1914—Of Coming of Age

The submarine has the smallest value of any vessel for the direct attack upon trade. She does not carry a crew which is capable of taking charge of a prize, she cannot remove passengers and other persons if she wishes to sink one.

—Memo written by the Royal Navy Assistant Director of Naval Operations, July 1914

On the eve of World War I, the Royal Navy's opinion of submarines was, well, not very high. As reflected in the British professional journals of 1914, submarines were:

—Useful as local defense vessels
—Playthings
—The weapon of the weaker power
—Weather-limited
—Too slow
—Unproven
—Too easily caught on the surface
—Too limited in range
—Probably a waste of money

Perhaps a submarine could serve as a picket, locate the enemy, and send a message via wireless radio (if not more than forty miles from a receiving station) or carrier pigeon, with coordinates, course, and speed. Under international law, she could not do much else.

The issues surrounding clandestine warfare, raised in Fulton's day and later, had never been addressed, but the establishment of the International Red Cross had led to a series of peace conferences during which the matter was considered.

From the first Geneva Peace Conference, 1864, came the "Geneva Convention" which set land warfare standards for the treatment of prisoners and of wounded, and outlawed weapons which caused "unnecessary suffering." Ratification was slow and steady, but was not completed until 1906.

In the meantime, twenty-six nations met to extend the Geneva Convention to war at sea. This first Hague Peace Conference, in 1899, considered, but did not adopt, the outlawing of submarine warfare. In 1899, there was hardly enough "submarine warfare" to make any difference.

The Second Hague Conference, held in 1907, was attended by forty-four nations and wrestled (somewhat inconclusively) with unrestricted right of capture, the use of mines and submarines, contraband, prize courts, and the rights and duties of neutrals. A follow-on meeting in London produced the London Declaration of 1908 which held submarines to the same Prize Rules which had traditionally applied to cruisers or other surface warships, the same rules which had been in effect at the time of the American Civil War.

A submarine could stop and search any merchant ship for contraband, weapons, or other materials of war or items deemed critical to the prosecution of war. If contraband was found, the ship could be sunk. *If,* that is, it was an enemy ship and the passengers and crew first were provided with the means necessary to reach safe haven. That did not mean only "sufficient lifeboats," but also provisions, navigational equipment, and environmental protection necessary to ensure safe passage and arrival. If the offending carrier was registered to a neutral flag, it could be escorted into port and interned.

The London Declaration also defined a new category of warship, the "auxiliary cruiser." It now was acceptable to give a merchantman defensive weapons, but the arming had to be done before the ship left port and the ship had to be designated as a naval auxiliary—thus, no longer protected by the rules applying to merchant ships, and subject to attack without warning. Just in case, many pre-World War I British and German merchant ships included gun foundations.

Fisher retired in 1910, at the age of sixty-nine, but didn't let age or retirement get in the way of his crusade for submarines. On May 8, 1914, he sent an exasperated note to Prime Minister Asquith: "*We had more submarines 4 years ago when I left the Admiralty than we have now!* . . . and the Germans *then* nil. *Now* they have more high sea submarines than ourselves. . . . Myself I should drop a Dreadnought secretly & go in for 20 submarines instead."

In a series of Lectures on International Law, a professor had insisted that no submarine would dare to touch a vessel without the proper visit and search being made. This prompted a May 14th memorandum from Fisher to Churchill:

Those who lecture on International Law say the civilized world would hold up its hands in horror at such acts of barbarism as a submarine sinking its prey, but yet an enemy can lay mines without outraging propriety! After all, submarines can exercise discretion—mines can't! . . . [the submarine] cannot capture the merchant ship; she has not spare hands to put a prize crew on board . . . she cannot convoy her into harbour; and, in fact, it is impossible for the submarine to deal with commerce in the light and provisions of accepted international law. . . . There is nothing else the submarine can do except sink her capture. . . . this submarine menace is a terrible one for British commerce and Great Britain alike, for no means can be suggested at present of meeting it except by reprisals.

Fisher sent this memo also to Prime Minister Herbert Asquith, suggesting that it be circulated among the government. The Prime Minister refused. Sink merchantmen without warning? The notion was unthinkable; he had it on the authority of Winston Churchill, who'd said just a few months earlier: "This would never be done by a civilized Power."

Admiral Percy Scott tried to send a "wake-up call." In a letter to the Daily *Mail,* June 2, 1914, he charged that ". . . as the motor has driven the horse from the road, so has the submarine driven the battleship from the sea." Three days later, another Scott letter was published in *The Times,* in which he reiterated that the day of the battleship was over: ". . . battleships are of no use either for defensive or offensive purposes, and, consequently, building any more in 1914 will be a misuse of money subscribed by the citizens for the defence of the Empire. . . . Submarines and aeroplanes have entirely revolutionized naval warfare; no fleet can hide from the aeroplane eye, and the submarine can deliver a deadly attack even in broad daylight."

Scott was attacked from all sides—by senior Naval officers, the government, and the conservative press. Five admirals took him on in print: His letter was a mischievous scare; he simply didn't understand; submarines were undeveloped and inaccurate; since submarines could *only* operate in the daytime, they were highly vulnerable and even a machine gun could put them out of action. One writer was astonished at Scott's presumption of knowledge on a subject of which he could not have any personal experience.

The *Hampshire Telegraph* offered a second-hand comment, that Lord Sydenham "regards Sir Percy Scott's theory as a fantastic dream." The *Manchester Courier* assigned his predictions "to the novel shelf." And the *Pall Mall Gazette* affirmed that his ideas "approached the boundaries of midsummer madness."

Scott was supported by some junior officers and the liberal press, which had long been campaigning against "bloated armaments" and thought they had found an ally. The Navy brushed him aside. A memo of July 1914 from the Assistant Director of Naval Operations blindly restated an earlier Admiralty position, as if someone had turned on the switch to play an old recording:

> The submarine has the smallest value of any vessel for the direct attack upon trade. She does not carry a crew which is capable of taking charge of a prize, she cannot remove passengers and other persons if she wishes to sink one.

We might note that, while the naval professionals seemed blind to the possibilities of the submarine, at least one professional writer had it right. Sir Arthur Conan Doyle's 1913 short story "Danger! A Story of England's Peril" told of the defeat of Britain at the hands "the navy of one of the smallest Powers in Europe." A navy whose fleet of eight submarines waged a campaign of unrestricted war against merchant shipping, cutting off Britain's lines of supply and thereby starving the nation into submission.

Professional intransigence aside, and thanks to Admiral Fisher, on the eve of World War I, Great Britain had the world's largest submarine fleet—

seventy-four in service, thirty-one under construction, and fourteen projected. While these large numbers reflected an underwater component of the naval arms race then in full stride, in larger degree they reflected improving technology. The British had not locked in any one style or model of submarine, nor had they contracted for any appreciable number of any given model. They'd simply ordered up another batch with each major improvement. By 1914, having followed the first Hollands with an improved "A" class, the British had reached "E."

Some comparisons:

France, with sixty-two boats in service and nine under construction, had the second largest fleet, although presenting a wide range of type and capability. Six were "oceangoing," fourteen had steam plants, thirty-three were of the loosely defined "Labeuf" class.

Russia was in third place. Having eagerly embraced the submarine after the Battle of Tsushima, the Russian Navy had acquired forty-eight boats variously from America (five Hollands and eight Lakes), Britain, France, and Germany, in service or under construction. Most of the Russian submarines were stationed in the Baltic and the Pacific; eleven were in the Black Sea. Russian strategic vision—if we can bend the language to call it a "vision"—saw submarines, indeed the entire Navy, as a defensive perimeter for the Army. Russian submarines had so little impact on World War I that we need not mention them again.

In numbers, Germany came next: twenty-eight boats in service, seventeen under construction. Of the twenty-eight, the oldest (*U-1* through *U-4*) were suitable only for training, and another four boats were in shipyard overhaul. The boats through *U-18* were limited by kerosene-burning engines, which left an easily spotted dusky trail on the water in daylight and sputtered forth a shower of sparks at night. The year 1911 saw the introduction of the diesel engine with *U-19,* and by 1912 the relatively advanced "30s" series (*U-31* to *U-41*) were under construction. These displaced 685 tons on the surface, 878 submerged, carried six torpedos and one 88mm deck gun, and had a surface speed of 16.4 knots, 9.7 knots submerged. Maximum range: 7,800 miles at eight knots.

Germany's submarine fleet may have been among the smallest, but the late start in joining the club allowed Germany to take advantage of lessons learned in other navies. The result was clear: When she went to war, Germany had the most capable underwater fleet. The German High Seas surface fleet also was smaller than the Grand Fleet of Great Britain, but—ship for ship—was notably superior, with better armor, higher speed, and a larger throw-weight of armament.

The United States had thirty submarines in service, ten under construction; Italy, twenty-one in commission, seven under construction; Japan, thirteen and three; Austria, six in service and two on the way. With the exception of some Austrian boats taken over by the Germans, few of the submarines of any of these nations saw meaningful service during the war.

The submarine safety record worldwide was a bit mixed, but surprisingly good for a system so inherently dangerous. The United States, with the second-longest experience with submarines (if you discount the Civil War), had the best record (one accident, two men killed). Germany, with the shortest experience, had one accident with three men killed. Japan had lost *H6* (fourteen killed), and Italy had also had one submarine lost (fourteen killed). The safety "records" in the other three submarine-operating navies were records at the other end of the scale: Great Britain, eight accidents, seventy-nine killed; France, eleven accidents, fifty-seven killed; Russia, five accidents, seventy killed. The accident rates for Great Britain, France, and Russia are roughly proportional to the number of boats in operation, and probably have no other significance. The high accident rate in the Royal Navy, however, forced the adoption of an unrealistic component in all training. Until 1910, a submarine on maneuvers was always to be accompanied by a surface ship flying a large red warning flag.

Thus, the total world submarine fleet was, give or take, 282 boats which ranged in age from twelve years to last week, and in size from the first Hollands of sixty-five tons to the 660-ton British E-boats and 685-ton *U-31*. Those are impressive numbers for a weapon which had only been in service a dozen years, but they are nothing compared with the numbers and capabilities which would enter service over the next four years.

On June 28, 1914, a Serbian nationalist assassinated the heir to the thrones of Austria and Hungary, rekindling European disputes which had been alternatively smoldering or in flames since the Middle Ages. It was time once again for flames.

Russia and Germany were at war by July 1st; France came in July 3rd; Great Britain, protesting the German violation of Belgium neutrality, issued an ultimatum; it expired, without response, at midnight August 4, 1914. War technically began for Great Britain at that moment; however, the Royal Navy's war started one hour earlier. Churchill sent his fleet to war at 11 p.m. London time because that was midnight Berlin time. He was taking no chances.

The German naval strategy was straightforward: Take out individual British ships whenever and wherever possible to whittle British strength down to a point where the fleets were more or less matched. Then, pick a fight.

1. Our object [wrote Kaiser Wilhelm] is to damage the British Fleet by means of offensive advances against the forces watching or blockading the German Bight, and also by means of a ruthless mining and, if possible, a submarine offensive, carried as far as the British coast.

2. When an equalization of forces has been obtained by these measures, all our forces are to be ready and concentrated, and an endeavor will be made to bring out Fleet into action under favorable conditions. If a favorable opportunity for an action occurs before this, advantage will be taken of it.

3. War is to be carried on against commerce as laid down in the Prize Regulations. The Commander in Chief High Seas Fleet will determine to what degree this warfare is to be carried on in home waters. The ships allocated for war against commerce in foreign waters are to proceed as soon as possible.

Note that the Kaiser wrote of "ships" assigned for commerce raiding; submarine attack on merchant shipping was not included in the war plans of any nation; under the rules of search and seizure, any attack would be too complicated and too dangerous. The submarine was to be used against the enemy's battle fleet (even if the men who laid down those doctrines had little faith in the ability of the submarine to accomplish anything significant).

The Germans assumed that the British—whose naval doctrine from time immemorial had been "attack"—immediately would send the Grand Fleet to take on the numerically inferior German High Seas Fleet at Heligoland Bight. Accordingly, as the war began, all available U-boats were on picket duty, on a line about thirty miles out but five miles behind a trip wire of patrolling destroyers. The U-boats waited on the surface, moored to anchored buoys but ready to slip the cable and submerge at a moment's notice. In theory, the destroyers would launch massive torpedo attacks against the approaching British Fleet, the submarines next would join in, and as the survivors kept coming, they would hit yet another defensive perimeter of torpedo boats. By the time the British reached the Germans (by then coming out to do battle) the numbers should be close to even and the greater effectiveness of the individual German ships would win the day. Finis, Grand Fleet.

The British confused things. They didn't come.

The Germans sent U-boats out to reconnoitre.

The first German boat was sunk the day after the war began—*U-15*—rammed by a cruiser while the crew desperately went through the steps to dive, which it took the early-model submarine five minutes to accomplish. Another U-boat sent out on patrol at the same time also was lost, to unknown causes.

These first patrols did nothing to enhance opinion of the submarine within the German Navy. The British blockade lines had not been found; the raid was a net plus for the British, with two submarines sunk and no British ships even damaged, although one torpedo had been fired against a battleship, but missed. The German Naval Staff came to the conclusion:

> The British Main Fleet and probably all war vessels which are worth attacking by U-boats are so far away from Germany that it is beyond the technical capacities of our U-boats to find them. Neither is it possible for U-boats to lie in wait off the bases for long periods owing to their distance from Germany. U-boat operations will therefore be abandoned for the present.

Most of the U-boats went back to the picket line, but several were on patrol. Then, a month into the war, the submarines of both nations began to make contact. September 5th: *U-21* hit the cruiser *Pathfinder* with one torpedo. A lookout on the cruiser spotted the periscope and gave warning, but *Pathfinder* had been following peacetime steaming orders: The economical cruising speed of six knots did not give enough momentum for a quick change of course. Not only were her options reduced by poor steerageway, but an incredulous throttle man in the engine room refused to believe—probably had never seen—the signal from the bridge for an emergency, accelerated turn: "Full Ahead—Port" and "Full Astern—Starboard."

U-21 Commander Otto Hersing carefully followed doctrine: observe the hit, evaluate the results (to aid the High Command in assessing future threat), dive to maximum depth (then 148 feet) and move quietly out of the area. Hersing estimated that total elapsed time, from torpedo launch to sinking, was three minutes. The British claimed it took four minutes. There were nine survivors, 259 lost. The Admiralty issued new orders, setting minimum patrol speed at fifteen knots.

A week later, the British *E.9*, commanded by Max Horton, sank the German light cruiser *Hela* with two torpedoes. On the same day, the Australian Navy's *AE.1*—a twin of *E.9*—was lost to unknown causes while searching for another German cruiser off New Guinea.

Then, on September 22, 1914, the submarine came of age; predictions which had been strung across three hundred years came to pass, and the submarine did indeed change the face of naval warfare.

On September 22, 1914, one small, virtually prehistoric German submarine sank three British cruisers. On the same day. Within slightly more than ninety minutes.

U-9 was commanded by Otto Weddigen. Taking a leaf from the book of Admiral Horatio Nelson, who drilled his gunners until they had cut the time to reload and fire to less than one-fourth that taken by their enemies, Weddigen drilled his crew in regaining the trim lost when a torpedo was fired, and then reloading for another shot. This was done by quickly shifting members of the crew forward, then aft, while the ballast was regulated, and then reloading.

Weddigen was aided by the innocence of the British captains who, each in turn, went to the aid of their stricken comrades. Initially, all thought the first ship had hit a mine. Even when the second ship was hit and it became obvious that a submarine was at work, the third ship nonetheless remained on scene.

Aboukir was sunk at 0720 (German time), *Hogue* at 0755. Cressy was first hit about 0820, and given the coup de grâce at 0855. A few weeks later, on October 15th, *U-9* took out another cruiser, *Hawke.*

In the sinking of just these four ships by one antiquated submarine, more British sailors were killed than were lost by Lord Nelson in all his battles put together. Reaction in Germany was jubilant; in England, funereal. Fleet Commander in Chief Admiral Jellicoe wrote to First Lord of the Admiralty Churchill:

It is suicidal to forego our advantageous position in the big ships by risking them in waters infested with submarines. The result might quite easily be such a weakening of our battle fleet and battle cruiser strength as seriously to jeopardize the future of the country by giving over to the Germans the command of the open seas.

One wonders if Admiral Jellicoe had ever heard of Citizen Fulton.

15 1914—The Proving Ground

They bear, in place of classic names,
Letters and numbers on their skin.
They play their grisly blindfold games
In little boxes made of tin.

—*Rudyard Kipling, "The Trade," 1916*

There was little privacy and little comfort in a submarine. There was no bath and only one lavatory for the use of all three officers and twenty-nine men aboard. Few shaved and no one changed their clothes from the beginning to the end of the voyage. The officers used eau-de-cologne to mask their body odour and the indescribable damp, oil-laden, stale smells of the sweating interior of the boat.

—*Geoffrey Clough, RN, Radio Operator*

When the war started, the art of submarine warfare was barely a dozen years old and no nation had submarine-qualified officers serving at the senior staff level. There were few officers at any level—except the submariners themselves—who cared. The ancient prejudices against submarines remained: First, they represented an unethical form of warfare best left alone, and second, they did not "fit" the classic structure of a navy.

Fleets were built of layers of ships of roughly equal capability, and with only one exception, no ship was expected to take on a superior class. Battleships fought battleships, battle cruisers and other types of cruisers could fight each other, but were faster than battleships and thus could get out of harm's way. Patrol boats patrolled. Perhaps a patrol boat, equipped with a torpedo, could do damage to a larger warship, but capital ships steamed in a formation protected by the antitorpedo boat destroyer. Admiral Fisher and a handful of other senior officers excepted, the naval professionals of the world simply did not understand that the submarine had broken the code. The smallest, cheapest warship could now take on the battleship, head to head, and win.

This minimal interest in submarines had two roughly opposite effects. On the one hand, the submariners themselves were pretty much left alone to develop their own tactics and operational procedures, which was good; on the other hand, submarine and weapon *design* and construction and the allocation of resources was dominated by the non-submariners, which means that the men who had to use the equipment had little influence over the equipment around which their tactics were being developed. This was especially grievous when it came to the habitability—rather, the almost complete *lack* of habitability—in the submarines of any nation.

Even under a peacetime regimen, life aboard a submarine was barely tolerable. After about ten hours submerged, there was so little oxygen that a lighted match would soon fizzle out and refuse to burn; a constant damp encouraged mildew and rot; odors of spoiled food and the smell of diesel fuel clung to everything and everyone; it was said that you could identify a submariner long before you could see him. As submariner Geoffrey Clough would later note:

> There was little privacy and little comfort in a submarine. There was no bath and only one lavatory for the use of all three officers and twenty-nine men aboard. Few shaved and no one changed their clothes from the beginning to the end of the voyage. The officers used eau-de-cologne to mask their body odour and the indescribable damp, oil-laden, stale smells of the sweating interior of the boat.

And that was for routine, fair-weather operations. Add bad food, gassing batteries, and bad weather to the mix: "Only those who have actually experienced the horrors of seasickness," another submariner was to caution, "can have any conception of the agony men who served in submarines suffered. . . . Imagine trying to work out problems in navigation when your stomach was in such revolt that you worked with a pail beside you and cold, clammy sweat, trickling down from your forehead and dripping off the end of your chin, smeared the pages of your work book. . . ."

Finally, there seemed always to be an overflowing toilet waiting for the opportune moment for "flushing"—never when an enemy was near. If you've ever flushed a modern marine toilet, where you must close and open certain valves in a certain order and then work the plunger of a pump while your face is uncomfortably close to the bowl, you may have some small appreciation of the flushing process in a 1914-vintage submarine cruising underwater at thirty feet. Close and open certain valves in a certain order, then blast the shot overboard with high-pressure air. If anything was not quite right, you were just as likely to blow the load back into your face.

In compensation, British submarine pay added 60 percent to an officer's base pay and more or less doubled that of a seaman. However, conditions were not much improved until many years later.

Despite—perhaps because of—such halfhearted efforts as the 1910 British Anti-Submarine Conference, no nation had developed any methods for detecting submarines, or antisubmarine weapons for use should a submarine be found. For lack of sufficient high-level interest, inventions which might have been put to practical use had pretty much stayed in the laboratory.

And many of the weapons to be used by the submariners themselves were not very effective. The 18-inch torpedo was too light to do serious damage to a modern armored warship. Two direct hits on Germany's *Prinz Adalbert* on July 2, 1915, put the cruiser out of action, but only for four months. In August, one hit on the German cruiser *Moltke* did little damage—although the fleet commander was so unnerved that he pulled his ships back into harbor and an invasion of Russia was forestalled.

On one of the first British attacks against a German cruiser, two well-aimed torpedoes ran under the target without effect. This frustration was repeated throughout the submarine fleet, before someone realized that the explosive warhead weighed forty pounds more than the peacetime practice load upon which the torpedo depth-settings had been based. This frustration was repeated throughout much of the next war for exactly the same reason. The study of history seems always to verge on a lost art.

The Germans laid antisubmarine minefields, but had not developed an antisubmarine mine. If a submerged boat entered a minefield, by choice or by chance, it stood a fair chance of coming out the other side—because the detonating horns on the German mines were arranged on top of the mine. Thus, even if a submarine snagged a mooring cable and pulled a mine down, the boat only came in contact with the bottom of the mine. If the cable was tightly caught, and if the submarine could surface, the crew would try to free the cable and mine without triggering the charge. Some did. Some did not.

The British laid antisubmarine minefields, but had not developed a mine—of any sort—that worked very well. Many U-boats passed through the minefields untouched. Some did not. There is no way of knowing exactly how many submarines on either side were lost to mines. The daybook entry "Overdue—presumed lost" covered status, but not cause.

No nation had considered the problem of "friend or foe" recognition for a submerged submarine, and few had initiated any useful training in the identification of enemy ships.

No one envisioned a long war; no nation had planned for a long war; all assumed that the war would be fought with the equipment already in place.

The war was to be a costly proving ground.

Following close behind the horrors of September came the sinking of *E.3* by *U-27* on October 17th, and ten days later the battleship *Audacious* hit a mine and sank in the Irish Sea. Some among the public were suspicious about the loyalty of the German-born First Sea Lord, Prince Louis Battenberg, and the series of naval disasters gave Churchill the lever to bring in a more forceful, dynamic, even ruthless replacment. Lord Louis resigned on October 29th (and soon enough had his German name translated into English—Battenberg became Mountbatten). On October 30, 1914, the seventy-four-year-old Jacky Fisher was brought out of retirement and back to duty as First Sea Lord.

Churchill was just at this time pushing hard to add more submarines to the fleet. On October 6th, Max Horton sank a destroyer, entered port flying the "Jolly Roger," and became a national hero—at a time when the nation needed some heroes. In an October 13th memo to the Admiralty staff, Churchill asked, "What measures can be adopted for increasing the number of submarines? Is it possible to let further contracts for submarines on a fifteen-month basis? It is indispensable that the whole possible plant for submarine construction should be kept at the fullest possible pressure day and night." Another memo, on October 28th: "You should exert every effort of ingenuity and organisation to secure the utmost possible delivery."

Ironically, the sinking of *Audacious* was to provide the necessary link for increased submarine production. The passenger liner *Olympic*, just completing a crossing from New York, had been nearby; *Olympic* rendered assistance but could not save the battleship. The Navy was worried about the public reaction to yet another disaster, and hoping to keep the sinking hidden from the Germans, Admiral Jellicoe had the passengers temporarily interned until they could be sworn to secrecy. One American passenger, Charles Schwab, claimed urgent business and asked to be allowed to get on up to London.

What sort of business? An appointment with the Secretary of State for War, Lord Kitchener, to discuss munitions. In the process of explaining himself to Jellicoe, Schwab mentioned that he was the head of the Bethlehem Steel Company which, among other things, operated the Union Iron Works and the Fore River Shipbuilding Company. Which built, among other things, submarines. Jellicoe urged Schwab to meet with Fisher and talk submarines.

As a result, in a meeting which Fisher was later to say lasted not more than five minutes, Schwab agreed to build twenty submarines for Britain in the shortest possible time at the highest possible price. The project would be expedited by cloning the "H" model boats he was already building for the U.S. Navy. Schwab planned to ship the boats in sections, for assembly in England. He was confident that there would be no problem in getting around the neutrality laws; there was, after all, the precedent set by Electric Boat in selling disassembled submarines to the Japanese. He was wrong.

At first, officials at the U.S. State Department gave an oral approval, and within a week of the meeting with Fisher, Bethlehem Steel and Electric Boat facilities all over the country had started work on the submarines. The next day, the *New York Times* reported that, in a contract worth $10,000,000, an unidentified belligerent government had ordered twenty submarines, to be shipped in parts for assembly overseas.

Schwab was soon to be reminded that there was another, stronger prece-

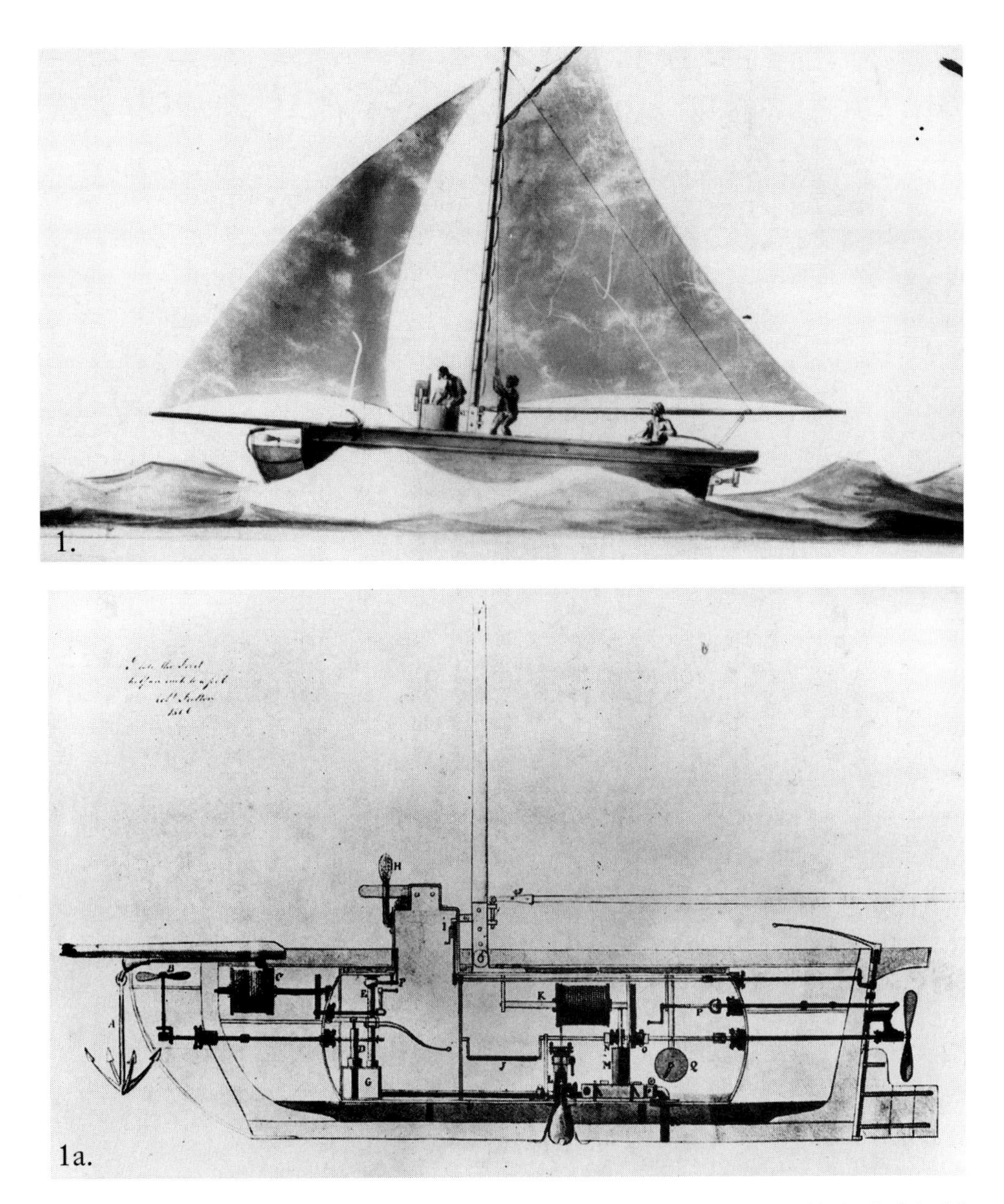

1, 1a. Robert Fulton's "improved"—but never built—*Nautilus,* 1804. (Compare with the "original," page 42.) The submarine now "shall have the property of sailing like an ordinary fishing boat." This is a sailing boat with a one-inch-thick pressure hull, comprised of a conning tower (to provide access and visual reference when running awash) and a central cylinder, six feet by twenty-four feet, "the ends forming a part of a sphere to resist the pressure of the water in all directions." Other changes from the 1798 version: a down-haul propeller has been added at the bow; there is a folding main propeller to reduce drag while under sail, and ballast control has been shifted from a hollow iron keel to tanks outside the pressure hull. Fulton has abandoned the *Turtle*-like "horn of the *Nautilus*" for attaching a bomb to an enemy vessel; the submarine is now intended only as a mine-layer. Note the two anchors: one forward for normal use, one amidships to adjust submerged depth when not moving. (William Barclary Parsons Collection, New York Public Library)

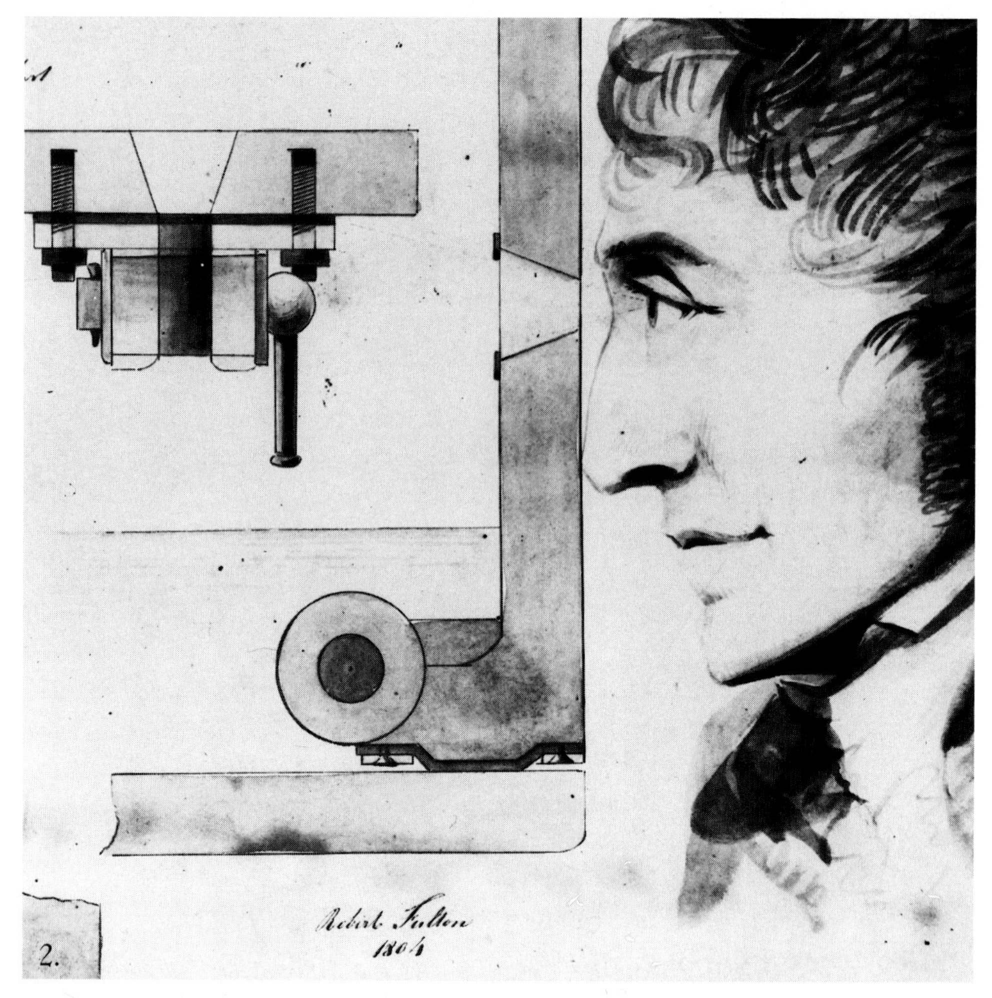

2. Among other drawings put in safekeeping with the American Counsel in London in 1806: this self-portrait of Robert Fulton peering through...something. Most authorities wrongly assume this to be a periscope, because Fulton called it a "perxscopt." As described by the inventor himself, it is a "conic glass window"—that is, a small deadlight—in the conning tower. (William Barclary Parsons Collection, New York Public Library)

3. What may—or may not—be the 1862 Confederate submarine *Pioneer,* being moved to public display at the Louisiana State Museum, New Orleans, 1957. This boat, discovered and salvaged in 1878, does not match, in some particulars, written descriptions of *Pioneer* but is nonetheless a Civil War–era submarine. (Courtesy *New Orleans Times-Picayune*)

4. Cotton broker—and submarine promoter—Horace L. Hunley (Courtesy Louisiana State Museum, New Orleans)

5. Confederate submarine designer and builder James R. McClintock (Courtesy Naval Historical Foundation)

5a, 5b. CSS *Hunley,* the first submarine to sink a warship in combat. The Conrad Wise pencil sketch (5a) was made while Confederate General P. G. T. Beauregard was contemplating the future of the *Hunley,* sunk the week before with the loss of all hands. The sketch became the engraving (5b) dated December 1863. Note that the engraving is reversed from the pencil sketch, left to right (see the discussion of the *Symons* boat, page 21), and adds a semi-military guard detail.

6. One of several Charleston-based steam-powered Confederate *Davids,* designed to run just awash and attack Union blockading forces with spar torpedo or by ramming.

7. The French submarine *Gustav Zede,* 1893. Early submarine crews (and passengers) seemed to delight in posing with this sort of awash nonchalance; not a life jacket, preserver or safety line in sight! (Farenholt Collection, Naval Historical Foundation)

8. Simon Lake's prominently wheeled *Argonaut*—under construction in the same graving dock in Baltimore as John P. Holland's never finished *Plunger*, rear. *Argonaut* made an open-ocean passage—on the surface—from Baltimore to New York, earning warm praise from French novelist Jules Verne and prompting his prediction: "The next war may be largely a contest between submarine boats."

9. *Holland* in drydock. Compare the hull form with that of the circa-1956-designed *Barbel* (photo 45). Holland seems intuitively to have grasped the essence of underwater bodies, but it was to be more than half a century before technology caught up with his concepts.

10. First crew of *Holland,* June 1901. Admiral Dewey's former aide, Lieutenant Harry Caldwell, commanding officer.

11, 12. Desperately in need of cash, Simon Lake pulled his *Protector* from the 1903 U.S. Navy procurement competition he had engineered—and sold it to Russia, then at war with Japan. The Electric Boat Company—desperately in need of cash—held on to the *Fulton* long enough to win the "competition"—and then sold the boat to Russia. (12) Here, *Fulton* is being prepared for a voyage as deck cargo on a Russian steamer. (Courtesy Floyd Houston)

13. The British *Holland No. 3,* which served from 1902–1913.

14. The British design *D-1,* 1908–1918. Note the marked shift from the *Holland* porpoise-like hull-shape to that of a surface ship—a shift common in all navies of the day (and one that would last until the end of World War II), as the designers came to acknowledge that submarines would necessarily spend most of their lives operating on the surface. The armored cruiser in the background is possibly *HMS Aboukir*—soon to be one of three warships sunk by the German U-boat *U-21* on the day the submarine came of age, September 22, 1914.

15. The Russian submarine *Akula*, under construction at St. Petersburg, 1908. The designer, Naval Constructor Bubnov, poses in the foreground. (Courtesy Boris V. Drashpil)

16. The French submarine *Ventose,* 1907–1920; a good illustration of extra-large limber holes (to facilitate the rapid escape of entrapped air when diving) and of an external torpedo launching system, a feature adopted by most navies of the day because it allowed a boat to carry more weapons. Adopted—but not perpetuated much beyond the start of World War II, since it soon enough became obvious that a torpedo carried outside the pressure hull could not be maintained in combat readiness and soon became useless.

17. The U.S. Navy—in part, attempting to break the hold of the Electric Boat Company on submarine design—purchased a set of plans from the Italian designer Laurenti. It was not a happy move. This October 1912 photo shows *G-4* (the 26th U.S. submarine) at the Cramp shipyard in Philadelphia a year after launching and with another yet to go before commissioning. The Laurentis had some advanced features—including double-hull construction—but proved to be difficult to build and operationally awkward.

18. The German civilian cargo submarine *Deutschland,* engaged in high-value transatlantic commerce, New London, Connecticut, November 1916; three months later, she had been converted and sent to war as *U-155*.

19. The crew of *U-53* poses on October 7, 1916, during a twenty-four-hour "courtesy" visit to Newport, Rhode Island. A few hours later and just after entering international waters, *U-53* sank one Norwegian, one Dutch, and three British ships. (Collection of Commander H. H. Lippincott, CHC, USN)

20. 1918, left to right: A French submarine (easily identified by the limber holes), German late-model coastal boat *UB-133,* and early model *UB-24*.

21. The "pattern" camouflage (shown on the World War I American troopship *Louisville*) was designed to confuse the U-boat's visual fire-control systems—making it difficult to judge range, size, speed, and course. Submarines themselves employed more "natural" schemes of camouflage, typically colors to blend in with operating conditions: white for arctic waters, different shades of gray for different parts of the world. Eventually, all navies adopted some version of the U.S. Navy's "haze gray" for surface ships, black for submarines. An exception: periscope housings continue to be painted with a dappled pattern similar to the shadows thrown down by sunlight on dancing surface waters. (National Archives)

22. Fulton would have approved: USS *R-14*, temporarily converted to a sailboat in Hawaiian waters, May 1921. Using the periscope as a mast and canvas covers sewn together for a sail, the crew brought her into port at Hilo five days after a power plant casualty.

23. Most navies have tried to use submarines as aircraft carriers—never with much success. Here is one American attempt: *S-1* (the 105th U.S. submarine) equipped with an on-deck hangar and the Martin MS-1 single-seat reconnaissance floatplane, October 1923. Twelve of the airplanes were built before reality caught up with wishful thinking. The MS-1 had to be assembled for operation, and disassembled for stowage, forcing the submarine to remain too long on the surface; in addition, launching and recovery were virtually impossible. (Courtesy Lieutenant Gustave J. Freret, USN [Ret])

24. Jackie Fisher's dream become operational nightmare: the British *M-1*, circa 1919, a K-boat with a 12-inch naval gun.

25. The French, never willing to be outclassed by the English, took the big-gun submarine to the next stage: twin 8-inchers aboard the 1929 *Surcouf* (shown here in 1941, shortly before being lost at sea, probably from collision with a merchantman).

26. Another *Nautilus*—the 168th American submarine—and another effort at putting big guns on submarines; in this case, twin 6-inch. One improvement over the British and French efforts: these guns could be trained and independently aimed. However, the shells were too heavy for safe and efficient handling and the V-class boat itself was too cumbersome for operation as an attack submarine. *Nautilus* was modified as a seaplane fueling station and used for amphibious operations in World War II.

27. Submarine pioneer Admiral Chester Nimitz assumes command of the U.S. Pacific Fleet on December 31, 1941—on board the only available undamaged warship, the submarine *Grayling*. The submarine was later lost with all hands, September 1943. Admiral Nimitz capped his career as the post-war Chief of Naval Operations.

28. *Hammerhead*, SS-364, October 24, 1943. Wisconsin's Manitowoc Shipyard developed this sideways technique to accommodate launching a boat into a narrow river.

29. Control room of *Batfish*, SS-310, 1945. (National Archives)

30. *Batfish* hoists trophy flags at the end of a war patrol. This boat is now ashore as a submarine memorial, Muskogee, Oklahoma. (National Archives)

31. The American flag flies over *U-505*—the first enemy warship captured on the high seas by the U.S. Navy since the War of 1812—as the jeep carrier *Guadalcanal* rides escort, June 6, 1944. The Type IX *U-505* is now on permanent display—ashore—at Chicago's Museum of Science and Industry.

32. A Type VIIC U-boat—the mainstay of the German World War II *unterseeboot* fleet—returning to base at Trondheim, Norway.

33a. A type IX U-boat in the bombproof pen at St. Nazaire, France, on May 13, 1945. This boat had just completed a 110-day voyage to Japanese waters. (U.S. Army Signal Corps)

33b. The 1943 experimental Type Wa-201 Walter boat, *U-793*, here partially dismantled at the end of the war.

34. Collapsible hydrogen peroxide bags being removed from the Type XVIIB Walter boat *U-1407* after the war. With this storage outside the pressure hull, fuel could be consumed without appreciable change in trim—seawater simply replaced the depleted volume.

35. To minimize the effect of Allied bombing, the late-war Type XXI boats were built in virtually complete sections at scattered locations, and transported by canal barge to assembly yards (here, the Deschimag Shipyard in Bremen). Note the "figure-8" cross-section of the pressure hull. The lower section was initially intended for hydrogen peroxide storage for a Walter power plant; it became, instead, the compartment for the enlarged battery capacity, which gave these boats the nickname "Electroboot." (Courtesy Captain Henry A. Arnold, USN [Ret])

35a. The *U-3008,* one of only two Type XXI U-boats to make a wartime operational patrol and later, one of two taken over by the U.S. Navy. These hastily built late-war boats were not very durable: *U-3008* developed leaks (because of faulty welding) and became a spare-parts locker for sister ship *U-2513,* which remained in U.S. service until the battery supply was depleted, July 1949. (National Archives)

36, 37. For some, the war ended too soon. (36) With more hope than sense, Germany had more than 1,000 increased-capability Type XXI (and more than 900 smaller Type XXIII) under construction or on order the last day of the European war. (37) With more desperation than hope, the Japanese launched a massive building program of suicide and midget submarines; here, eighty-four *Koryu* boats, of four different designs, are huddled in drydock at the Kure Naval Base, October 1945. (U.S. Army Signal Corps; National Archives)

38. Post-war, the U.S. Navy applied lessons-learned—many, learned from the Germans. *Sabalo* SS-302 was given a "fleet-snorkel" conversion.

39. The greatest post-war advance was nuclear power: an American development (which grew out of a German theory). The first nuclear-powered submarine—another *Nautilus*—borrowed a hull-form from...guess where? This post-retirement photo SSN-571 *Nautilus,* taken while en route to a memorial berth at New London, Connecticut, clearly shows a Type XXI family tree.

40. The 1955 U.S. midget submarine *X-1*—the only American boat designed and built by an aircraft manufacturer (Fairchild) rather than a shipbuilder. The fifty-foot-long *X-1* began life with a three-way power plant: diesel on the surface, a closed-circuit diesel-hydrogen peroxide plant while submerged, batteries for slow-running endurance. The H_2O_2 system exploded in May 1957, and was removed. *X-1* continued in service until 1973, and is now ashore as a memorial in Annapolis.

41. Another German legacy: here, the Loon guided missile aboard *Cusk* SSG-348, January 1948. Loon was a copy of the German V-1 buzz bomb, and difficulties of setup and launch matched those of the 1923 Martin MS-1 seaplane—although, in the case of Loon, recovery and disassembly were not an issue. (National Archives)

42. The next generation submarine-launched guided missile: Regulus aboard *Grayback* SSG-574. The hangar (to hold two missiles) is now better integrated into the hull; when the *next* generation missile forced retirement of Regulus, the hangar became a compartment for clandestine amphibious assault troops.

43, 44. The German V-2 rocket became the American Jupiter ballistic missile; although exceedingly large, at least one scheme was set forth to mount four Jupiters in a submarine. Timely development of the much smaller Polaris missile permitted sixteen on the boat. Here, firing tube hatches open on *Sam Rayburn* SSBN-635.

45. The next-to-last American diesel boat: *Barbel,* SS-580 (1959–1989), showing the aerodynamic hull shape tested in *Albacore* and soon adapted to all new construction. *Barbel* is now a memorial at Portland, Oregon.

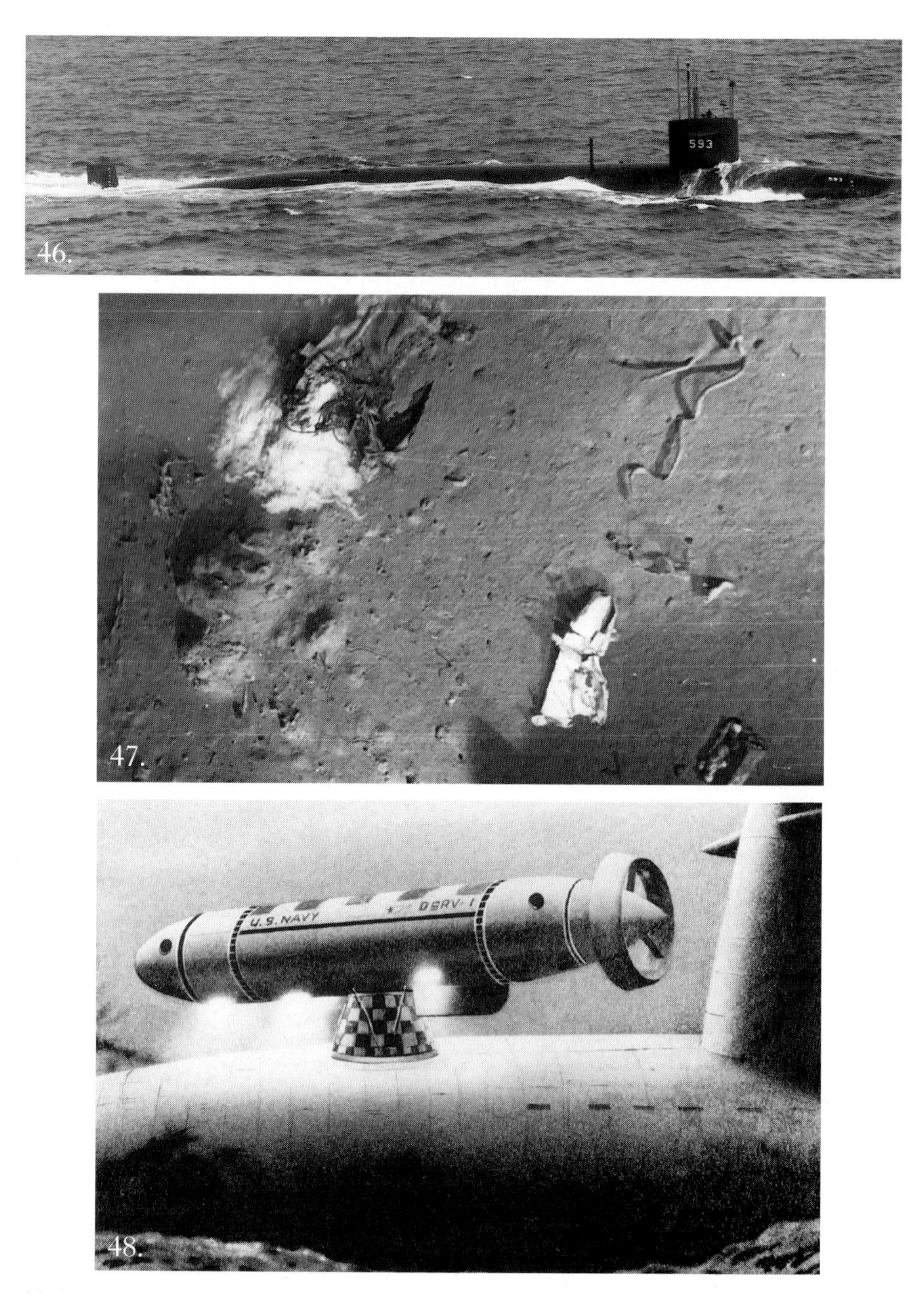

46, 47, 48. The ill-fated *Thresher* (SSN-593). Intended as the first of a new class of attack boats, *Thresher* was lost to accident with all hands April 10, 1963. A Court of Inquiry uncovered design shortcomings that probably contributed to the loss. Redesign delayed but improved the safety of follow-on boats, redesignated the *Permit* (SSN-594) class. (47) *Thresher* wreckage, photographed at 8,300 feet. (48) The Deep Submergence Rescue Vehicle (DSRV), developed following the loss of *Thresher,* to assist any submarine that bottomed short of crush depth.

49, 50, 50a. Two recent Russian boats: (49) a Victor III attack submarine and (50) a Delta III ballistic missile submarine, both photographed in the summer of 1994; (50a) The Strategic Arms Reduction Treaty (START) at work at the Zvezdochka Shipyard, Severodvinsk, Russia, October 1996; the final days of a circa-1967 Yankee-class nuclear-powered ballistic missile submarine. (U.S. Navy; Department of Defense)

51. *Seawolf* (SSN-21—the "submarine for the twenty-first century") on sea trials, July 1995, hailed as "...the most powerful warship in the world; not just the most powerful submarine."

dent than any set during the Russo-Japanese War. During the American Civil War, British firms had built several Confederate raiders, including the infamous *Alabama,* and the U.S. Government had later placed a claim against the British government for damages. A neutral, it was argued, is obligated to exercise due diligence to prevent the fitting out and departure of vessels of war for a belligerent. Even though the armaments were not installed until the ships had reached the Azores, the purpose for which the ships were being built was well known; Great Britain accepted responsibility, and a Geneva tribunal awarded $15,500,000 to the United States.

Based on *that* precedent (and bolstered by his own personal animus to big business, especially big munitions business) Secretary of State William Jennings Bryan overruled his staff. An export license would be denied. Schwab appealed to the Joint State and Navy Neutrality Board, which held that the contract, as framed, was within international law and custom. President Woodrow Wilson overruled the Board: no submarines, in whole or in part, to be shipped to England.

Therefore, Schwab arranged for the first ten submarines to be built in Canada. He persuaded the Admiralty to lease the Canadian Vickers shipyard in Montreal (through an amicable arrangement with the parent Vickers Corporation), then put his own management team and skilled workers on site and began shipping submarine parts from his American operations to Canadian Vickers in the guise of "structual steel for bridges to replace those destoyed in Europe." Because of the increased complexity, he also negotiated a 20 percent increase in price, to $600,000 per boat, plus incentives for timely delivery.

The American press, especially the *New York Times* and the *Wall Street Journal,* followed the story with close and reasonably accurate detail. The German and Austrian ambassadors lodged a query with the State Department. Were the stories true that "the Bethlehem plant sends secretly the component parts of submarines to Canada whence they are to be shipped to England . . . ?"

Schwab assured friendly officials at the State Department (he did *not* deal with the Secretary) that while it was true that his subsidiary companies were providing "materials" to Canadian Vickers, these required additional fabrication before they could be considered components for a submarine. This, in fact, was the case for everything except the engines. These were being built by a subsidiary of Electric Boat and were just, well, engines, suitable for any maritime application. The State Department therefore told the ambassadors on February 7, 1915, that "no component parts are being built by the Bethlehem Steel Works and being sent to Canada."

The first of ten submarines was launched April 18th, the last on May 26th. They sailed to England under their own power, and were integrated into the fleet. However, by this time, the Admiralty had come to realize that Great Britain did not really need many more submarines than she already had or were on order, and no effort was made to acquire the second ten.

The Germans had gone into the war without any submarine strategy beyond a hopeful "send the U-boats out and see what happens." On October 20, 1914, "what" began with a small and not immediately appreciated act. Operating fourteen miles off the Norwegian coast, *U-17* became the first submarine to sink a merchantman—the small British steamer *Glitra.* The attack was carried out under strict observence of the prize rules. Once the boarding party from *U-17* had verified the British registry of the ship, they ordered the crew into lifeboats and then—saving ammunition—opened the sea cocks to flood and sink the ship. Then the submarine hitched the lifeboats to a towline and hauled them closer inshore to a safer position.

Too bad that *U-17* didn't set a precedent. Six days later and without warning, *U-24* torpedoed *Amiral Gauteaume,* an unarmed ship loaded with Belgian refugees. U-boat Commander Rudolph Schneider assumed that the people he saw on deck were soldiers, which would have made the ship a legitimate target. *Assumed,* in the heat of the moment and perhaps the dream of glory, but did not verify. This was not to be the last time in the war when a young, inexperienced, overzealous U-boat commander would break the rules, such as they were. The ship survived, but forty passengers were killed.

The Germans marked another seminal event not quite a month later when *U-18* did the unthinkable—insofar as British thinking was concerned—and on November 23rd, penetrated the fleet anchorage at Scapa Flow. Fortunately (for the British), the Grand Fleet was at sea; unfortunately (for *U-18*), the boat was spotted, rammed by two ships, became unmanageable, got caught in the tidal flow, and put up on the rocks. Fortunately (for the British) the German High Command assumed that *U-18* had fallen victim to man-made, not natural antisubmarine defenses and did not take advantage of this weakness.

The impact on the Royal Navy, however, was electric. Scapa Flow was 475 miles from the U-boat's home base—so much farther than the range of the current British submarines that no one had bothered to put up any defensive measures. This one small boat forced the most powerful battle fleet in history to abandon for a time its main base and shift to a harbor on the other side of Scotland. And next—after an inconclusive sighting of perhaps another U-boat—the fleet moved for a time to an even more remote location in Ireland.

The British weren't totally asleep at the defense switch, and almost immediately began moves to block U-boat passage through the Strait of Dover—only twenty two miles wide. Presumably feasible, the operation was to prove quite difficult. Between October 2nd and the middle of February, the British laid 7,154 mines—of which more than half sank to the bottom or drifted away without result. Of the remainder—well, no navy at that time seems to have learned how to make an effective mine. At low tide, these were clearly visible, floating at the surface; when the occasional U-boat did make contact, the mines often failed to explode.

Other defensive measures in the Strait included an arrangement of net-holding booms to be stretched across the Channel, and a flotilla of trawlers towing three-hundred-foot-long wire-mesh nets. The nets were fitted with indicators to reveal when a U-boat was caught, in which event patrol craft would converge and drag an explosive sweep line over the spot. This was not very useful, accounting for only one U-boat kill.

The weather proved so unsuitable—what with strong tides, heavy currents, and frequent gales—that the boom line was abandoned before half finished. The trawlers slogged through.

The true value of the Dover patrol was exactly the same as that of the U-boat which reached Scapa Flow: intimidation. On April 10, 1915, the High Seas Fleet Flotilla declared the English Channel too dangerous and off-limits for U-boat operations. This interdiction was to continue until December 1916.

On November 2nd, Great Britain declared the whole of the North Sea a war zone and set a blockade of Germany. All ships, of whatever flag, were subject to search; the British were looking for contraband destined for Germany—whether directly, or by way of one of the neutral ports in Holland or Scandinavia. Contraband cargo was subject to seizure, and questionable cargo to forced purchase by the British.

The nature of "blockade" had not changed since the American Civil War. A *blockade* must be limited to enemy territory, confined to specified limits, and effective. Merely "declaring" a blockade does not establish a blockade. There are separate categories of contraband: *absolute* (munitions), *conditional* (food, clothing, fuel, etc. which had dual civilian-military use) and *free goods* (cotton, wool, leather, raw materials)—the latter usually free of restraint.

The London Declaration of 1908 determined that blockade would apply only to absolute contraband. However, the Declaration of London was never ratified by Great Britain, and thus she felt no obligation to allow the shipment of any goods to her enemy. The British did not limit their searches to absolute contraband, but included *anything* which they felt could help keep Germany in the war—including food, fodder, fertilizer, fuel, and raw materials.

Would this matter? On the books, Germany was 80 percent self-sufficient. However, German soil tends to be poor and sandy; without fertilizer—imported—German farmers grew poor crops, increasingly poorer as the soil nutrients were depleted. The country produced limited fodder in the best of years, and usually imported most of the nation's requirements. Fodder was another fertilizer-dependent crop, and without adequate fodder cattle farming was handicapped.

Then too, the concept of "80 percent self-sufficiency" did not address the powerful German economy: industrial production $3 billion, compared with $4 billion for Great Britain (and $7 billion for the United States), all heavily dependent on imports and exports. In just twenty years, the German merchant marine had trebled in size, and now had 12 percent of total world tonnage, second only to that of Great Britain (at 48 percent). Her foreign commerce also was second only to that of Great Britain. Expansion of the German Navy was in part predicated on the need to protect her overseas commerce.

There were some efforts by neutrals to import materials for their own use, and export homegrown equivalents to Germany, but this did not have much impact overall. The impact of the blockade, however, was immediate. As General von Freytag-Loringhoven was later to write, in rather formal jargon:

The consequences of the blockade showed themselves at once. Although we succeeded in establishing our war economics by internal strength, yet the unfavorable state of the world economic situation was felt by us throughout the war. That alone explains why our enemies found ever fresh possibilities of resistance, because the sea stood open to them, and why victories which would otherwise have been absolutely decisive, and the conquest of whole kingdoms, did not bring us nearer peace.

16 1915—Poking at Sleeping Giants

. . . a U-boat cannot spare the crews of steamers, but must send them to the bottom with their ships. The shipping world can be warned of these consequences, and it can be pointed out that ships which attempt to make British ports run the risk of being destroyed with their crews. This warning that the lives of steamers' crews will be endangered will be one good reason why all shipping trade with England will cease within a short space of time.

—Admiral Freiderich von Ingenohl, Commander in Chief, Imperial German Navy, November 1914

By the end of the year, U-boats had sunk seven warships and ten merchant ships—and Germany had lost five submarines: Two hit mines, two were rammed, one was lost to causes unknown. The British sank two small warships, but lost four submarines: one to a mine, one torpedoed by a U-boat, two to causes unknown. Call it an ambiguous baptism of fire.

But the British Navy had learned the lesson of *Aboukir* and companions: No ships went to the aid of the antique battleship *Formidable,* sunk on New

Year's Day with the loss of 550 of a crew of eight hundred. And at the Battle of the Dogger Bank, January 24, 1915, British fleet units in hot pursuit of German battle cruisers made a ninety-degree turn upon a lookout's report of a periscope sighting, which gave the Germans an insurmountable lead. Admiral Beatty, in his post-battle dispatch, noted that he "personally observed the wash of a periscope two points on the starboard bow." The nearest U-boat was sixty miles away. Lieutenant Nimitz had not been very far from the mark with his scarecrow-periscope scheme.

But the skirmishing was about to end, and the real battle—the battle for which the submarine was uniquely fitted—was soon to be joined.

A few months before the war, a member of the German Naval Staff was musing about the possibility of a U-boat war against British commerce. He may have been almost the only person in Germany to have been thinking thus, and we cannot help but wonder if he was not, perhaps, privy to the "warnings" of the British 1912 Admiralty Report on war risk insurance. He asked a junior officer to take a look. The junior officer estimated that it would take a force of 222 U-boats to isolate Great Britain. His study ended up where most staff studies end up: filed away for historians. Germany's operational fleet only reached *U-38* by the end of 1914; there clearly had been no long-range plan to take on any significant portion of the merchant fleet.

But now it was November, a blockade had been declared, and in addition, Britain was playing a bit loose with the rules. In reaction to the twin sinkings of commercial vessels, the British government was suggesting that British ships fly neutral flags. Commander in Chief of the High Seas Fleet Admiral Freiderich von Ingenohl urged Chief of Naval Staff Admiral Hugo von Pohl to propose an immediate counter-blockade, using submarines:

> As England is trying to destroy our trade it is only fair if we retaliate by carrying on the campaign against her trade by all possible means. Further, as England completely disregards international law in her actions, there is not the least reason why we should exercise any restraint in our conduct of the war. We can wound England most seriously by injuring her trade. By means of the U-boat we should be able to inflict the greatest injury. We must therefore make use of this weapon, and do so, moreover, in the way most suited to its peculiarities. . . . Consequently a U-boat cannot spare the crews of steamers, but must send them to the

> bottom with their ships. The shipping world can be warned of these consequences, and it can be pointed out that ships which attempt to make British ports run the risk of being destroyed with their crews. This warning that the lives of steamers' crews will be endangered will be one good reason why all shipping trade with England will cease within a short space of time. . . . The gravity of the situation demands that we should free ourselves from all scruples which certainly no longer have any justification.

Pohl submitted a formal proposal on November 7th; however, the rest of the Admiralty was not yet convinced, and more important, Chancellor Bethmann-Hollweg and Foreign Secretary von Jagow thought the scheme too risky. There were American sensitivities to be considered.

Pohl persisted, suggesting that intimidation would do the job, the mere announcement would frighten the neutrals away from British trade, and Germany wouldn't have to worry about provoking a wider war. But it all had to be done, he further suggested, before the South American grain harvest was due to arrive in British waters.

About the same time, Admiral von Tirpitz gave a notoriously optimistic interview to a newspaperman in which he made it all but certain that a U-boat war against British commerce would quickly settle the war in Germany's favor. The public couldn't wait!

Tirpitz was quoted in the *London Times* on December 24, 1914:

> We have learnt a great deal about submarines in this war. We thought they would not be able to remain much longer than three days away from their base, as the crew would then necessarily be exhausted. But we soon learned that the larger type of these boats can navigate around the whole of England and can remain absent as long as a fortnight.
>
> All that is necessary is that the crew gets an opportunity of resting and recuperating; and this opportunity can be afforded the men by taking the boat into shallow waters, where it can rest on the bottom and remain still in order that the crew can have a good sleep.

To this, the journalist added, "In Admiral Tirpitz's opinion a submarine war against British merchant ships would be more effective even than an invasion of England by means of zeppelins."

And so, first the German Admiralty (in January) and then the Chancellor (February 1st) agreed; Kaiser Wilhelm ratified the decision on February 4th. One day later, Pohl's proclamation was issued declaring "all the waters around Great Britain and Ireland, including the whole of the English Channel" a war zone, and warning that, from February 18th, all British merchant shipping in those waters would be destroyed. Further, the proclamation carefully noted, it would not always be possible "in view of the misuse of neutral flags ordained by the British Government . . . and owing to the hazards of naval warfare . . . to prevent the attacks meant for hostile ships from being directed against neutral ships."

Flags, indeed, had been "misused." The Cunard liner *Lusitania* sailed through Irish waters under an American flag on January 31st—on the specious grounds that she had some American passengers. Typically, violations of the laws of war have opened the door to reprisal. The form of reprisal was usually response in kind, but we must assume that German ships were not interested in hoisting the American flag.

To permit innocent neutral ships to reach neutral (and, hopefully, German) ports, a narrow corridor passing north of Scotland and running along the Dutch coast was declared safe from attack.

The first unrestricted campaign began on February 22nd; there was only one effective U-boat at sea, which sank five British steamers. However, as a modest beginning, the submarine blockade soon enough was taking out perhaps 1 percent of the monthly traffic.

Very soon, expediency dictated the need for larger numbers of less capable boats which could be brought more quickly into service, especially for use in the North Sea area where range and endurance were not so critical. Two classes were developed in 1915—the coastal-type UB-boats, which were prefabricated in sections in Germany and assembled in Belgium, and the UC minelayers, carrying only twelve mines.

Initially, the UBs were 127/142 tons, speed 6.5 knots on the surface, 5.5 knots submerged. Expediency? Since UBs were only one-fourth the size of the newer U-boats, they could be built in about one-seventh the time: seventy-five days for construction of the major sections, fourteen days for final assembly, compared with eighteen months. The first UBs entered service in April. Soon, the class was improved, with more than double the displacement, more torpedoes, 50 percent higher surface speed. The UBs and UCs made up the Flanders Fleet; some operated out of the Adriatic ports.

In the meantime . . . the prototypical symbol of the U-boat war had just gone down—literally—in history.

Lusitania was scheduled to leave New York on May 1, 1915, and in the normal practice of the day, the owner-operator placed sailing announcements in the New York newspapers. In a most abnormal procedure, the German Embassy to the United States placed advertisements in the same New York newspapers, adjacent to the Cunard Line notices, to remind the neutral Americans of the state of war. The ads warned, without equivocation, that "Travellers sailing in the war zone . . . do so at their own risk."

Just at noon, the Cunard liner was under way for Liverpool. At 9:15 p.m. six days later, *Lusitania* was hit by a single torpedo and sank within eighteen minutes. Newspaper headlines which proclaimed "TWICE TORPEDOED" were in error; there was a second explosion, but it was either from coal dust or the sympathetic detonation of munitions in the cargo. No matter; *Lusitania* went down, taking 1,198 men, women, and children to the bottom. These included 128 Americans.

The civilized world was horrified; former President Roosevelt called this act "piracy on a greater scale than any old-time pirate ever practiced." Lieutenant Commander Walter Schweiger, commanding *U-20,* saw *Lusitania* as a stroke of good luck. He was completing a successful patrol—three ships sunk on the previous two days—and was about to go home when he spotted the liner and rushed to take firing position.

In the immediate aftermath, and indeed at times in the more than eighty years since, a number of arguments have been offered to explain or excuse the sinking. For example: Was *Lusitania* operating as a naval auxiliary and therefore fair game? The Cunard Line and Schweiger had legitimately different opinions. Construction of the ship had been subsidized by a grant from the government, the decks had been built to a standard which would accommodate the installation of naval guns, and the owners received an annual stipend, all to make the ship available for conversion to government service in wartime—as provided for in the 1908 London Declaration. Copies of *Jane's Fighting Ships* and *The Naval Annual* carried aboard *U-20* and consulted before the attack described the ship as an Armed Merchant Cruiser. At the time of the sinking, however, the liner was in civilian, not military service and deck guns had not been installed.

Was *Lusitania* carrying contraband cargo?

President Wilson discounted reports of munitions aboard the liner and accused the Germans of trying to becloud the issue. When Senator Robert LaFollette of Wisconsin, solidly identified as the leader of a peace faction, attempted to introduce the subject of "munitions" into the Senate, he was threatened with expulsion; he was saved, more or less, by the Collector of the Port of New York, who announced that he would willingly testify to the facts of the matter—that munitions were indeed not only on the manifest but in the hold. The manifest listed 173 tons of ammunition, and in later years divers to the wreck identified quantities of military stores, so that answer is "yes."

Finally, did *Lusitania*'s Captain William Turner take all appropriate steps to avoid danger? Not really. The day before, he had been warned of submarine activity in the area; his response that night was to increase the lookouts, darken the ship, and close the watertight doors. However, on the fatal day, he was moving along his chosen course, off the south coast of Ireland, without zigzag or deception, a course which any U-boat navigator could predict with hardly a glance at the chart. Once Schweiger had spotted the big ship and realized that he could reach an attack position, the game was over.

Schweiger knew that the ship was listed as a naval auxiliary, and of course the *Lusitania*'s captain made his job easier by following a predictable course. However, the ship was clearly identifiable as a passenger liner. Schweiger did nothing to verify her "status" and fired his torpedo without warning and without hesitation.

Prior to the sinking of *Lusitania,* the American public was concerned more about the loss of trade with Germany than with other aspects of the war; afterwards, public opinion began a steady shift against the Germans. President Woodrow Wilson began to feel some pressure to *do* something. The American Ambassador in London, Walter Page, warned him that the United States "must declare war or forfeit European respect."

President Wilson offered a cautious, awkwardly phrased diplomatic response to the sinking: "The objection to the present method of attack against trade lies in the practical impossibility of employing U-boats in the destruction of commerce without disregarding those rules of fairness, reason, justice and humanity, which all modern opinion regards as imperative."

Admiral Scheer—in an internal staff memo, not passed to President Wilson—asserted that "U-boat warfare upon commerce is a deliverance; it has put British predominance at sea in question."

The first period of unrestricted U-boat war lasted from February through August 1915; free to attack any shipping in the declared war zone, the U-boats did so with great success. In August alone, they sank forty-two British merchant ships totaling 135,000 tons. However, on August 19th, about fifty miles from *Lusitania*'s grave, another civilian liner was sunk, *Arabic,* with minor loss of life but with three Americans among the casualties. The sinking was without warning and at the hand of Rudolph Schneider, who had put *Amiral Gauteame* in the history books as the first merchant ship sunk without warning. Again, there was great international indignation. Great pressure was brought upon the U.S. government to intervene to "protect American property and citizens."

It was apparent that further provocation could bring the United States into the war, bringing with it such immense natural and manpower resources that German victory would be in doubt. The German government had a debate.

Chief of the Naval Staff Bachmann was highly in favor of continuing the U-boat campaign; he was fully supported by German public opinion.

Foreign Minister Gottlieb von Jagow protested that continuation of unrestricted warfare would bring all of the neutral states into the war: "Germany will be treated like a mad dog against which everybody combines."

Interior Minister Karl Helfferich warned of America's strength, and added, "I can see in the employment of the weapon nothing but catastrophe."

With his own nervous eye on the industrial giant across the Atlantic, Chancellor Theodore von Bethmann-Hollweg called unrestricted warfare "an act of desperation." He settled the argument by replacing Admiral Bachmann with Admiral Henning von Holtzendorff, a close personal friend.

Unrestricted warfare was suspended. To the surprise of the government, the Naval Staff raised no further objection. However, the interests of the Naval Staff were more temporal than political: they looked at the operating tempo—the boats of the High Seas Fleet were dangerously overextended and overdue for overhaul—and they checked the calendar—winter was approaching and operating conditions in northern waters were about to become "impossible." The German war effort would not seriously be impeded by any temporary suspension.

The Naval planners also had been running some numbers. Admiral von Holtzendorff easily was persuaded that, "If after the winter season, that is to say under suitable weather conditions, the economic war by submarines be begun again with every means available and without restrictions which from the outset must cripple its effectiveness, a definite prospect may be that, judged by previous experience, British resistance will be broken in six months at the outside."

In the meantime, U-boat commanders were once again required to observe prize rules. Since this could so easily expose them to fatal jeopardy, they exercised extreme caution and did not attack any ship of which the identity or character was in doubt. This caution, and the weather, brought a sharp reduction in sinkings around the British Isles; by the end of September, the German focus was clearly in the Mediterranean, where the weather was good, there were targets galore, and not many Americans.

17 1915—Blind Faith

It is an omen—an Australian submarine has done the finest feat in submarine history and is going to torpedo all the ships bringing reinforcements, supplies and ammunition to Gallipoli.
—Commodore Roger Keyes, aboard the British Flagship Queen Elizabeth, *April 26, 1915*

In the meantime, on November 1, 1914, Turkey had entered the war as an ally of Germany. This put a choke on the Dardanelles—the maritime route connecting the Black Sea and the Mediterranean. In 1914, 90 percent of Russia's international trade moved along this route and thence to the rest of the world. A large part of that traffic was grain for Great Britain. Three warships—two Germans which had sought refuge in Constantinople while Turkey was still neutral, and one Turkish ship—were blocking the strait and occasionally roaming the Black Sea, shelling Russian port cities.

The British had stationed five "B" class submarines in the Mediterranean since 1910, Now, along with three French submarines, they set up a new base on the island of Lemnos to counter the counter-blockade of the Dardanelles. By the middle of December, the British submarine commander in the Mediterranean had decided to send one boat on a strike mission through the Dardanelles and into the Sea of Marmara.

Easy to say, harder to do. The strait is thirty-five miles long, and only one mile wide at the narrowest point. Five minefields had been laid across the channel. To welcome any submarine forced to the surface, searchlights swept the area all night and lookouts continued their vigil all day, the shoreline along both sides was lined with gun emplacements both new and old, and the gunners had mapped out their fields of fire. Finally, the currents running out of the Dardanelles were swift and erratic.

About 4 a.m. on December 13th, the British submarine *B.11*—chosen above her sisters because a brand-new battery bank had just been installed—began the underwater passage, against the current, through the minefields, and into the Dardanelles. By 10 a.m. she had sunk the Turkish cruiser *Messoudieh* with the first torpedo her skipper had ever fired, and by late afternoon was back at home base.

Well, it was not quite as simple as that might sound; the going in was easy enough, involving only one scrape against a mine, and the shells aimed at her periscope by *Messoudieh* all missed. However, on the way back out, the current swept the boat aground in shallow waters—shallow enough, that is, that the conning tower was above the surface and the boat came under fire from the shore batteries. By pushing the electric motors to the maximum, the commanding officer, Lieutenant Norman Holbrook, was able to slide back off this mud bank and reach deeper waters. When finally able to surface, he found that the air inside the boat had grown so foul that his gasoline engines wouldn't start until the boat had been ventilated for more than half an hour.

It was the sort of exploit the public loved. The government was happy enough to award Lieutenant Holbrook the Victoria Cross.

The British War Council decided that the Dardanelles must be put in Allied hands, and that "the Admiralty should prepare for a naval expedition in February to bombard and take the Gallipoli Peninsula with Constantinople as its objective."

The War Council did not explain by what means a Naval force could "take" a heavily defended position ashore. They might have done well to spend a few moments in a review of American military history. While it is true that Naval units under Admiral Farragut had captured the defenseless Confederate city of New Orleans, Admiral Farragut quickly needed the assistance of Army units to hold the ground. Gallipoli was far from defenseless. The local Turkish and German forces had taken their role in the counter-blockade seriously, by moving naval guns ashore, hardening emplacements, bringing in howitzers, laying minefields.

Senior British Naval officers did not object to the War Council's plan, apparently in the belief that the whole operation was to be a test; if things didn't go well, it would be dropped. Things didn't go well at all, but rather than cancellation, the operation was ratcheted up to become one of the bloodiest unsuccessful campaigns in the history of warfare.

The naval bombardments began in normal fashion in mid-February and continued through March 18th; then, the years of theory linking Gianibelli, Bushnell, Fulton, and Colt, and given a useful trial more than half a century earlier by Maury, came crashing into modern naval reality. At 1:54 p.m., the French battleship *Bouvet* struck a mine and sank within two minutes, with the loss of more than six hundred men. At 4:06, the British battle cruiser *Inflexible* hit a mine, with serious but not fatal result; at 4:16 the British battleship *Irresistable* found her mine, followed by another British battleship, *Ocean,* at 6:00. Both sank during the night, although the crews had been safely removed. In a period of slightly more than four hours, four major warships were taken out by a single line of twenty moored mines.

The British—more specifically, Winston Churchill—now frustrated by the apparent lack of progress from the naval bombardments, decided to take the Gallipoli Peninsula (and the Dardanelles) by force. Ironically, the bombardments actually had been effective; the defenses ashore now were barely functional and—historic hindsight has proclaimed—had an invasion been launched at or near that moment, it would certainly have succeeded. However, the Army wasn't ready. The Navy wasn't ready. It took five weeks to get ready, and then they still weren't ready; among other blunders, once the troops were on the beach and under fire from enemies who had used the same five weeks to get ready, it was discovered that the transports had been loaded upside down, with the guns and munitions buried under tents and clothing.

In the meantime—while everyone was getting ready—the British Naval Command had been so pleased with the exploit of *B.11* that seven new-model E-boats and the sister-ship Australian *AE.2* were sent to Lemnos. Then, in mid-April, in trying to be the first "modern" submarine through the Dardanelles, *E.15* was swept hopelessly aground by an exceptionally strong current, right under the guns of a Turkish fort. The captain was killed in the shelling, the crew was forced to surrender and taken ashore, and a series of British attempts to destroy the submarine became increasingly desperate. First, *B.6* fired two torpedoes, which missed. Two destroyers went in after dark, but were easily picked out by searchlights and turned back by the guns. Seaplanes came in at dawn to drop bombs, with no luck. *B.11* was next, but ran into fog and couldn't locate the "target." Two battleships made the next attempt, but couldn't get closer than six miles—too far for their own gunnery to be accurate.

The Turks moved more guns into position and sent two tugs to try to pull *E.15* free. In a final desperate effort, two British torpedo-equipped patrol boats, manned by volunteers, made the run in after dark and, despite heavy opposition, managed to get two torpedoes into *E.15* before one boat was taken out by an enemy shell. The other swung by, picked up the crew, and got away with only one casualty—a sailor who later died of his wounds.

This saga, however, was not over. The Naval Command wanted to be certain that *E.15* was completely destroyed, so *B.6* was dispatched to examine the site. *B.6* was caught in the same treacherous current which had put *E.15* aground, and also became stuck in the shallows, not one hundred yards from that wreck—in full view of the gunners ashore. She managed, however, to back clear and get away.

The invasion finally was launched on April 25, 1915; within hours, casualties were overwhelming and the military commanders, working from the battleship *Queen Elizabeth,* were stunned. The Australian boat *AE.2* was ordered into the Sea of Marmara on the morning of the 26th, with a clear mission: Interdict the shipping that was bringing supplies to the defenders at Gallipoli.

AE.2 made it through the Strait and sent an "arrival" message back to the flagship. Aboard *Queen Elizabeth,* Chief of Staff Roger Keyes read the message aloud to assembled senior staff, and announced: "It is an omen—an Australian submarine has done the finest feat in submarine history and is going to torpedo all the ships bringing reinforcements, supplies and ammunition to Gallipoli."

On the strength of this one report and Admiral Keyes's incredibly naive confidence in the submarine, the Naval Command decided not to cancel the operation and pull the troops back off the beach. By the time the affair was finally ended, eight months later, the casualty list was to include more than 25,000 killed, more than twelve thousand missing, more than 75,000 wounded—just on the Allied side. The enemy lost 66,000 killed. A strange "tribute" to the power of the submarine.

On the 27th, *E.14* was sent to join *AE.2* and, after the usual exciting passage, met the Australian boat to discover that although *AE.2* had taken on several targets, every torpedo had run too deep and she was now down to one. That information would have been included in the "arrival" message; however, transmission was cut short by the appearance of a Turkish destroyer. So much for Admiral Keyes's optimism.

The next day, *AE.2* ran into trouble with a Turkish torpedo boat, was damaged too badly to permit further diving, and was scuttled by her crew to prevent capture. Most crew members were taken prisoner; nine disappeared with the boat. On May 1st, the French submarine *Joule* tried to make the passage, but hit a mine and was lost with all hands.

In the meantime, *E.14* had been playing hunter/hunted with Turkish gunboats, sinking one, but otherwise there was not much success on either side. Then, on May 10th, *E.14* made a spectacular hit with her last torpedo on a troop-laden transport, which sank with a loss of six thousand men.

However, the whole affair at Gallipoli already was acknowledged to be a disaster, made even worse on May 11th when the British battleship *Goliath* was torpedoed by a Turkish destroyer (operating with a German crew). The loss of 570 sailors precipitated a crisis back home. Churchill wanted to send in more Naval units; Admiral Fisher called that throwing good money after bad and resigned. Then Churchill, under fire from all sides, also resigned. The slaughter continued.

To counter the Allied operations around the Dardanelles, *U-21,* Otto Hersing commanding, was sent into the Mediterranean. She arrived on station off Gallipoli on May 24th, and found her first victim the next day, the battleship *Triumph,* torpedoed at the range of three hundred yards. But there was a problem: British destoyers swung to the attack, heading for the spot from which the torpedo had been fired. Solution: *U-21* followed the trail of her own torpedo and dove under the sinking warship and out the other side, like a fox escaping from a secret exit while the hounds were worrying the entrance to the lair.

The next day *U-21* sank another battleship, *Majestic,* at a range of six hundred yards. A few hours later, the British ordered the flagship, queen of the fleet *Queen Elizabeth,* out of the war zone. *U-21,* maneuvering through a passage opened just for the purpose, threaded the Darndanelles to a new home berth in Constantinople. To confuse enemy spies ashore, Hersing changed his hull number to "*U-51.*"

The story of the Dardanelles continued: British submarines patrolled the Sea of Marmara for the next eight months, looking for troop and ammunition ships but taking on any target of opportunity. Operations in the Marmara were geographically constrained—the sea is about one hundred miles long by fifty miles at the widest point, small enough that a submarine could make a run from one end to the other in a day and small enough that she often could be spotted from the shore.

The most notable British submarine commander to operate in the Marmara—indeed, one of the most notable of the war—was Martin Nasmith, commanding *E.11,* who made three patrols in the Sea between May and December 1915. Nasmith exhibited an exceptional degree of both initiative and pluck. Before his first passage through the Dardanelles, he arranged for an "orientation flight" in a British seaplane. He had guards installed around his propellers and hydroplanes, to fend off the mooring lines of mines and the mesh of the antisubmarine nets. He took aboard an overload of twelve torpedoes—giving up his bunk for storage and sleeping on the floor.

On at least one occasion, once in the Marmara, Nasmith used a Turkish sailing ship as camouflage while he recharged his batteries and sent a lookout up the tallest mast in an effort (unsuccessful, as it turned out) to broaden the search for targets. He torpedoed one ammunition ship moored to a pier. Putting a team of sharpshooters on his bridge, he forced another ship to drive herself up on the beach.

In what was more of a psychological than military coup, he penetrated Constantinople harbor, where he took the first-ever photograph through a periscope, surfaced, fired two torpedoes at a transport loading out at a pier—one of which hit—then escaped. He did so under heavy enemy fire and after almost being hit by his other torpedo, which had gone astray and, as he later wrote, "travelled twice around the main harbor and up under the Galata bridge, turned hard left and sank a ship alongside the wall."

Nasmith was rigorous in his routine: Sunday was inspection and all hands to prayer. He tried to arrange some time each day for a portion of the crew to go overboard for a brief bath.

Running short on torpedoes, and vowing to sink one ship with each of the five remaining, Nasmith had them set to float rather than sink at the end of an unsuccessful shot. This was a violation of international law, but who was to challenge him? He went in the water himself to disarm the first one which had missed a target. One of his officers did the same for another recovered torpedo—which was brought back aboard the submarine through the simple expedient of sliding it nose-first into a torpedo tube, then bringing it inside for inspection, repair, and return to normal firing position. Overall, one of the final five had missed and disappeared; each of the other four eventually found a target.

Martin Nasmith was to become the leading British ace of the war, sinking more surface warships than any other commander—four, of the fourteen sunk by all British submarines throughout the war—and posting a total score of 122 ships.

Nasmith made a valuable contribution to the boats which would follow him into the Marmara: He determined that the difficult currents in the Darndanelles were casued by a steady outflow of fresh water from inland rivers, which in some places formed a top layer about sixty feet deep, while below that a layer of salt water extended to the bottom. In some places, the currents in the two layers was actually running in different directions, twisting a boat abruptly off course. A boat trimmed for Mediterranean salt water but hitting a patch of fresh water, might suddenly and unwittingly sink until once again in the denser water. Nasmith calculated just the right trim to allow his boat to float on top of the deeper layer, an alternative to spending the day resting on a bottom which in most places was too deep for *E.11*.

The Turks made constant efforts to close the channel. First, they installed a heavy steel-mesh net, suspended from floating buoys. Gunners ashore would watch for bobbing floats, and thus alerted, would be waiting for the trapped submarine to pop to the surface. The first boat to hit the net managed, by repeating the sequence of full astern, full ahead, full astern, to seesaw through before running out of power, although one of the motors was burned out in the effort and the submarine had to abandon the patrol.

The next boat to force the passage hit the net at full speed and broke through—to discover that two offshore torpedo launchers had been installed, triggered by a movement of the net. The torpedoes missed. On July 25th, a French submarine, *Mariotte,* was caught by the net and forced to the surface and heavy shelling; the captain had no choice but to scuttle the boat and surrender the crew.

The next obstacle to be installed was a heavy wire cable strung across the narrows just below the surface, intended to snag and damage periscopes. These efforts, of course, were all in addition to the minefields which claimed at least three Allied submarines. They may have been running on the surface and therefore, as far as the mines were concerned, would have been the "surface ships" against which they had been designed.

One submarine fell victim, in a manner of speaking, to a rowboat. *E.7* became trapped in a net and her propeller was fouled by a broken wire. Over a period of some five hours, during which several primitive depth charges were dropped from a patrol boat—without effect—*E.7* alternately tried to break free and rest.

The captain and the cook of *UB-15,* moored not far away, came over in a rowboat. They took soundings with a lead line until they located the trapped submarine, then set the fuse on a mine, lowered it down, and rowed like mad to clear away from the area. The submarine was fatally damaged, having no choice but to surface while it was still possible and surrender.

The submarine patrols were ended in January 1916, when the eight-month bloody mistake at Gallipoli was finally canceled, and all Allied troops were evacuated in one of the better-planned operations of the war, without additional casualties. The patrols had had no appreciable influence on the outcome—or rather, the *non*-outcome—of the battle.

The box score in the Marmara: for the Allies, seven warships, nine transports, seven supply ships, thirty steamers and two hundred sailing vessels. For the Turks and Germans: three French and four British submarines.

Nearby, the French had assumed responsibility for operations in the Adriatic, but the clear waters provided no cover for a submarine. German air patrols easily spotted lurking French boats, and six were lost in the first few months of the war. One French submarine, *Curie,* was caught in antisubmarine nets and scuttled. The Austrians easily salvaged and reconditioned the boat and put her into service as *U-XIV.* By way of differentiation from their ally's U-boats, the Austrians used Roman numerals.

The Allies attempted to close the entrance to the Adriatic with a "barrage"—from the French word for "barrier"—a series of nets strung across the forty-four-mile mouth of the sea, from Greece on the east to Otranto on the west. The effort came to be known as the "Otranto Barrage." The barrage was not very effective, only claiming one victim in the whole war—*U- VI,* on May 12, 1916.

Two months later, *E.41*—under the command of Norman Holbrook—was rammed by *E.4* in home waters; Holbrook and most of the crew managed to get free, but the boat sank with seven men trapped inside. They gathered in the control room and waited as the water compressed the remaining air enough to allow them to pop open the conning tower hatch. They had to wait until the water was up to their necks, but patience paid off for four of them, who were swept to the surface with the suddenly released air bubble. Two more men managed to swim out before the hatch closed again. One man remained below, unable to open the heavy hatch. All in the dark, he swam through several compartments, releasing air until he managed to bring air pressure up sufficiently so that he too could escape. *E.41* was salvaged and returned to service.

The other area for British submarine operations was the Baltic—minor, almost a footnote to the war, but an interesting footnote nonetheless.

In October 1914, a three-boat delegation was ordered to the Baltic; mechanical problems forced one to turn back. (That was Nasmith's *E.11,* thus "saving" him for the Dardanelles.) It was a naive move—the submarines had no assigned base, no arrangements had been made for supplies; the mission was based on the assumption that they would be welcomed and supported by the Russians, although the Russians appear not to have been consulted. In any event, Russian support was forthcoming—more or less.

It was naive, and limited, but the small force had an immediate impact on German operations. First, patrolling German forces pulled back to base. Next, the Germans were forced to set up convoys for iron ore shipments across from Sweden. This was two years before the Allies understood that they too must do the same. *E.1*'s torpedo into the battleship *Moltke* caused withdrawal of the Baltic units to Kiel—one torpedo perhaps saving Russia from German conquest!

The force was augmented in 1915 by four more E-class and four C- boats. *E.13* did not complete the trip; having run aground on a Danish mud bank, she was destroyed by her crew to prevent imminent capture by circling German forces. The C-boats, too small to make the passage on their own, were first stripped of batteries and torpedoes to reduce weight, towed to

Archangel, then loaded on barges and sent south through the system of rivers and canals. At two hundred tons, each still offered a challenge to barge-handlers and stevedores, but all arrived safely in St. Petersburg. However, the ship carrying the batteries was sunk by a U-boat, so service was delayed.

On October 11, 1915, *E.19,* Lieutenant Commander F.N.A. Cromie commanding, captured and sank—all under prize rules—*five* merchant ships. On October 23, 1915, *E.8* found and sank the restored *Prinz Adalbert* when one torpedo made a lucky hit on a forward magazine.

Max Horton, skipper of *E.9,* was rotated back to England in January 1916—but not before an adventure ashore with a German spy. Horton had a reputation as a lady-killer, and the Germans picked out a particularly attractive lady in a move to turn the tables, so to speak. However, she apparently was so taken by Horton's charm that she warned him of the plot; her employers were captured and executed and the lady—never publicly identified by Horton—eventually settled in England.

By the summer of 1916, German defensive measures had greatly reduced the opportunities for the British, but the box score was impressive: only two submarines lost, *E.13* while en route, and one C-boat to accident. The British continued their patrols, but with minimal success, and life became rather routine—for a while.

At one point, having run out of rum—for the traditional daily ration of the watered-down concoction known as "grog"—the British asked the Russians for a supply of vodka to help alleviate the chill after a winter's patrol. It was not forthcoming, even though the request eventually made its way all the way to the Tsar, who, it was reported, offered a 20th century naval equivalent of Marie Antoinette's infamous "If there is no bread, let them eat cake."

"If they are so cold," he said, "why can't they wear two shirts?"

Not too many months later the attitude exemplified by that remark exploded into the Revolution of 1917. The uprising began on November 6th; by the 17th, the Bolsheviks had concluded an armistice with Germany—one condition of which was the surrender of the British submarines. The British submariners declined that invitation and shifted their boats to Helsingfors (today's Helsinki). There they remained until April, with Russia in chaos and the German Army steadily moving up the coast; the senior British officer, Cromie, induced a Russian icebreaker to clear a path to open water, then led his remaining fleet of seven out of the harbor. Once in deep water, with explosive charges set, the crews transferred to a waiting tug to watch as their submarines blew themselves up and sank. The crews returned to England. Cromie remained behind as Naval Attaché in St. Petersburg; he was killed defending the Embassy against a band of looters.

The Revolution had other consequences: by this time, the United States was in the war and Allied military supplies worth $50,000,000 had accumulated in the Port of Archangel, where they now were vulnerable to seizure by the Germans. The American cruiser *Olympia* (Admiral Dewey's flagship at the Battle of Manila Bay, and now a public museum in Philadelphia) was sent in, along with a contingent of British troops, to stand guard. At about this time, some Russian Naval officers, now prevented by treaty from active participation in the war, had pooled their resources to mount an expedition against seals—sealskins, you understand, being somewhat in demand on the open market. When a U-boat sank their sealing ship, with the loss of all the cargo and most of the crew, they got even. They turned over their force of three destroyers to Allied control: one to the British, one to the French, and one to the Americans on *Olympia*. Thus it was that a warship with the exotic name *Karitan Yurasovsky* served as a unit of the U.S. Navy, manned half by Russian and half by American sailors.

Churchill was suceeded in May 1915 as First Lord of the Admiralty by Arthur Balfour, who established a Board of Invention and Research to serve as a central clearinghouse for technical and scientific innovations, especially regarding antisubmarine warfare. Admiral Fisher was appointed chairman—partly in sympathy, partly out of the hope that his strong leadership would energize the effort—and the BIR began operations on July 5th.

The BIR solicited advice from the general public, and got more than fourteen thousand typically enthusiastic but largely naive responses. Many centered on disabling the periscope—gum it up with heavy goo, festoon it with an explosive necklace, smash it with a hammer. None were useful. Trained animals were suggested—sea lions to find submarines, sea gulls to spot periscopes.

Fisher himself proposed development of a monster submarine, hauling a 16-inch gun which could be fired while submerged. Balfour seemed interested, but told Fisher, "At this moment I would give up all new submarines in the world if the Inventions Board could discover a method of destroying the old ones!"

More serious scientific efforts were pursued, but Fisher's abrasive and at times irrational personality drove such a large wedge between the scientific community and the Navy that little was accomplished. Behind his back, BIR soon became known as the Board of Intrigue and Revenge. It was disbanded after two unproductive years.

Not to be outdone—or perhaps, to do the outdoing—American Secretary of the Navy Josephus Daniels created a Naval Consulting Board. In his letter of July 7, 1915, inviting Thomas Edison to be the chairman, the Secretary noted, "One of the imperative needs of the Navy, in my judgement, is machinery and facilities for utilizing the natural inventive genius of Americans to meet the new conditions of warfare. . . . we are confronted with a new and terrible engine of warfare in the submarine . . . and I feel sure that with the practical knowledge of the officers of the Navy, with a department composed of the keenest and most inventive minds that we can gather together, and with your own wonderful brain to aide us, the United States will be able, as in the past, to meet this new danger with new devices. . . ."

The board reviewed more than 100,000 suggestions, 80 percent of them concerning submarines. There was a throwback to Bushnell: floating torpedoes—"launch them in channels when the tide was going in, let them float into the German harbors and blow up everything afloat."

There was the "porcupine" defense: hundreds—thousands—of spikes sticking out from the hull of a ship. As the inventor noted, "Torpedoes can't sink a ship unless they hit her, and if you put these long spikes all along the side . . . the torpedoes are stuck before they hit the boat." There was the "mattress" armor—sheathe all ships with a layer of thick cotton batting. There was the "cowcatcher" bow, for scooping up and diverting incoming torpedoes. And, of course, there was at least one "unsinkable ship"—filled with lumber. All were in good company with the British antiperiscope prescriptions of molasses and hammers.

The major contribution of the Consulting Board, however, was a thorough survey of U.S. industrial capabilities—a key element in the rapid wartime buildup two years later.

18 1916—Germany

The German Government cannot but reiterate its regret that sentiments of humanity, which the Government of the United States extends with such favour to the unhappy victims of submarine warfare, have not been extended with the same feeling to the many millions of women and children who, according to the avowed intention of the British Government, are to be starved. . . .

—*German government response to a formal complaint from President Wilson, May 4, 1916*

By the end of 1915, British patrols had intercepted 734 neutral ships carrying supplies to Germany; this was three times the cargo loss imposed on the British by the U-boats. In January 1916, the German Naval Staff made a comprehensive review of their own plans and prospects.

1. Our war aim, apart from destroying the English Fleet as the principal means by which Britain controls its Empire, is to reduce its total economy in the quickest possible time, bringing Great Britain to sue for unconditional peace. To achieve this it will be necessary:

   a. To cut off all trade routes to and from the British Isles.

b. To cripple, in all the seven seas, all ships flying under the British flag and all ships under neutral flags plying to and from Great Britain.

c. To destroy military and economic resources and, by means of air attack, disrupt the trade and commerce in the British Isles, showing its population quite mercilessly the stark realities of war.

2. The shutting-off of the British Isles from all incoming and outgoing passenger and mail supplies in such a way that the British Isles are encircled by blockade and forbidden to neutral shipping: any ship attempting to breach the blockade will be destroyed. . . .

In a belated, perhaps temporary, but no less sincere acknowledgment of the new stature of the submarine, the Navy Staff cautioned, "It is not advised that surface ships be used for this blockade on account of danger from English submarines and other warships." To support extended U-boat operations, overseas bases were needed; the Faeroes, the Azores, and the Spanish coast were suggested. And ominously, although the current war was not yet two years old, they were already looking forward to the "next war":

> One cannot tell at this point in time whether, when peace is declared, the Faeroes and the Azores may be acquired and whether in the next war it will be possible to obtain the use of Spanish ports for our purposes: all this will depend completely on future political alignments. . . . we must consider neutral intervention against us in the event of England collapsing at our onslaught. Reorganization of U-boat fleets must take place in the shortest time so that we have an effective weapon to back up our policies. This building-up must be done in, at most, five years.

Five years . . .

The Germans had continued to make improvements to the longer-range submarines—but in small classes, which allowed new features and capabilities easily to be introduced. Eight boats of the *U-43* type, built in 1915–1916, were slight improvements over the "30s." *U-81*s, *U-93*s, and *U-99*s each brought greater range or firepower. The longer-range *U-81*s (810/950 tons displacement) could take ten torpedoes on a 7,630-mile cruise; the larger *U-93*s carried sixteen torpedoes but had a limited range of 3,800 miles; the smaller *U-99*s carried twelve torpedoes with an endurance of 4,080 miles.

But now, the focus shifted to long-range planning: Pick the most effective of the models, and shift all production. Yes, there was a hazard in a long-term building program: "The possible disadvantage in the building of a large number of similar U- boats, with the attendant feature that developing technology may overtake them, is outweighed by the need in which we find ourselves." The Navy ordered forty-seven U-cruisers, designated UA—long-range raiders, 1,930/2,483 tons, which could make 15.3 knots on the surface and had a range of 12,630 miles at eight knots. Those UA-boats were well designed for the "cruiser" mission, with two 150mm (5.9-inch) deck guns, storage for one thousand rounds of ammunition, an armor-plated conning station, nineteen torpedoes, and living space for twenty men beyond the crew of fifty-six to provide prize crews. Forty-seven were ordered; but only nine made it into service before the Armistice.

A Naval Staff study of British merchant shipping showed that by this point in the war, about 20 percent of British tonnage had been diverted to military use; military manpower requirements had taken skilled workers out of the shipyards; shipbuilding and repair priorities had been shifted to warships. For the whole of 1916, only 650,000 tons of new shipping could be delivered. A fleet of thirty-five U-boats had destroyed on average eighty thousand tons of British shipping per month; double the fleet to seventy boats, and take out 160,000 tons a month for a yearly total of almost two million tons, and the British would never catch up. The Naval Staff recommended a return to unrestricted warfare.

The Chancellor remained concerned about the American factor; the Kaiser agreed to a "restricted" unrestricted warfare, beginning February 29, 1916. Enemy armed vessels could be attacked anywhere, and enemy merchant ships in the war zone could be attacked without warning; passenger ships, under any national flag including British, were not to be accosted. Von Tirpitz resented the limitations and resigned, effective March 12th. The more moderate Admiral von Capelle took his place.

On March 18th, Weddigen—now promoted and in command of the "modern" *U-29*—was lost when his boat was rammed and sunk by *Dreadnought. U-29* was the only warship ever sunk by this seminal vessel.

On March 28th, the master of the British merchantman *Brussels* tried to ram a U-boat which had his ship under attack. Neither were successful. The

British Admiralty rewarded Captain Fyratt with a gold watch; the Germans—possibly reacting to the loss of Weddigen—put him on their "most wanted" list. On June 22, 1916, *Brussels* was surrounded by a flotilla of destroyers and Captain Fryatt was captured. He was tried as a *franc-tireur*—a "free shooter" or guerilla of the sea—convicted, and executed on July 29th. The law in this case was a bit murky. Some authorities held that civilians were supposed to flee from armed attack, not fight back; others (perhaps in the majority) asserted that a merchant ship had every right to self-defense. In any event, the Germans did not repeat the judicial process when they captured another would-be merchant rammer, and the judges who pronounced sentence on Captain Fyratt were placed on the British postwar list of war criminals. They were not brought to trial; see page 243.

The new rules lasted about three weeks—until March 24th, when the French cross-channel ferry *Sussex* was torpedoed "without summons or warning." The ship survived, but twenty five Americans were among the eighty five killed. There was another flurry of diplomatic wrangling. The German Ambassador to the United States asserted that *Sussex* had hit a mine. Inspection of the damage revealed torpedo fragments embedded in the hull. The U-boat commander, Pustkuchen, claimed that he had seen men in uniform on deck and assumed *Sussex* to be a troopship and therefore fair game. On April 18th the U.S. Government lodged a formal complaint and a warning:

> If it is still the purpose of the Imperial Government to prosecute relentless and indiscriminate warfare against vessels of commerce by the use of U-boats without regard to what the Government of the United States must consider the sacred and indisputable rules of international law and universally recognized dictates of humanity, the Government of the United States is at last forced to the conclusion that there is but one course it can pursue.
>
> Unless the Imperial Government should now immediately declare and effect an abandonment of its present methods of submarine warfare against passenger and freight-carrying vessels, the Government of the United States can have no choice but to sever diplomatic relations with the German Empire altogether.

The German response of May 4th was both argumentative and conciliatory:

> The German Government cannot but reiterate its regret that sentiments of humanity, which the Government of the United States extends with such favour to the unhappy victims of submarine warfare, have not been extended with the same feeling to the many millions of women and children who, according to the avowed intention of the British Government, are to be starved, and who by suffering are to force the victorious armies of the Central Powers into an ignominious capitulation.

However,

> The German Government is prepared to do its utmost to confine the operations of war for the rest of its duration to the fighting forces of the belligerents, thereby also insuring the freedom of the seas. . . . the German naval forces have received the following orders: In accordance with the general principles of visit and search and destruction of merchant vessels recognized by international law, such vessels, both within and without the area declared a war zone, shall not be sunk without warning and without saving human lives, unless these ships attempt to escape or offer resistance.

"But," the response continued, shifting back toward belligerence, "neutrals cannot expect that Germany, forced to fight for her existence, shall, for the sake of neutral interest, restrict the use of an effective weapon if her enemy is permitted to continue to apply at will methods of warfare violating the rules of international law." In other words, we'll pull back for now but reserve the right to do whatever we must do to survive, whether or not Americans get in the way.

Chancellor Bethmann-Hollweg had forced the Navy to back down and observe prize rules, effective from April 24th. American intervention was again forestalled. Admiral Reinhard Scheer, who had relieved the cancer-stricken Pohl in January, recalled all High Seas Fleet and Flanders boats, asserting that the "new rules" made their life too precarious. His true motive may have been to force the resignation of the Chancellor, hoping perhaps for a public outcry against Bethmann-Hollweg's wavering support for a popular cause; in any event, public response was muted and the Chancellor stood his ground.

By the spring of 1916, the U-boat fleet was greatly enlarged and vastly more powerful than that which had ended the first full year of the war. The British Grand Fleet had been growing also, and now counted almost twice as many dreadnoughts and battle cruisers as the High Seas Fleet. But the German strategy had not changed. Find a way to redress the balance, then take on the fleet.

Scheer had a plan. Set a trap, send U-boats to lie in wait off the main British bases and across transit routes and lay mines at harbor entrances—and then send the High Seas Fleet out on a raid. This would most certainly draw the British out to sea and into the waiting arms of the U-boats. Seventeen U-boats of the High Seas Fleet took up assigned stations, each under orders to maintain radio silence unless and until the enemy had put to sea.

All forces would be alerted to the commencement of operations by the rather transparent message "Take into account that enemy's forces may be putting to sea."

But the sortie of the German fleet was delayed for eight days, the U-boats were almost worried to death by British patrols, and when the "signal" was finally sent on May 30th, it was picked up by only four of the boats. Two of these actually sighted units of the Grand Fleet; one got a chance at a target, and missed.

Another eight boats of the Flanders Flotilla took up patrols on May 30th; one sighted British destroyers off in the distance, too far for an attack.

Thus, when the High Seas Fleet caught up with the Grand Fleet, it found that it was still at full strength. The result is known to history as the Battle of Jutland—actually, a series of four actions spread over twelve hours involving 254 ships. The British lost fourteen, with twenty damaged; the Germans lost eleven, with twenty-seven damaged. The battle was a tactical victory for the Germans. The British, under Admiral Jellicoe, were overcautious to a fault and the Germans, maneuvering boldly and aided by the weather and approaching darkness, were able to duck back into the safety of their own harbors. Two U-boats, waiting back off the British coast, each attacked a returning battleship. Both missed.

Scheer planned another U-boat ambush. Twenty-four boats were to be in position by August 20, 1916; intercepted signals gave warning to the Grand Fleet, which sailed five hours before the Germans put to sea. Scheer—based on a report from a zeppelin which mistook a group of destroyers (a force he could handle) for a detached squadron of battleships, went off chasing phantoms. Then, upon an accurate report that the Grand Fleet itself was closing in from the north, Scheer once again scurried for home. The submarine ambush was slightly more profitable. Two light cruisers were sunk.

But the Battle of Jutland now had become a total strategic victory for the British—by default. The High Seas Fleet, battered and still outgunned, was not again to leave port until steaming out for surrender two and half years later. Another consequence of Jutland, a plus for the British, was the shift of German shipyard facilities—and shipyard workers—from the U-boat program to battle fleet repairs.

On July 9, 1916, the world's first commercial submarine made a sixteen-day passage across the Atlantic, submerging at the beginning and end of the journey as protection against any roaming British patrols.

Deutschland—at 230 feet and 1440/1820 tons, just about the same size as Jules Verne's *Nautilus*—was built to move high-value low-volume cargo through the blockade. Where international law forbade a warship from taking cargo in and out of neutral ports, *Deutschland* only *looked* like a warship; she was unarmed and was operated by North German Lloyd Lines. *Deutschland* had exceptional range: more than thirteen thousand miles at nine knots, 25,000 miles at 5.5 knots, and an underwater endurance of sixty-five miles at three knots—for twenty-four hours, if the air held out.

On this first voyage, she carried mail, dyestuff (worth about $1.4 million), and gemstones. On arrival in Baltimore, the crew was given a gala reception, complete with a German-American band and swarms of newsreel cameramen.

Simon Lake had charged the Germans with patent infringment, and wanted *Deutschland* seized and held hostage. The German Embassy promised him a lot of business, after the war, if he would back off. He did, even going aboard *Deutschland* to welcome Captain Paul Koening and his crew and making little speeches for the press about the marvelous commercial potential of the submarine. The boat could carry seven hundred tons of cargo, and while that is not much when compared with a surface ship, we should note that it is roughly seven times the cargo-carrying capacity of the largest airlifters of the 1990s. Lake envisioned commercial submarines of 11,500 tons carrying 7,500 tons of cargo, ideal for allowing belligerents to evade blockade.

> In my judgment, [he wrote a year or so later] the only way that any nation will be able ultimately to continue its commerce with any degree of safety or certainty when blockaded by submarines will be by the construction of large merchant submarines which will be able to evade the enemy U-boats successfully.
>
> I have pointed out . . . that "submarines cannot fight submarines," because they cannot see or locate each other. It is this very thing which will enable the cargo-carrying submarine to evade the military submarine. They are also able to evade all surface craft. . . . Captain Paul Koening, of the *Deutschland,* told me that most of his journey in the *Deutschland* was upon the surface. He stated that her low visibility enabled him to see all approaching ships before they could see her, and that it was only necessary for him to submerge and rest until the surface ship had passed on her way.

Deutschland made the return trip to Germany with a cargo of nickel, tin, and rubber (much of that carried outside the pressure hull; the rubber alone was reported to be worth more than $17.5 million), and made one more round trip, therefore evading the British blockade four times. However, a sister ship, *Bremen,* disappeared without a trace on her first trans-Atlantic effort.

In February 1917, as part of the UA building program, *Deutschland* was converted as *U-155*—with six external torpedo tubes and two deck guns. Six other converted "merchant" hulls and two purpose-built "U-cruisers" made it into wartime service, five on operations along the U.S. East Coast.

The Battle of Jutland forced one issue; in his post-battle report to the Kaiser, July 4, 1916, and hard though it was for a professional naval officer to admit, Scheer acknowledged that there was simply no way that the High Seas Fleet or any combination of offensive moves against the Royal Navy could force England to the peace table:

> If we are not finally to be bled to death we must make full use of the U-boat as a means of war, so as to grip England's vital nerve. . . . A victorious end to the war within a reasonable time can only be achieved through the defeat of British economic life—that is, by using the U-boats against British trade. In this connection, I feel it is my duty to again strongly advise Your Majesty against the adoption of any

> half-measures, not only because these would contradict the nature of the weapon and would produce commensurate results, but also because in British waters, where American interests are strong, it is impossible to avoid incidents, however conscientious our commanding officers may be. . . .

"However conscientious our commanding officers may be . . ." The fate of Germany—in some sense, of the civilized world—was in the hands of a tribe of under-thirty zealots, most of whom wouldn't know a "treaty" from a "treat" and who had never been schooled in the niceties of diplomacy, international relations, or even warfare. Schweiger, who sunk the *Lusitania* and *Hesperian* (the first ambiguous, the second clearly against operating instructions); Schnieder, who sunk *Amiral Gauteaume* and *Arabic* (the first the first-ever unannounced sinking, the second clearly against operating instructions); Pustkuchen, who sunk *Sussex* (overzealous at the least and disingenuous after the fact). Aging men at the height of their profession, military or diplomatic, now had to make excuses for (or bluff their way through) the grotesque mistakes of men at the bottom end of the scale of training, experience, and sense.

The impact of the Allied blockade was being felt—not quietly, but with anger. "If we were to starve like rats in a trap, then surely it was our sacred right to cut off the enemy's supplies as well," wrote Claus Bergen, a civilian artist who went on patrol with *U-53.* He was incensed by the richness of foodstuff found in the larder of the lowliest British fishing trawler—cocoa, coffee, tea, butter, ham, beef, marmalade, bacon, bread:

> All these things, which were now completely strange to us, we had removed from a few paltry enemy fishing boats, while in Germany the women and children were starving and dying of empty stomachs or supporting life on vile, injurious, almost inedible food-substitutes. The poor in Germany thought of the old days as they sat over their watery turnips while, in the cabins of these trawlers which we happened to have sunk just at dinner time, were plates piled with, what seemed to us, lavish helpings of good fresh roast meat and potatoes, such as we only saw in dreams.

A submarine petty officer salved his conscience in much the same way:

The British ships deserved sinking because each "was certainly directly helping to destroy Germany and to carry on a system of war that thrust into the hands of innocent German children a slice of raw onion for their supper . . ." When on leave, he "so often saw children whose angel-souls shone through their pale starved bodies, or soldiers, themselves but skin and bone, carrying home their last loaf to their wives whose hour was nearly come" and was "seized with fury against this inhuman enemy who had cut off Germany's food imports."

He did not mention the efforts to make cloth out of nettles or paper; the ersatz coffee made from toasted sawdust; the shortages of beer and tobacco—perhaps not so compelling as propaganda, but certainly important to the Germans.

At a meeting on August 31st, the members of the High Command who had never willingly accepted unrestricted U-boat warfare—Field Marshal Paul von Hindenberg and General Erich Ludendorff—started coming around. The nation was in danger and had to employ every possible weapon and technique; a return to unrestricted U-boat warfare was therefore inevitable and they better get on with it.

Foreign Minister Jagow remained opposed, because of the probable effect on the United States and other neutrals, but the Chancellor too began to fall in line. He finally agreed with the proposition but reserved to himself the time of execution.

In the interim, all parties agreed to resume organized U-boat operations against merchant shipping in British waters in October, under prize rules. In the interim, some ships had been attacked in those waters by U-boats proceeding to and from other operating zones.

Also, as a warning to American politicians that their waters were as easily reached by U-boats as by cargo-carriers, *U-53* surfaced off Newport, Rhode Island, on October 7th for a "courtesy visit" and twenty-four-hour refueling stop, as permitted by international law. Immediately upon departure, *U-53* sank five steamers in the vicinity of the Nantucket lightship—three British, one Norwegian and one Dutch—and headed for home.

19 1917—Toward "World" War

Strictly secret—B -35640-I, Berlin December 22, 1916: The beginning and the declaration of the unrestricted U-boat war must follow so quickly one upon the other that there is no time for negotiations, especially between England and the neutrals. The wholesome terror will exercise in this case upon enemy and neutral alike.

—*German Naval Staff document*

Despite the tightened rules of engagement, with three times the number of U-boats operational as they had the year before, the Germans were taking out about 300,000 tons a month. In addition, anxiety and intimidation had worked to some considerable degree: Sailings were delayed by U-boat warnings, reducing the effective carrying power of the British merchant fleet by about 30 percent, and neutral traffic had fallen from about 3.4 million tons to just under one million tons—in six months. And if submarines could not risk the jeopardy of open-ocean visit-and-search—neither, as it developed, could surface warships because of the danger from submarines! At first, the British ordered all intercepted ships into the nearest British port for the search, then later issued a standing order that *all* ships transiting British-controlled waters first must "voluntarily" stop at a British port.

Admiral Jellicoe, appointed First Sea Lord in December, to First Lord of the Admiralty Arthur Balfour:

> There appears to be a serious danger that our losses in merchant ships, combined with the losses in neutral merchant ships, may by early summer of 1917 have such a serious effect upon the import of food and other necessaries into the Allied countries as to force us into accepting peace terms, which the military position on the Continent would not justify and which would fall short of our desires.

Admiral Jellicoe recited the reasons for "our present want of success" in combating the menace: the unexpectedly long range and endurance of the U-boats, failures of the various antisubmarine efforts, the increased firepower of the newer boats "which makes them more than a match for our smaller patrol craft," and the increased sophistication of the U-boat commanders, who, "having become aware of the methods hitherto in use, are more or less able to avoid destruction by these means."

He had two examples in mind. First, was the decoy or "Q" ship—the "Q" standing for nothing in particular unless derived from the Queenstown base where many were stationed—whereby a seemingly defenseless merchantman under attack metamorphosed into a heavily armed warship. At least 225 of these decoys were put into service, and while it seemed like a good idea at the time, it didn't take long before the U-boats caught on and, as a result, more merchant ships were hit by torpedo, without warning.

One of those, the armed merchantman *Noralina,* became an accidental "Q" ship on June 4, 1917, when the first mate ordered the crew to abandon ship after a torpedo had hit, but not really damaged, the ship. When the U-boat saw the crew taking to lifeboats, she started to move in for the kill—only to be taken under fire by the armed guard which had not yet left the ship.

Another ruse: An old C-boat was taken under tow by an armed trawler while it lurked just below the surface with a telephone connection. When a U-boat would surface to attack, the British submarine would break away and take her own position for attack. The scheme claimed two victims: *U-40* and *U-23*. The German High Command learned of the ruse when survivors of *U- 23* passed the word to German civilian prisoners in England who were being repatriated. *U-23* was the last to be tricked.

Jellicoe went on to say that, in short, the rub of the matter was that Britain would be defeated by the U-boats unless "new methods of attack be devised and put into execution at the earliest possible moment."

Jellicoe created an "Anti-Submarine Division" in the Naval Staff; a bit late, but not yet *too* late. He saw only three ways of dealing with U-boats and did not necessarily agree with the methods implicit in all of them: Prevent them from getting to sea; sink them once they were at sea; protect shipping from attack. Admiral David Beatty—Jellicoe's successor as Commander in Chief of the Grand Fleet—seized on the first, and suggested sowing a vast minefield, eighty thousand units, around the exits from the U-boat bases. Jellicoe disabused Beatty of this notion. Mines weren't all that effective against submarines, he noted, "proved by the ease with which submarines pass the Dover barrage." He probably didn't mention that in any event, the country had no more than 1,100 mines on hand.

As for protecting shipping from attack, Jellicoe believed that the problems of convoy were insurmountable. Therefore, the Anti- Submarine Division concentrated on the business of finding and sinking—*trying* to find and sink—U-boats.

The destroyer-type had been created to combat the torpedo boat, and logically enough it became the natural enemy of the submarine. However, without any weapons specifically designed for use underwater, the destroyer's principal weapon was speed—thirty knots-plus, giving it a surface advantage of at least three to one over most U-boats. Speed *could* bring a destroyer quickly into range for a few shots from a deck gun and into position for ramming. Unfortunately, speed was of marginal value in fighting a submarine, which could slip out of sight (or at least, out of lethal range) in thirty or forty seconds. In 1914, a destroyer skipper found himself sitting above a U-boat which he could see, but not touch. "What we need," a staff officer mused, "is some sort of bomb to drop in the water."

Thus was launched development of the depth charge (called the "water bomb" by the Germans). However, a satisfactory version was not available until 1916—two versions, actually. The primary weapon, Type D, had a three-hundred-pound charge of TNT. This was powerful enough to be useful against submarines, but was a hazard to escort ships smaller than destroyers, slow-moving and lightly built, which could themselves be caught in the blast. Therefore, an alternative 120-pound version, Type D*, was created; however, 120 pounds was not enough to cause serious injury to a U-boat. Type D* was soon enough retired, and the self-inflicted injury problem was resolved by installing hydrostatic fuses which could be set for detonation at one hundred, 150, and two hundred feet. This, of course, protected the smaller surface ship, which would have steamed far enough ahead before the blast, but might well have protected a shallow-running U-boat as well.

The first U-boat sunk by a depth charge was *U-68* on March 22, 1916. By January 1917, depth charges had claimed only one more victim. So was 1917 then to be the year of the depth charge? Not really; because of production delays, each ASW ship was limited to two charges of each type through much of the year. In fact, it was not until 1918 that a sufficient quantity were on hand to permit stocking to the planned level of thirty five units per ship.

The depth charge inspired confidence in the surface forces and provoked fear among the U-boat crews, but it was not effective unless dropped with great accuracy. Tests conducted some years later (1940) demonstrated that, to be destructive, a charge had to go off within fourteen feet of the target; up to twenty-eight feet, the boat would be disabled and probably be forced to surface. Out to sixty feet, the crew would be terrorized but the ship relatively untouched.

Without any useful means of determining the U-boat's position, the near-term solution was to lay down a pattern of charges, some rolled off the stern, some shot out to the sides from a "Y" gun. It was not by itself a very effective solution.

Two other ASW weapons were put into service in the first two years of the war: an explosive sweep with eighty-pound charges hanging from a dragline, which sank one U-boat during the entire war, and a paravane—essentially, an underwater kite towed by the ASW vessel—which packed a four-hundred-pound charge but only accounted for one additional kill. There were two problems common to explosive sweeps and paravanes. They had to actually make contact with the U-boat, and the towline was too easily wrapped around the screws of the towing ship.

In December 1916, the U-boats of the High Seas Flotilla returned to the Dover Strait for a most practical reason: to shorten the voyage to the Western Approaches by about 1,400 miles. They did so, with almost boring success. A document recovered from a U-boat sunk later in 1917 detailed 190 safe passages between December 23, 1916, and June 6, 1917. During that period, the British had detected only sixteen submarine contacts at the barrage.

Also in December, Chief of Naval Staff Henning von Holtzendorff presented the mirror image of Jellicoe's dire prediction. By his latest calculations, a monthly rate of sinking of 600,000 tons would bring about Britain's collapse within six months, before U.S. intervention could make any difference.

> I arrive at the conclusion that an unrestricted war, started at the proper time, will bring about peace before the harvesting period of the 1917 summer, that is, before 1 August; the break with America must be accepted; we have no choice but to do so. In spite of the danger of a breach with America, unrestricted submarine war, started soon, is the proper, and indeed the only way to end the war with victory.

The quality of life in Germany continued to slide. There was little food—one historian documents daily rations limited to "five slices of bread, half a small cutlet, half a tumbler of milk, two thimblefuls of fat, a few potatoes and egg cup of sugar." Houses went unheated, he noted, adding: "The majority of the urban population were either cold or wet or hungry for the greater part of the day."

The Naval Staff announced the inevitable:

> Strictly secret—B -35640-I, Berlin December 22, 1916: The beginning and the declaration of the unrestricted U-boat war must follow so quickly one upon the other that there is no time for negotiations, especially between England and the neutrals. The wholesome terror will exercise in this case upon enemy and neutral alike.

On the possibility of American entry into the war:

> As regards tonnage this influence would be negligible. It is not to be expected that more than a small fraction of the tonnage of the Central Powers lying in American and many other neutral harbors could then be enlisted for the traffic to England. For the far greatest part of this shipping can be damaged in such a way that it cannot sail in the decisive

> time of the first months. Preparations to this effect have been made. There would also be no crews to be found for them.
> Just as little effect can be ascribed to any considerable extent to American troops, which, in the first place, cannot be brought over through lack of tonnage.

Still, Germany hoped to find some accommodation with the United States, and was encouraged by the fact that Woodrow Wilson had campaigned and won his 1916 re-election to the Presidency on the slogan "He kept us out of war." But the United States was a nation divided, more so than at any time since the Civil War. One faction called for war with Germany anytime, anyplace; another was against war on general principles but was willing to stand up for "national honor." Some people were irritated with Great Britain for cutting off trade with Germany; there were some—Irish-Americans sharing age-old hatreds and appalled by the more recent British supression of the Easter Rebellion—so opposed to Great Britain that they would support any opposition. Some were appalled at the indiscriminate German U-boat campaign; some, through kinship or philosophy, supported the German cause.

That German cause was not helped by several heavy-handed acts of sabotage to which the German military and naval attachés were linked. One involved a shipping terminal in New Jersey, blown up July 30, 1916.

Another meeting of the High Command was held on January 9, 1917. No one objected to a resumption of unrestricted sumbarine warfare on February 1st.

Another act of sabotage involved a foundry, also in New Jersey, destroyed on January 11, 1917. And then came one of the strangest blunders of the war: a telegram from German Foreign Minister Alfred Zimmerman to the German Minister in Mexico, suggesting an alliance with Mexico in a war against the United States and hinting at a similar arrangement for Japan. When it was passed along by British intelligence and made public in America at the beginning of March, it became perhaps the single most critical factor in rallying divergent factions and bringing Americans together on the war.

January 19, 1917

Dr. Alfred Zimmerman to Minister to Mexico Heinrich von Eckhardt:

On the first of February we intend to begin submarine warfare unrestricted. In spite of this it is our intention to endeavor to keep neutral the United States of America.

If this attempt is not successful, we propose an alliance on the following basis with Mexico:

That we shall make war together and together make peace. We shall give general financial support and it is understood that Mexico is to reconquer lost territory in New Mexico, Texas and Arizona. The details are left to you for settlement.

You are instructed to inform the President of Mexico of the above in the greatest confidence as soon as it is certain that there will be an outbreak of war with the United States, and suggest that the President of Mexico, on his own initiative, should communicate with Japan suggesting adherence at once to this plan; at the same time offer to mediate between Germany and Japan.

Please call to the attention of the President of Mexico that the employment of ruthless submarine warfare now promises to compel England to make peace in a few months.

We might note, in passing, that the crux of the Zimmerman plan—to bring Mexico into action against the United States—was contingent on the United States entering the war against Germany. The grand irony: the Zimmerman telegram was the direct instrument of that entry.

A letter announcing resumption of unrestricted operations on February 1st was prepared, and scheduled for delivery to the U.S. Government on January 31st. A set of orders was transmitted to all U-boat commanders on January 27th:

. . . to force England to make peace and thereby decide the whole war. *Energetic* action is required, and above all *rapidity* of action. The campaign is to be prosecuted with the utmost vigour. No vessel must remain afloat. . . .

Here, in a slightly edited format for ease of reading, are excerpts from the Orders of January 27, 1917:

- Leave is only possible hitherto when repairs to the boat are not affected by it.
- For this reason, food allowances of U-boats are increased.
- The sole aim is that each boat shall fire her entire supply of ammunition as often as possible.
- The standard of achievement is not to be judged by each separate enterprise, but on the total result over any given period. Therefore short cruises, with short visits to the dockyard, result in a considerable curtailment of effort and are to be avoided. The best form of practice is for the U-boat to remain at sea for fourteen days in each month with the object of firing her full supply of torpedoes and gun ammunition.
- During periods of overhaul, only what is absolutely necessary is to be done.
- The crew and reserve personnel are to be made the utmost use of.
- Venereal diseases represent an avoidable loss from a military standpoint—often a very serious loss—which must be eliminated in the future.
- Our object is to cut England off from traffic by sea, and not to achieve occasional results at far distant points. As far as possible, therefore, stations must be taken up near the English coast, where routes converge and where divergence becomes impossible.
- When the weather is so bad that weapons cannot be used, better weather conditions and chance successes are not to be sought at far distant points at sea, but the boat is to remain on her station, if necessary submerging to avoid damage. An improvement in the weather will be perceived immediately and can at once be utilized without any delay in reaching the spot.
- The principal advantageous methods of prosecuting ruthless U-boat warfare are:

(1) Attack submerged all ships which are armed or suspect, whenever the U-boat is in a position for attacking submerged, or is able by means of her speed to reach such a position.

(2) Utilize all chances of attack by night.

(3) Open immediate effective firing when a ship is stopped by gunfire, without wasting time on warning shots.

(4) No boat-to-ship communication.

- Way should always be kept on the U-boat to render a counterattack more difficult, particularly that of an enemy submarine, and in order that the boat herself may remain completely under control, so as to avoid a torpedo, if necessary.
- When a ship, abandoned by her crew, is to be sunk by gunfire, she should be approached from aft; she is then not in a position either to ram or to open fire, as the U-boat traps with hidden armament [Q-ships], reported to date, in every case had their guns on the broadside. A trap will endeavor without exciting notice to keep her beam on to the U-boat and will turn accordingly. Beware of this!
- When approaching keep one bow torpedo ready for firing, with the tube flooded, and keep the enemy under fire; have the boat ready for diving and no men on deck except those actually required.
- As a rule, expend only one torpedo on each ship stopped. . . . she should be finished off with gunfire, if possible.
- A sharp lookout must be kept for the approach of any further vessels, particularly any which lie hidden under the lee of the ship attacked. An approach from astern permits observation on both sides.

Then, with all in readiness, the message was delivered.

Washington D.C., January 31, 1917
Mr. Secretary of State:

Your excellency was good enough to transmit to the Imperial Government a copy of the message which the President of the United States of America addressed to the Senate on the 22nd inst. The Imperial Government has given it the earnest consideration which the

President's statements deserve, inspired, as they are, by a deep sentiment of responsibility.

It is highly gratifying to the Imperial Government to ascertain that the main tendencies of this important statement correspond largely to the desires and principles professed by Germany. These principles especially include self-government and equality of rights for all nations. Germany would be sincerely glad if, in recognition of this principle, countries like Ireland and India, which do not enjoy the benefits of political independence, should now obtain their freedom. . . . The freedom of the seas, being a preliminary condition of the free existence of nations and the peaceful intercourse between them as well as the open door for the commerce of all nations, has always formed part of the leading principles of Germany's political program. All the more the Imperial Government regrets that the attitude of her enemies, who are so entirely opposed to peace, makes it impossible for the world at present to bring about the realization of these lofty ideals.

. . . The attempt of the four allied powers to bring about peace has failed, owing to the lust of conquest of their enemies, who desired to dictate the conditions of peace. Under the pretense of following the principle of nationality, our enemies have disclosed their real aims in this way, *viz.*, to dismember and dishonor Germany, Austria-Hungary, Turkey and Bulgaria. To the wish for reconcilliation they oppose the will of destruction. They desire to fight to the bitter end.

A new situation has thus been created which forces Germany to new decisions. Since two years and a half England is using her naval power for a criminal attempt to force Germany into submission by starvation. In brutal contempt of international law, the group of powers led by England not only curtail the legitimate trade of their opponents, but they also, by ruthless pressure, compel neutral countries either to altogether forego every trade not agreeable to the Entente Powers, or to limit it according to their arbitrary decrees.

The American Government knows the steps which have been taken to cause England and her allies to return to the rules of international law and to respect the freedom of the seas. The English Government, however, insists upon continuing its war of starvation, which does not at all affect the military power of its opponents, but compels women and children, the sick and the aged, to suffer for their country pains and

> privations which endanger the vitality of the nation. Thus British tyranny mercilessly increases the sufferings of the world, indifferent to the laws of humanity, indifferent to the protests of the neutrals whom they severely harm, indifferent even to the silent longing for peace among England's own allies. Each day of the terrible struggle causes new destruction, new sufferings. Each day shortening the war will, on both sides, preserve the lives of thousands of brave soldiers and be a benefit to mankind.
>
> The Imperial Government could not justify before its own conscience, before the German people, and before history the neglect of any means destined to bring about the end of the war. Like the President of the United States, the Imperial Government had hoped to reach this goal by negotiations. Since the attempts to come to an understanding with the Entente Powers have been answered by the latter with the announcement of an intensifed continuation of the war, the Imperial Government—in order to serve the welfare of mankind in a higher sense and not to wrong its own people—is now compelled to continue the fight for existence, again forced upon it, with the full employment of all the weapons which are at its disposal.

Thenceforth, from "February 1, 1917, sea traffic will be stopped with every available weapon and without further notice in the [designated] blockade zones around Great Britain, France, Italy and in the Eastern Mediterranean."

Neutrals already at sea, or in a blockaded port, were to "be spared during a sufficiently long period" to clear out of the zone. The fate of Americans already at sea on enemy freighters was in the hands of the shipping companies, "as the enemy shipping firms can prevent such ships in time from entering the zone." Germany allowed for the passage of one well-marked American passenger ship, each way, each week, between the United States and the port of Falmouth, England.

In a memorandum which accompanied the formal note, the German Ambassador reemphasized "the situation thus forced upon Germany by the Entente Allies' brutal methods of war and by their determination to destroy the Central Powers" which "gives back to Germany the freedom of action which she reserved in her note addressed to the Government of the United States on May 4, 1916."

Ambassador Bernstorff had the text of the January 31st message in hand on the same day as the coded Zimmerman telegram passed through Washington on its way to Mexico. Before presenting that message, Bernstorff executed the "preparations to this effect" promised by von Holtzendorff in December, ordering the destruction of "the engines of all German ships lying in American harbors." There were more than one hundred German and Austrian ships which had elected voluntary internment when the European war began.

The sabotage was not very effective. At least one-third of the ships soon were repaired and put into American service—most notably, the world's largest ship, *Vaterland,* renamed *Leviathan,* which could carry almost ten thousand troops at a time. In fact, recycled German ships carried more than half of all troops sent overseas under the American flag.

20 1917—The Giant, Aroused

At the present rate of destruction more than four million tons will be sunk before the summer is gone. Such is this dire submarine danger. The English thought that they controlled the sea; the Germans, that they were invincible on land. Each side is losing where it thought itself strongest.

—*Ambassador Walter Page in London to President Wilson, May 4, 1917*

To no one's surprise, President Wilson invoked the "freedom of action" he had reserved in his note of April 18, 1916, and on February 3rd, severed diplomatic relations, recalled the American Ambassador to Berlin, and told the German diplomatic mission in the U.S. to go home. At almost the same time that the President was announcing these moves, in one of those strange parallels laid down from time to time in history, another *Housatonic* was sunk by an enemy submarine—this time, an American merchant steamer and the attacker survived.

Events moved quickly. On February 12th, the schooner *Lyman M. Law* was sunk. On February 26th, President Wilson asked the Congress to authorize the arming of merchant ships with Naval guns and crews; the legislation easily passed in the House, but was stalled in the Senate by a filibuster led by pacifist Robert La Follette of Wisconsin. He was branded by Theodore Roosevelt as "the most sinister enemy of democracy in the United States . . ." He was certainly out of step with the rest of the nation, but he was not an enemy; LaFollette believed that the salvation of democracy began at home, in Washington, not on a foreign battlefield.

In the interim, four more American ships were sunk by U-boats.

The Congress adjourned on March 4th; the President used executive authority to accomplish what the legislature would not approve, and by March 14th, the first armed guard units were ready for service. In a gesture more clearly tuned to public relations than the needs of the service, the crew of the Secretary of the Navy's official yacht *Dolphin* was among the first to volunteer, to a man, to man the guns. One, John T. Eopolucci, would become, on April 1st, the first American serviceman killed in action against Germany when his armed merchantman, *Aztec,* was torpedoed.

Meanwhile, on March 18th, three American merchant ships were sunk by the U-boats. Great Britain was down to about six weeks' food supply, and on April 2nd President Wilson called the Congress "into extraordinary session because," in his words, "there are serious, very serious choices of policy to be made."

> Since April of last year the Imperial Government has somewhat restrained the commanders of its undersea craft. . . . The precautions taken were meager and haphazard enough, as was proved in distressing instance after instance in the progress of the cruel and unmanly business, but a certain restraint was observed. The new policy has swept every restriction aside. . . .
>
> I was for a little while unable to believe that such things would in fact be done by any Government that had hitherto subscribed to humane practices of civilized nations. International law had its origin in the attempt to set up some law which would be respected and observed upon the seas, where no nation had right of dominion and where lay the free highways of the world. By painful stage after stage has that law been built up, with meager enough results, indeed . . . but always with a clear view, at least, of what the heart and conscience of mankind demanded. This minimum of right the German Government has swept aside, under the plea of retaliation and necessity. . . . I am not now thinking of the loss of property involved, immense and serious as that is. . . . Property can be paid for; the lives of peaceful and innocent people cannot be. The

> present German submarine warfare against commerce is a warfare against mankind.

He reminded the members of his hope, expressed in his last address to the Congress on February 26th, that

> . . . it would suffice to assert our neutral rights with arms, our right to use the seas against unlawful interference, our right to keep our people safe against unlawful violence. But armed neutrality, it now appears, is impractical. Because submarines are in effect outlaws when used as the German submarines have been used against merchant shipping, it is impossible to defend ships against their attacks as the law of nations has assumed that merchantmen would defend themselves against privateers or cruisers, visible craft giving chase upon the open sea. . . .
>
> With a profound sense of the solemn and even tragical character of the step I am taking and of the grave responsibilities which it involves, but in unhesitating obedience to what I deem my constitutional duty, I advise that the Congress declare the recent course of the Imperial German Government to be in fact nothing less than war against the Government and people of the United States; that it formally accept the status of belligerent which has thus been thrust upon it; and that it take immediate steps not only to put the country in a more thorough state of defense, but also to exert all its power and employ all its resources to bring the Government of the German Empire to terms and end the war.

"The world must be made safe for democracy," he affirmed. "Its peace must be planted upon the tested foundations of political liberty." The President called for full mobilization of "all the material resources of the country," for supplying the Navy "with the best means of dealing with the enemy's submarines," and for the addition of "at least five hundred thousand men, who should, in my opinion, be chosen upon the principle of universal liability to service."

Discussion in the Congress was heated, raising all the issues which had been rehearsed in the election and in recent Congressional debate. Some saw the push for war as a ploy by bankers and businessmen: "We are going into war on the command of gold," one senator charged during floor debate, adding, ". . . war madness has taken possession of the financial and political powers of our country. . . . you shall not coin into gold the lifeblood of my brethren [or] put the dollar sign upon the American flag!"

Nonetheless, the declaration of war was effective April 6, 1917. Eventually, more than two million men would enlist or be drafted into the armed forces. Theodore Roosevelt was not one of them. His request to lead a specially trained "elite" division was ignored.

The President had already sent the president of the Navy War College, Rear Admiral William S. Sims, to establish naval liaison with the British. The President asked Sims to urge two actions upon the British: "First, that every effort should be made to prevent the submarines getting into the Atlantic—that they ought to be shut up in their own coasts, or some method should be found to prevent their ingress and egress; second, that all ships ought to be convoyed."

Sims was not unknown in England. During a visit in 1910, he had given a speech at the London Guildhall in which he had rather boldly promised, "If the time ever comes when the British Empire is seriously menaced by an external enemy, it is my opinion that you can count upon every man, every dollar, every drop of blood of your kindred across the sea." Management back home was not pleased, but let him off with a mild rebuke.

But here's a strange wrinkle: Just before he was to leave on this new mission, Sims was admonished by recently appointed Chief of Naval Operations William Benson, "Don't let the British pull the wool over your eyes. It is none of our business pulling their chestnuts out of the fire. We would as soon fight the British as the Germans." It was not only among members of the Congress that a certain ambiguity prevailed.

Sims later asserted that he gave the remark no particular significance, ascribing it to "a personal idiosyncrasy of the Admiral. I had known," he added, "the general opinion that he was intensely anti-British, but it did not affect me particularly." When this all came to light in 1920, Benson admitted that he may have used "very forcible language" to impress upon Sims the seriousness of the situation and the importance of making certain that "his feelings toward the British did not lead him into any indiscretion."

Sims and his aide embarked on the liner *New York* while the nation was still at peace, and therefore traveled incognito. They arrived to a formally declared war and a spectacular welcome: *New York* hit a mine in the outer harbor of Liverpool. Sims also was welcomed, although perhaps not as spectacularly, by the Admiralty. On his first report back to Washington, he noted that the submarine issue was

> . . . very much more serious than the people realize in America. The recent success of operations and the rapidity of construction constitute the real crisis of the war. The morale of the enemy submarines is not broken, only about fifty-four are known to have been captured or sunk and no voluntary surrenders have been recorded. Supplies and communications of forces on all fronts, including the Russian, are threatened and control of the sea actually imperilled. German submarines are constantly extending their operations into the Atlantic, increasing areas and the difficulty of patrolling. . . . the amount of British, neutral and Allied shipping lost in February was 536,000 tons, in March 571,000 tons, and in the first ten days of April 205,000 tons. With short nights and better weather these losses will be increasing.

Sims asked Admiral Jellicoe: "Is there no solution for the problem?" Admiral Jellicoe responded, according to Sims, "Absolutely none that we can see now."

Sims brought up the subject of "convoy." He was advised that, "after trying various methods of controlling shipping, the Admiralty now believes the best policy to be one of dispersion. They use about six relatively large avenues or arcs of approach to the United Kingdom and Channels, changing their limits or area responsibility if necessity demands."

Jellicoe assured Sims that "the Admiralty have had frequent conferences with merchant masters and sought their advice. Their most unanimous demand is: 'Give us a gun and let us look out for ourselves.' They are also insistent that it is impractical for merchant vessels to proceed in formation, at least in any considerable numbers, due principally to difficulty in controlling their speed and to the inexperience of their subordinate officers."

The arguments against convoy—raised and retired in 1910—had not changed:

There were too many ships. The Admiralty counted 2,500 arriving and departing British ports each week.

Port facilities were already strained by the numbers. Bringing a convoy's worth of ships into one port all at the same time would create impossible conditions.

Coordination within a convoy would be difficult if not impossible because their crews presented a mixed lot of experience and skill, and many did not speak English.

Most ships were fast enough to outrun a U-boat; the only time they were really exposed to danger was in the approaches, and more patrols would do the job.

Only a bare handful of sinkings occurred in mid-ocean—indeed, few more than one hundred miles from land—where ships sailed on widely scattered tracks. But those tracks converged as they neared the British Isles, and it was in those "western approaches" that most attacks took place.

Finally, as explained in an Admiralty report in January 1917:

> The system of several ships sailing in company, as in a convoy, is not recommended in any area where submarine attack is a possibility. It is evident that the larger number of ships forming the convoy, the greater is the chance of a submarine being enabled to attack successfully, and the greater the difficulty of the escort in preventing such an attack.
>
> In the case of defensively armed merchant vessels, it is preferable that they should sail singly rather than that they should be formed into a convoy with several other vessels. A submarine could remain at a distance and fire her torpedo into the middle of a convoy—with every chance of success. A defensively armed merchant vessel of good speed should rarely, if ever, be captured. If the submarine comes to the surface to overtake and attack with her gun, the merchant vessel's gun will always make the submarine dive, in which case the preponderance of speed will allow of the merchant vessel escaping.

Well, perhaps. One duel between *J.L. Luckenbach* and a submarine lasted an unbelievable four hours; the steamer fired 202 rounds, the U-boat 225 (scoring a dozen or fewer hits, although one shell disabled *Luckenbach*'s engine). The U- boat was finally driven off by the appearance of a destroyer, arriving in response to frantic messages from *Luckenbach*. However, the basic point was correct: U-boats usually chose to attack armed ships while submerged, expending from their limited supply of torpedoes.

* * *

It is bizarre to note, but at this time the British were involved in convoy with overwhelming success in two quite separate arenas, but had not yet translated that success to the Atlantic crossing.

In one, all troop movements to the European battlefields across the English Channel were escorted; not a single soldier was lost to the U-boats in the entire war.

In the other, there was a monthly traffic of about eight hundred small colliers carrying vital coal to the French ports. French coal fields had been occupied by the Germans, and the nation's industrial production and heat now depended upon the import of about two million tons a month from Great Britain. Losses had been mounting to the point of French concern, so in February, the Admiralty instituted "controlled sailings" whereby groups of colliers were given an escort of small armed trawlers. The perjorative term "convoy" seems not to have been used, but "convoy" it was. By the end of August, 8,871 transits had been made—with the loss of sixteen ships, 0.18 percent.

In April, the Admiralty office which coordinated the collier convoys reviewed the dreaded "2,500 ships a week" and discovered that most were small coasters and cross-channel ferries, not the vital trans-oceanic merchant fleet. The actual number of those was between 120 and 140 per week.

The Anti-Submarine Division ran an operational analysis, and advised Jellicoe on April 25th that:

> The larger the convoy passing through any given danger zone, provided it is moderately protected, the less the loss to the merchant services; that is, for instance, were it feasible to convoy the entire volume of trade which enter the UK *per diem* in one large group, the submarine as now working would be limited to one attack, which, with a destroyer escort, would result in negligible losses compared with those now being experienced.

Jellicoe was not yet persuaded; in a memo of his own to the First Lord on April 27th, he was most direct:

> The real fact of the matter is this. We are carrying on the war at the present time as if we had the absolute command of the sea. . . . It is quite true that we are masters of the situation as far as surface ships are

> concerned but it must be realised, and realised at once, that such supremacy will be quite useless if the enemy's submarines paralyse us as they do now our lines of communication.

Jellicoe understood the situation, but not the solution. He proposed reducing the "lines of communication"—the number of merchant ships entering British waters.

However, he did agree to a trial program: one convoy from Gibraltar to home waters, another from Scotland to Norway. These would demonstrate that many of the objections to convoy were without merit—or at least, easily enough handled. The Admiralty would begin the process of full-scale convoy, but it would be some time before it would have much effect.

Nonetheless, the move was none too soon. As stated in a memo from Ambassador Walter Page in London to President Wilson, May 4, 1917:

> At the present rate of destruction more than four million tons will be sunk before the summer is gone. Such is this dire submarine danger. The English thought that they controlled the sea; the Germans, that they were invincible on land. Each side is losing where it thought itself strongest.

Finally, on June 20th, Jellicoe himself admitted to the War Policy Committee that because of the great shortage of shipping caused by the U-boats, "There is no good discussing plans for next Spring—we cannot go on." Losses that month were 687,507 tons—the second highest monthly total of the war.

There was some distance to go between approving convoy and making it happen, however. There were problems with independent-minded masters, who would be damned if they'd let some Naval officer tell them what course to steer and at what speed. They would jump ahead on their own and thereby, some of them, become the U-boat's favorite sort of target. And be damned.

The typical merchant ship in peacetime commerce, once clear of the harbor, would set engine orders at "Full Ahead" and not change them until nearing her destination. In convoy, frequent changes were necessary to keep in station; frequent changes put an unaccustomed load on the machinery, which too often broke down creating stragglers. Another U-boat favorite.

There were other growing pains. The earliest convoys were too small and too heavily escorted, putting a strain on scarce resources. There were no standard operating procedures.

Most of all, convoys needed escorts and there weren't very many available. In May 1917, the British ordered 1,108 new antisubmarine craft, from destroyers on through sloops and trawlers to submarines, and expected immediate support from the United States.

However, the British had been keeping their losses secret—from their own people and from the Americans. Sims was astonished when he learned the facts; the Navy Department was skeptical. Besides, U.S. Navy planners were transfixed by the visits of *U-53* and *Deutschland* to U.S. shores. They insisted on keeping scarce escort ships close to home, where as a practical matter the U-boat threat was minimal to insignificant, and resisted Admiral Sims's urgent pleas to send them to the war zone, where they were desperately needed.

The annual U.S. building program for escorts had long been pegged to the building program of the capital fleet; what, after all, was an escort but a companion for the battleship? The *battle fleet* must be protected against submarines, not the merchant fleet; as the great Mahan had noted, commerce raiding was an insignificant naval exercise. It was not until some other Naval leaders toured the war zone for a first-hand look that Sims began to get the support he needed.

The American industrial prowess, airily dismissed by von Capelle, came thundering down. Work on capital ships just begun was suspended, and the shipyard assets were shifted to the construction of more than 340 submarine chasers and 267 destroyers. The keel for "Liberty Destroyer #139" was laid at the Mare Island Naval Shipyard on May 15th, 1918; the USS *Ward* was launched eighteen days later and placed in commission after seventy days. The prewar norm for a destroyer was more than a year. An entire shipyard was built from scratch on the mud flats of the Delaware River at Hog Island, just below Philadelphia; the war ended before one shipbuilding milestone was reached, the launching of five cargo ships in the same yard on the same day.

In the meantime, even though shipping losses were averaging almost the 600,000 tons prescribed by von Holztendorff, August 1st was fast approaching without any sign of the promised collapse of the British government "within six months." One German ally, Austria-Hungary, already had offered a heretical proposition:

> All the information we receive about England combines to prove that a collapse of our most powerful and dangerous adversary is simply out of the question. Submarine war would damage but not ruin her; would it not, then, be better to abandon the idea that the campaign would be an instrument of final, decisive victory, and to make a serious effort to begin peace negotiations?

The German government began asking much the same question of itself, but the High Command would not allow the question to be answered. On July 2nd, General von Hindenberg restated the proposition, "The war is won for us if we can withstand the enemy attacks until the submarine has done its work." Chancellor Bethmann-Hollweg was forced to resign. The High Command, in effect, imposed a military dictatorship on the country and began ordering U-boats in astonishing numbers: 435 of various types, of which only five entered service in time to be surrendered after the Armistice.

The tempo of U-boat operations was kept at the highest possible level, with a minimum of R&R—which for the crews meant "rest and recreation," but for operations planners was more likely "refit and reload"—and only the most necessary shipyard repairs. Refresher training? The crews were too busy operating to have time, or the need, for training.

But "tempo" was not necessarily proportional to "success." Spotting a convoy was one thing; getting the U-boat into a firing position was another and major problem. If the convoy escorts were in sight, the U-boat could not dash ahead on the surface. The U-boat could not attack on the surface; it almost had to be in a lucky position to begin with. An attack, successful or not, would reveal her presence to the escorts.

Otto Hersing, now in the Atlantic with *U-21,* inspired by his success in diving under the sinking *Triumph* and remaining where he would least be expected, lingered in the middle of any convoy he had just attacked. Even if he was discovered, he reasoned, the enemy destroyers could not safely drop depth charges on him without possibly causing damage to the merchant ships among which he was hiding. The fox was hiding in the chicken coop.

21 1917–1918—The Amateur Plays the End Game

I will back the amateur against the professional every time, because the professional does it out of the book and the amateur does it with his eyes open upon a new world and with a new set of circumstances. He knows so little about it that he is fool enough to try the right thing.

—*President Woodrow Wilson, address to officers of the U. S. Atlantic Fleet, August 11, 1917*

The head of the German government had been forced out because he was leaning toward peace and would not support the continued activities of the U-boats; the head of the American government had been returned to office because he was committed to peace, but now was stepping up to the unaccustomed role the U-boats had thrust upon him: Commander in Chief.

> July 4, 1917: Strictly confidential. From the President for Admiral Sims. From the beginning of the war, I have been greatly surprised at the failure of the British Admiralty to use Great Britain's great naval superiority in an effective way. In the presence of the present submarine

emergency they are helpless to the point of panic. Every plan we suggest they reject for some reason of prudence. In my view this is not a time for prudence but for boldness, even at the risk of great losses.

And on August 11, 1917, at Yorktown, Virginia, a few hundred yards from the spot where Lord Cornwallis surrendered his force to General Washington, the President shared his thoughts with the assembled officers of the Atlantic Fleet:

> We are hunting hornets all over the farm and letting the nest alone. . . . I despair of hunting for hornets all over the sea when I know where the nest is and know that the nest is breeding hornets as fast as I can find them.
>
> Now . . . I take it for granted that nothing that I say here will be repeated and therefore I am going to say this: every time we have suggested anything to the British Admiralty the reply has come back that virtually amounted to this, that it had never been done that way, and I felt like saying, "Well, nothing was ever done so systematically as nothing is being done now."
>
> . . . America has always boasted that she could find men to do anything. She is the prize amateur nation of the world. Germany is the prize professional nation of the world. Now, when it comes to doing new things and doing them well, I will back the amateur against the professional every time, because the professional does it out of the book and the amateur does it with his eyes open upon a new world and with a new set of circumstances. He knows so little about it that he is fool enough to try the right thing.

At an Allied Naval conference in September, Admiral Jellicoe (perhaps tipped off to the President's hornet analogy) proposed to seal off all German ports. He was easily dissuaded by more practical Naval officers; in his postwar memoir, he noted that sealing off the ports seemed

> . . . most attractive in theory and appealed strongly to those who looked at the question superficially . . . the idea of sealing the exits from submarine bases was urged by so many people on both sides of the Atlantic.

As a substitute, a massive minefield was to be laid across the whole of the North Sea. Massive, indeed, to cover a strip 250 miles wide and up to nine hundred feet deep. There were not enough mines in the world, nor the capacity to produce them, to sow such a field with the contact-detonated designs then in use.

However, the Americans developed a mine with a long copper floating antenna. Being touched by a ship's hull set off an electrical charge sufficient to detonate the mine. Thus, fewer mines were required. One mine moored at one depth could do the duty which would have required several to cover the two-hundred-foot column of water through which a U-boat might pass. Minelaying would begin in February.

The main and continuing ASW problem was detecting and locating submarines; humorist Will Rogers proposed a simple solution: drain the Atlantic Ocean. When asked how that could be done, he replied, "Well, that is a detail. I am not a detail man."

However, as far as detection was concerned, a solution was ready at hand. Submarines under way made noise, and sound travels far better in water than in air.

A hydrophone is essentially an underwater microphone, connected to a binaural listening device. By 1911, hydrophones had been installed on forty-five British surface ships and one submarine—but for navigation, not hunting. Various navigational aids—buoys, lightships, markers on sandbars and wrecks—were equipped with underwater bells, rung by the constantly moving sea, and the hydrophones were mounted in pairs, one on each side of the bow. By turning the ship this way and that, until the volume in each ear was about the same, the operator knew that the bow was pointed at the bell, and thus could lay down a line of position.

Adapting the system to finding submarines seemed to have promise, but was not so easily accomplished. Among other problems, if the listening ship was moving, its own noise invariably blotted out any sound from the submarine. If it was not moving, it couldn't turn this way and that to get a bearing, and could only determine that a U-boat was somewhere in the vicinity.

One part of the solution was to mount a trainable U-shaped tube, parallel to the surface but under the keel of the ship. The operator could turn the tube to localize the sound. He would then know the direction to the U-boat, but not distance or depth. Several escorts, operating in concert, could get a rough fix on the U-boat's position and track its direction of movement. One of them could then run over the spot and drop a depth-charge pattern. Without a method for establishing depth, however, setting the depth charges was often a matter of trial and error.

The other part of the solution was to find some way to differentiate the U-boat's noise from that of the escorts, and attenuate the latter with filters. This was not very effective, especially if the escort was moving at any appreciable speed.

An active listening device was waiting in the wings, but not brought on stage until too late to be of value during the war. Spurred by the sinking of *Titanic* by an iceberg in 1912, R.A. Fessenden established the American Submarine Signal Company to market an electromagnetic oscillator which sent a signal underwater, to bounce off a solid object and send an echo back to a listener aboard the sending ship. In April 1914, this device detected an iceberg at a range of two miles, but for many years was used only for underwater signaling.

For the most part, and for most of the war, the usual method of detection was visual—catching the submarine on the surface, or by looking down into the water (if the water was clear enough). Operating manuals of the day reflected the naiveté of the day. A 1917 guidebook for ASW asserted that "the submarine menace can be checked by present-day aircraft."

"Present-day aircraft" played an important role in patrolling shipping lanes and escorting convoys, principally by forcing the U-boats to submerge before they could get into attack position and to stay submerged, draining battery power and therefore reducing combat effectiveness. I suppose that counts as a "check," as in chess, but the "present-day aircraft" were almost totally ineffective when it came to sinking U-boats. Almost. An airplane was credited with one kill, one only, *UB-32,* on September 22, 1917.

The airship made an ideal spotting platform, offering range well above one thousand miles, high endurance (almost fifty hours), a great load-carrying ability, room for wireless equipment (which in that day was too heavy and cumbersome to be carried in an airplane), and slow speed—or no speed—facilitating the tracking of a nine-knot submerged submarine. On the other hand, its slow speed made the airship a poor attack vehicle. The airship was more visible to a surfaced submarine than the other way around, and the U-boat would be deep and gone before the airship would arrive over the spot. However, until the invention of the "sky scope"—a periscope which provided an overhead range of view—more than one submarine skipper found, to his distress, that he had surfaced just beneath an airship.

An early defense against submarines was to cover the hulls of surface ships with wild painted patterns, intended to confuse a submarine skipper about course, speed—even the type of ship. Some merchant ships had phantom destroyers painted alongside; at least one German submarine was painted to look half its size (and thus, presumably, present a less inviting target). The razzle-dazzle worked. In the words of one U.S. skipper, "In a quick look through the periscope, you couldn't tell which way a ship was headed!" It worked, at least until advancing technology provided adequate target data without recourse to periscopic investigation.

Submarines themselves were painted in varying schemes, depending on the operating area. In northern waters, dingy white; in other areas, dappled gray. Eventually, the basic U.S. Navy haze gray was adopted for surface ships and flat black for submarines.

The military intelligence services played a major role in the ASW effort. German code books were recovered from shipwrecks and found on the bodies of crewmen, and over time many—if not most—radio messages were intercepted and decoded; virtually any German radio message—from land or sea—was an open letter.

The effectiveness of the German radio equipment was superior to any available to Britain and her allies—a blessing and a curse. The blessing: Messages from the main radio in Berlin could be picked up as far away as the United States and China. U-boats at sea could, when conditions were right, send and receive messages over distances greater than seven hundred miles. The curse: The Germans used this ability to excess, and radio direction finders were able to track a U-boat across the North Sea.

Remarkably, U-boat commanders realized that their transmissions could be detected, but seemed not to care. They assumed that they would be long gone before a destroyer could be vectored to their boat's position. What they didn't understand was that knowing where *they* were, even if out of attack range, helped the British reroute the *convoys* away from the danger zone.

For the year 1917, 5,090 ships were convoyed; sixty-three were lost, a rate of 1.23 percent. For the first five months of convoy, in any convoy attacked, only one and never more than two ships were sunk. For the whole of the year, only *one* ship was lost from a convoy under air support. The continued shipping losses were among the independent sailers—over the first six months of convoy, that amounted to 577 sinkings, almost 90 percent of those who did not sail in convoy.

Another measure: In March 1917, the oceangoing boats of the High Seas Fleet were averaging about one ship sunk every other day of patrol; by June 1918, they were scoring one ship every two weeks. By the end of the war, more than 88,000 ships had been convoyed, of which only 436 were sunk.

Convoy worked, but not because the escorts were sinking many U-boats—they were not. But convoy escorts and air support forced the U-boats to attack submerged, an almost impossible task unless the submarine just happened to be in the right position. In May 1918, the U-boat Command tried an experiment: About a dozen U-boats were sent out in a group in hopes that one or another would spot a convoy and alert the others, by radio, in time for all or most to run ahead, on the surface, and take position for attack. The two-week effort was not very successful. Four ships were sunk, one was damaged, while 293 ships had been safely convoyed through the zone in which they were operating.

Toward the end of the year, the British began a major effort to improve the Dover barrage; they brought in an upgraded (and tested) mine, and laid down a tight field of parallel rows, anchored at depths from thirty feet down to one hundred feet below the average low tidal level, across the entire Strait from Folkestone to Cape Gris-Nez. To block any surface transit and force the U-boats down into the minefield, as many as one hundred boats equipped with illumination flares and searchlights moved across the waters, day and night.

The improvements were basically in place by February 1918, although adjustments continued to the end of the war. By that time, 9,573 mines had been laid in twenty parallel rows in a band about six miles wide. Approved traffic was funneled through guarded gates inshore along the English and French coasts. U-boats which had returned to using the Channel on an "optional" basis were once again forced into the long northern passage; the barrage claimed twelve confirmed kills by the end of the war, and one possible.

About the time these improvements were completed, the effort shifted to what had been called the largest engineering effort in history—the Northern Barrage. A combined British-American effort set 71,126 units in one area, beginning on March 3rd. Sowing of the broader field began June 8, 1918. On that day, 3,400 mines were laid in a string forty-seven miles long; one minelayer dropped a mine every 11.5 seconds for more than two hours, without a break. The Northern Barrage was completed by October 26th, very late in a war which was to end two weeks later.

How shall we evaluate the Northern Barrage? Read the literature and you can find any evaluation you want: total failure, expensive experiment, useless example of American arrogance, marginal success. The magnetic mines proved to be hypersensitive and would often detonate for no discernible reason, setting off whole strings of their brethren. Think of explosive dominoes.

However, the barrage is credited with the sinking of between four and seven U-boats (the difference remaining in the "probable" category) just in the month of September; another three boats were damaged. Discounting the psychological effect on U-boat crews (U.S. Secretary of the Navy Daniels wrote of "mutinies" among crews "forced" to make the transit), four to seven boats in one month is a pretty high score in a four-year war in which two hundred U-boats were lost (or "destroyed"—I suppose it depends on whether sympathies were with the Germans or Allies).

The barrage cost $80,000,000, a lot of money in anybody's day, but only slightly more than the average one-month loss of ships and supplies throughout most of the war. However, laying the barrage was the easy part; with the war over, the mines had to be recovered. That operation occupied six months from April to September 1919; twenty-three minesweepers were damaged and eleven men killed in the process.

In the spring and summer of 1918, the success of convoy had pushed U-boat operations well out into other areas—most notably, deploying U-cruisers to the eastern seaboard of the United States. First out, in May, was *U-151.* In a ninety-four-day, eleven-thousand-mile cruise, she laid some mines in the Chesapeake and Delaware Bays, cut or damaged underwater telegraph cables, sank twenty-seven ships (four were victims of the mines), and brought back a load of copper which had been liberated from a freighter. To provide the cargo space, she jettisoned a bunch of iron ballast.

In July, three U-cruisers (*U117, U140,* and *U156*) laid mines which sank one cruiser and damaged a battleship; they sank fourteen ships with torpedos, twenty eight by gunfire or planted charges, and picked off more than fifty smaller boats.

In all, six U-cruisers operated in North American waters. They proved that it was possible for submarines to conduct operations three thousand miles from home base, and sink some ships—in the event, mostly fishing craft and coastal cargo-carriers not equipped with radios and which therefore could neither be warned nor give warning. The U-cruisers failed in their primary mission: to block the movement of troops and supplies to Europe. Not one convoy seems to have been threatened, let alone attacked.

The U-cruisers did succeed, however, in provoking the usual panic among the coastal-dwelling civilian population. In the twenty-four hours after the first reported attacks, the Navy Department logged in more than five thousand telegrams, telephone calls, and letters, all demanding that the Navy "do something."

By full summer, conditions in Germany had grown so bad that much of the nation was in rebellion—which threatened to turn into a Soviet-style revolution. The game was up, and the government knew it. The Kaiser appointed a new Chancellor (Prince Max of Baden) with authority to negotiate an armistice. He began peace negotiations on October 3rd.

Some in the government service were not very cooperative. The U-boats had put the United States into the war; now, they came close to ending any hopes of a negotiated peace. On October 4th, the passenger ship *Hiramo Maru* was torpedoed and sunk off the Irish coast with the loss of 292 of 320 on board; on the 10th, an Irish mail boat was torpedoed with the loss of 527 lives. In a formal note of October 14th, President Wilson told the German government that an armistice was impossible

> . . . so long as the armed forces of Germany continue the illegal and inhuman practices which they still persist in. At the very time that the German Government approaches the Government of the United States with proposals of peace, its submarines are engaged in sinking passenger ships at sea—and not the ships alone but the very boats in which their passengers and crews seek to make their way to safety.

Opinion was divided within the High Command; give in to Wilson and give up any leverage, or keep continued pressure on the Allies. This was Scheer's position: ". . . to comply with this condition would be tantamount to a complete cessation of U-boat warfare," he complained. "In so doing we should lay aside our chief weapon, while the enemy could continue hostilities and drag out the negotiations as long as he pleased."

The German response of October 20th protested the charge of "illegal and inhuman practices" and denied that lifeboats had been attacked; in that, they may have been correct as neither of these sinkings were included in a British postwar list of incidents considered to have been war crimes.

North Sea U-boats were recalled on October 21st—but not necessarily as a gesture of defeat. The leaders of the German Navy planned one last—as in "suicide"—mission. This was an echo of the first operation of the war: With all available U-boats sent out in ambush, the High Seas Fleet was to appear at sea, luring the Grand Fleet over the waiting U-boats and then into one grand final battle. The latest Fleet Commander, Admiral Franz Hipper, believed that "an honorable battle by the fleet—even if it should be a fight to the death" would "sow the seed of a new German fleet of the future. . . . [there could] be no future for a fleet fettered by a dishonorable peace."

Twenty-two U-boats went into position. One, *UB-116,* became the last U-boat casualty of the war, destroyed by shore-controlled mines while trying to get into the fleet anchorage at Scapa Flow on October 28th.

Under Hipper's plan, the High Seas Fleet was to sortie on October 30th, but the sailors of the fleet had their own ideas about ending the war. They did not include suicide. The sailors began taking control of the ships; some, emulating their counterparts in Russia the year before, were hoisting red flags as a sign of revolution.

On October 31st, *U-135* was assigned one of the more bizarre missions of any war: to hold in check and be prepared to attack, with gun and torpedoes, the German battleship *Helgoland.* On November 6th, the British intercepted the following order sent to Baltic-based U-boats:

> . . . fire without warning on ships with the red flag. The whole of Kiel is hostile. Occupy exit from Kiel harbour. . . . Use all means to break down resistance.

The U-boat crews remained loyal, and U-boats continued the war to the bitter end. Two days before the negotiated Armistice of November 11th was to take effect, *UB-50* put a torpedo into and sank the aging battleship *Britannia.* It was defiance to the last, and defiant many would remain over the ensuing twenty-one years of "peace." The world had not heard the last of the U-boat.

22 Interregnum: Europe

International law is somewhat ephemeral, at best; and whenever relatively new weapons like the submarine are introduced, nations concoct their own international law to suit. It prevails if they can make it prevail.

—War Propaganda and the United States, *by Harold Lavine and James Wechsler, Yale University, 1940*

On November 14, 1918, Korvettenkapitän Lothar von Arnauld de la Perière nosed *U-139* into berth in Kiel for the last time, saw that the lines were properly secured, changed into civilian clothes, and went ashore incognito. He was the most successful U-boat commander of the war—indeed, the most successful submarine commander of any nation, any war. From April 1916 to the end, he sank 454,000 tons of enemy shipping—two warships, five troopships, and 187 merchantmen. During one patrol he sank better than two ships a day, to a total of fifty-four.

His adherence to prize rules was punctilious—he relied on deck guns and fair warning to non-combatants. During the entire war, he fired only four torpedoes (one of which missed). Apparently—during the entire war—he did not cause injury to even one person aboard any unarmed vessel. One of his patrols became the heart of a 1917 film documentary, *Der Magische Gürtel*

("The Magic Girdle"); all ships were politely stopped and searched, all crews shifted to lifeboats before any sinking began.

We should resist, however, treating him as some sort of avenging saint. He was successful because he was persistent, intelligent, a thoroughly professional Naval officer. He preferred artillery to torpedoes because it contributed to a more efficient patrol—his U-boat could carry only six torpedoes, but perhaps as many as one thousand shells. He was to be killed in an air raid in 1941, serving as a vice admiral in the next war.

At the other end of the scale was Max Valentiner, commander of *U-38*. He too was very successful, the third-ranking U-boat ace, credited with 141 merchant ships and one warship, total 300,000 tons. But while von Arnauld de la Perière and some other U-boat commanders may be enshrouded in the myth of the chivalrous warrior, Valentiner is not among of them. Among other questionable exploits, he sank two Italian liners in November 1914, even though Germany and Italy were not then at war with each other. His action was not in error. He went on the attack flying the Austrian flag, knowing that German ally Austria was at war with Italy. In the ensuing uproar, the Germans arranged to have *U-38* taken into the Austrian Navy with an effective date two weeks before the first Italian sinking.

In between de la Perière and Valentiner there were more than four hundred other wartime U-boat commanders. Most were not vicious. Despite numerous Allied stories of atrocities against helpless survivors, that record remains clouded by wartime propaganda. An American book on *The German Submarine War 1914–1918* published in 1931 certified only one case where U-boats fired on lifeboats: that of Helmut Patzig, commander of *U-86*. Having sunk the hospital ship *Llandovery Castle* on June 27, 1918, Patzig wanted to bury the evidence. He was not successful, as twenty four surivors escaped to tell the tale. In an interview with the *New York Tribune,* April 15, 1923, Admiral Sims affirmed *Llandovery Castle* to be the only such instance of which he was aware, adding, "The press accounts of the 'terrible atrocities' were nothing but propaganda. The British naval records and our own are filled with reports showing that German U-boat commanders aided in the rescue of crews and passengers of ships they sank."

The issue of postwar retribution was debated by the Allies, off and on, for two years before the war ended, and for another two years afterwards. Some held that the authors of the war must be accountable for all excesses of the war—

the Kaiser and other senior members of the government and the military. Precedents were cited: Confederate President Jefferson Davis was charged as a co-conspirator in the prisoner-of-war atrocities at the Andersonville camp. However, he was never brought to trial. Then there was the matter of war crimes committed by military personnel: Where should the line be drawn? Should subordinates be excused for carrying out the orders of superiors? Experts on international law decided that a violation of the law should be punished, orders or no orders.

The Allies drew up accusations against some eight hundred Germans (including the government, and along with a handful of Austrians, Turks, and Bulgarians). But then came the matter of jurisdiction. Where should the trials be conducted, and by whom? The Wilson Administration urged moderation, noting that the German government was already under siege by revolutionaries and radicals of every stripe and that forcing so many recent "heroes" to trial might trigger a disaster.

Eventually, the issue was resolved in favor of not doing much, and letting German civil courts do it. Forty-six cases were readied—none involving senior officials—but by the time the trial opened on May 23, 1921, the number had been reduced to twelve: three for the *Llandovery Castle,* eight for prisoner-of-war offenses, and one for cruelty to French children.

Of the total, half were found guilty with sentences ranging from six months to four years. *Llandovery Castle* was the last case, tried July 17th. The U-boat commander, Patzig, had fled the scene, but two of his officers were tried as accomplices. The court did not accept the plea of Lieutenants Dittmar and Boldt that they were "only following orders," nor did it give any credence to a somewhat contradictory defense ploy that the accused believed they had been firing on an approaching enemy ship, not at lifeboats. They were each sentenced to four years in prison. The verdict was not popular in Germany. Lieutenant Dittmar "escaped" from prison after only four months; Lieutenant Boldt made it out two months later.

We must note that no Allied personnel were charged with war crimes, perhaps in tribute to the purity of the cause. In the Yale University 1940 publication *War Propaganda and the United States,* authors Harold Lavine and James Wechsler noted:

> International law is somewhat ephemeral, at best; and whenever relatively new weapons like the submarine are introduced, nations

> concoct their own international law to suit. It prevails if they can make it prevail.

Such one-sided "justice," even though the Germans were barely touched, was another contribution to smoldering resentments which would contribute, within a generation, to another war.

The subject of "unrestricted submarine warfare" had dominated the strategic debate, particularly among politicians and the public press. After wartime passions had cooled, some people acknowledged that most warfare was, and always had been, "unrestricted." A 1936 article by British General Sir Henry F. Thuiller—"Can Methods of Warfare Be Restricted?"—answered its own question:

> Submarines cannot take off passengers and crew, and there were occasions in the late war when these were set adrift in open boats. There were also times when they were not even given that option, but were sunk without attempt at rescue. This was certainly very inhumane. But is it more so than to bombard a town with heavy artillery regardless of the civilians and the women and children in it?—a practice which has prevailed for centuries and no one makes any protest against it.

General Thuiller included the blockade of Germany in his list of inhumane but accepted methods of warfare, "cutting off the food supply of the whole of Germany and Austria, knowing full well that these countries could not produce sufficient food and milk for their population . . ." and continued;

> All war is terribly inhumane. It is very splendid of our navy to have kept up its chivalrous custom of ensuring the safety of civilians at sea right into the XXth century, but their less sensitive comrades on land have for long been in the habit of firing at railway trains without asking any questions about who are in them.

Overall, in four years of war, 5,708 Allied ships were sunk, more than four thousand of them by U-boats. In total, more than eleven million tons, one-fourth of the world's supply. Half of those sunk were British. Twenty-two of the four hundred U-boat commanders were responsible for more than 60 percent of all sinkings.

In the most critical period, 1915 through mid-1917, 104 German submarines sank more than 3.5 million tons of shipping, two million of which were British. It was estimated then—and later as well—that at that pace the war would have been over in another six weeks. However, two things happened just at that time: America entered the war, and convoy was instituted. From the spring of 1917, 83,959 ships sailed in convoy; U-boats sank 257 while at the same time sinking 2,616 independent sailers. Of those 83,959 ships, 16,070 sailed in ocean convoys (Atlantic and Mediterranean); ninety-six were sunk. 67,888 ships were convoyed in northern local waters (coastal trade and colliers, for example); 161 were sunk.

The British Empire sent 5,399,563 officers and men across the Channel; the Americans sent 1,876,000 of their own across the Atlantic. All were convoyed. Not one—none—was lost to enemy action while at sea.

Hard to believe, but the Admiralty's Anti-Submarine Division seems never to have run a statistical analysis of the survival rate of convoy versus independent sailers. To the end of the war, the Division held to the dogma that warships should be out on the offensive, and while convoys certainly were providing some measure of safety to the merchant fleet, "attack" was still the best, really the *only,* defense. In March 1918, the Division even suggested reducing escorts by 30 percent and shifting the ships to independent patrol and U-boat hunting. They seem not to have noted Mahan's "needle in the haystack" analogy. At the end of the war, with more than five thousand warships in commission, the British still had only 257 assigned as convoy escorts.

On the other side—according to figures published in 1967 by R.M. Grant—Germany started with twenty-six U-boats in service at the end of 1914, built 390 U-boats by the end of the war, and lost 178. At war's end, 171 boats were in the water (although not all were in service) and another 149 were under construction.

Of the total lost, nineteen were to accident, nineteen to unknown causes, and the rest to enemy action. Of those, mines claimed at least forty-eight, depth charges took thirty, twenty were sunk by gunfire, fourteen were rammed by warships and five by merchant ships. British submarines destroyed seventeen; the American submarine force claimed, but Grant did not confirm, a tenuous one. Aircraft—one kill. Sixteen U-boats were destroyed by their own crews at war's end. Another eight, which had already surrendered, mysteriously "sank" en route to designated British ports. Their crews, we may be assured, were not in the sort of jeopardy as had been their missing compatriots. For the war, U-boat personnel losses were 515 officers and 4,894 enlisted men, perhaps 30 percent of assigned personnel.

A postwar article by Admiral Sims listed the relative effectiveness of antisubmarine vessels. His figures put yachts and patrol craft at the bottom of the list: The more than three thousand in service were credited with sinking thirty one U-boats. Destroyers: five hundred in service, score thirty-four U-boats. By comparison, one hundred Allied submarines probably sank twenty U-boats.

Thus, in spite of Simon Lake's belief that submarines could not fight submarines, the submarine itself proved to be the most effective antisubmarine weapon, although issues of identification were never fully resolved. Because of the distances involved in their extended operations, the Germans had to spend as much time as possible on the surface whereas the Allied boats, in defensive positions closer to home, could lie submerged in ambush.

A hopeful sign for future ASW was the success of the depth charge. Through the end of 1917, only eight boats had fallen victim to this infant weapon; in the ten wartime months of 1918, the more mature depth charge took out twenty-two U-boats. An airdropped version was just coming along—three hundred pounds of explosive in a 520-pound package, with a dual contact and pressure-sensitive fuse. If the bomb did not make a direct hit, it would explode at forty feet.

Even more important—for the future, because the present was still handicapped by technology—was air patrol. The lifting capacity of most airplanes was yet too limited to carry suitable bombs, so their role was simple enough: one aircraft patrolling the outer fringes to prevent a submarine from maneuvering into position on the surface; another flying over the convoy looking for periscopes and, even more important, torpedo tracks—those arrows in the water pointing directly back to the enemy. U-boat commanders knew all of this; U-boat commanders tended not to bother with any convoy with an air patrol.

ASW research and development showed promising progress. An echo-ranging system which grew out of the Fessenden iceberg project, now called ASDIC, had been introduced too late to have any impact in the war, but was installed on all British destroyers by the mid-1930s. No one seems to know what the letters ASDIC really stood for, if anything. The long-accepted solution was "Allied Submarine Detection and Investigation Committee," alleged to have been a joint effort of 1918; however, one historian made a detailed search and was unable to find that any such committee ever existed. If anything, the initials may be related to the work of the "Anti-Submarine Division" of the Admiralty. ASDIC was the British version of SONAR (*those* letters stand for "SOund Navigation And Ranging") whereby a burst of sound is reflected from a target and back to the sending ship. Direction to the target could be determined by the direction in which the sound head was pointed; range was computed from the time it took for the signal to make the round trip. There was, however, no method for determining the depth of the target (unless it was directly below the sending unit). Nor was sonar effective if the target was on the surface. Another disadvantage: The "ping" would also alert the target to the fact that it had probably been found—usually well before the attacking unit could arrive on-scene.

Antisubmarine bombs were developed in the 1920s and were put into "service" in 1934—without testing. In *actual* service, some detonated prematurely, some took erratic directions once in the water, and unless scoring a direct hit, none were likely to cause serious damage. The air forces soon enough took up the depth charge instead.

During the Great War, the British brought forth the K-class submarine, three times the size of any earlier class and with a steam engine to push it at twenty-four-knot battle-fleet speed. The omniscient Jacky Fisher had dismissed the idea in 1913—"the most fatal error imaginable would be to put steam engines in a submarine." However, intelligence reports that Germany was then building a boat capable of twenty-two knots forced the issue. The reports were in error. So were the K-boats.

A U.S. Navy observer reported in 1918 that a dive by *K.8* took eleven minutes. First, the boilers were shut down (turn off blowers, shut off fuel to the burners, lower the funnels, close the ventilators, throw out the clutches, and secure the turbines). Once under the water, the temperature in the boiler room reached 160°F; no one stayed in the boiler room, but it was checked every thirty minutes to make sure that nothing was going wrong. Temperature in the engine room was a bit more tolerable at 90°F, but since the engines were not running, no one needed to remain there either, so no one did. Surfacing was almost as slow as diving: ten minutes until the boat was on the surface with the engines at full power ahead.

The Naval planners were not concerned about dive time—they assumed that the crew would spot the masts of any approaching enemy ship in plenty of time to dive before themselves being spotted. The Naval planners seem to have forgotten the recent developments of airplanes and airships.

The heavy Ks had a lot of forward momentum when they dived, and the planing effect of the large flat deck overwhelmed the diving planes. They were not stable, and suffered so many accidents, minor and major, that the class was thought to be jinxed. On one trip—with the future King George VI aboard—the first of the K-boats took a sudden plunge in 150 feet of water, getting her nose stuck in the mud while her stern, propellers spinning wildly, stuck out at the surface. This ignominy lasted but a few minutes, but set the tone.

The numerous problems with the Ks are not worth noting here, except to mention the most spectacular accident in submarine history, in February 1918. Like a pileup on a fog-shrouded expressway, nine K boats speeding through the night at nineteen knots ran into each other and some surface escorts, with the result that three submarines and two surface ships were damaged, two submarines sunk. The crew of *K.17* had abandoned ship, but surface ships churning through the water left only eight survivors; all hands were lost aboard *K.4.*

Despite a rather dismal record, six more Ks were ordered in June 1918, although the war soon ended and only one, *K.26,* was commissioned. At the same time, four Ks still under construction were changed from steam to diesel power and were reconfigured as "submarine monitors." This "M" class went along with Jacky Fisher's BIR idea, to give a submarine the firepower to sink a battleship. In theory, the 12-inch gun could be fired while the boat was underwater; in practice, the boat had to surface each time to reload the gun. Only one of the class was equipped with the gun: *M.1.* Her first commander was Max Horton. She sank in 1925, the victim of a collision.

M.2 was fitted with a hangar for a small seaplane and finished in 1920 as an aircraft carrier. The experiment was not successful. The aircraft had an endurance of only two hours and recovery at sea proved difficult; *M.2* herself sank when the hangar door was opened by mistake before the boat was fully on the surface.

The third "M" boat was finished as a minelayer; where the normal submarine carried eighteen mines, *M.3* could carry one hundred.

In 1925, the British tested another large submarine—the three-thousand-ton *X.1* armed with four 5.2-inch guns and six 21-inch torpedo tubes. It didn't work very well and was scrapped. Not to be outdone, France came out with the 1927 *Surcouf*—twin 8-inch guns and an airplane. *Surcouf* was probably rammed and sunk—an accident—by an American freighter in the Caribbean on February 18, 1942. Both nations were trying to create the underwater version of a cruiser. They had not succeeded.

On April 28, 1915, former First Lord of the Admiralty Fisher had posed this question to Admiral Jellicoe:

> Supposing that there is another war in a generation hence and that Germany then has 500 submarines of immense power and able to keep the sea for weeks, what is going to be our success? At the present moment there is no known way of surely protecting our mercantile marine from such a fleet of submarines. . . .

"At the present moment" the Germans had perhaps sixty-eight U-boats in service, and never more than 170 in service at any time during the war. Allied success with convoy, air patrol, and new weapons had saved the day—aided in no small measure by German strategic flip-flops and tactical mistakes. One would think, therefore, that the first postwar order of business would be to prepare for the next war by following the time-honored military practice of planning to refight the last war. Not so.

The end of the war brought absolute complacency. What to do next? The world was at peace, and there were problems more pressing than "the next war." There was the matter of the war debt—Great Britain's National Debt had risen more than tenfold between 1914 and 1918. The British Government made "peace" a matter of policy. "It should be assumed," the Cabinet proclaimed in August 1919, "that for framing revised Estimates [the annual military budget], that the British Empire will not be engaged in any great war during the next ten years." That became a rolling policy, renewed every year and covering not only the period through 1929, but all the way to 1938.

Some among the British drifted into the interregnum with a pious idealism or a specious take on the war. In a February 1928 lecture at the Royal Naval War College, submarine veteran Captain Charles G. Brodie asserted that "barbarous and forbidden methods" were counterproductive, as they too much demoralized the perpetrators! Further, he contributed to a sense of security that was to prove altogether false when he proclaimed:

> . . . the pendulum has swung very markedly against [submarines]. Detection devices have vastly improved, and the complete immunity enjoyed by a well-handled submarine against all attack, unaided by mishap or remarkable luck, will never recur. . . . I consider the submarine menace against our commerce definitely dead as a decisive factor.

In *his* postwar summation, Winston Churchill took a giant step further into fantasy-land: ". . . the seafaring resources of Great Britain were in fact and in the circumstances always superior to the U-boat attack."

Thus, we have one participant claiming that the problem had been solved, and his wartime superior claiming that there never really had been a problem at all!

Those assurance asides, the government did set up a Shipping Defence Advisory Committee in 1937, which developed training programs for merchant marine officers, published a handbook on *The Defense of Merchant Shipping,* worked through signaling procedures, and urged that merchant ships be prepared for the mounting of naval guns. The Committee discovered that there weren't enough antiaircraft guns in stock for maritime air defence, nor did Britain have the industrial capability to make up the shortage. Amid reluctance to "buy foreign," the Swiss Oerlikon company was given a contract for five hundred guns, and the British were given a license to build more at home.

Notwithstanding any of the above, and not linked to any evidence from the real world, First Lord of the Admiralty (again!) Winston Churchill told Prime Minister Neville Chamberlain in March 1939: "The submarine has been mastered."

23 Interregnum: America

It is inconceivable that submarines of our service would ever be used against merchant ships as was done during the World War.
—Member of the U. S. Navy General Board, 1926

In a letter entitled "Certain Naval Lessons of the Great War," under a date of January 7, 1920, the senior American officer in the war zone, Admiral Sims, took the civilian head of the Navy, Secretary Daniels, to the metaphorical woodshed. The letter had been intended as an official communication with the Navy Department, but an advance copy somehow reached a member of the Senate just then engaged in a hearing over a relatively minor issue: the Secretary's arbitrary method of recommending favorite personnel for high-level medals and awards. The letter raised the stakes.

In essence it said that the Navy had not been prepared for a war that was patently inevitable; the Navy had made no plans for mobilization, and had had no plans for antisubmarine and transport operations, even though it was obvious that these would be of primary importance should the United States enter the war.

These were not idle complaints; while the Secretary played public relations by creating the Edison Board, he was, in fact, blocking any moves to increase personnel or fleet readiness. On the day war was declared, 67 percent of all Navy ships were in need of repairs; instead of quickly going into wartime service, they went into the shipyards for an average stay of fifty-six days.

The Assistant Secretary of the Navy—the other Roosevelt, Franklin D., who served in that post for five years—had some, but not much success in working around the Secretary. But Daniels's boss, the President, had campaigned on the slogan "He kept us out of war," and the Secretary would do nothing that was contrary.

The Congress did what Congresses usually do when some members perceive a political advantage—it held another round of hearings. Nothing happened.

The primary wartime contribution of the U.S. Navy was convoy duty, providing 80 percent of all trans-Atlantic escorts. The first American destroyers arrived on station on May 3, 1917; by the armistice, seventy were operating in European waters. An uncounted number of civilian yachts, donated for the duration and given makeshift status as patrol boats, bounced around in the Bay of Biscay. Ten submarines made a remarkable mid-winter passage across the Atlantic to take up station off Ireland and in the Azores. They played a bit part in the drama of the war, but like apprentices working backstage, the officers and men watched the featured players, and practiced, and learned.

One who watched was Lieutenant Chester Nimitz, assigned to the staff of the Submarine Force, Atlantic Fleet. Nimitz toured British facilities and helped organize the U.S. submarine effort in the war. Immediately after the war, now a lieutenant commander, he became the Senior Member, Board of Submarine Design.

Among other things, American submariners soon learned that their crews were too small for wartime operations; more people had to instantly be ready for a dive, rather than relying on some who were on the off watch and sleeping. They learned that all dives were crash dives, to be completed in one minute—or less. They adopted the European practice of "riding the vents." The bottom tank valves were left open all the time; the vents at the top of the tank were kept closed, and the air in the tank kept the water out until the vents were opened. When flooding, the American superstructure tended to trap air, slowing the dive; the solution was to cut in more limber holes, allowing air to escape more quickly.

* * *

The unfinished U-boats were broken up in the shipyards; the surrendered U-boats were divvied up between the victorious Allies: Britain, 105; France, forty-six; Italy, ten; Japan, seven; the United States, six; Belgium, two. Most were sold for scrap.

But a few found their way into service, at the least as templates for the next generation of boats. In a report to the Office of Naval Intelligence dated July 15, 1919, the U.S. Naval Attaché to Japan noted some of the more significant features of those U-boats sent to Japan, features not widespread in the boats of any other navy:

1. Excepting the *UC-90,* they have a mast for radio, which can be erected or dropped down from the inside of the submarine by an electrical apparatus.

2. Excepting the *U-46,* they are all provided with two sawlike torpedo net cutters, one on deck and the other on the bottom.

3. In order to prevent the projecting parts [from] getting entangled with lines or cords, while sailing under water, steel wires are stretched from the top of the conning tower to the bow and two from the conning tower to the stern. These wires can, by special device, be utilized for radio signaling to short distances. The vertical and horizontal rudders, and propellers are protected by strong frames around them.

4. In order to purify the air inside the submarine, there are provided apparatuses which purify the air by means of chemicals, and others that give out compressed oxygen.

5. In order to show the site of the submarine when stuck to the bottom, they have buoys which are attached to the submarine by chains, which can be floated from the submarine. There are also rubber tubes through which the salvaging vessel can send down air or liquid food. They are also equipped with apparatus for telephone communication.

The six U-boats brought to the U.S. were put on public display, part of a postwar Victory Bond fund-raising tour (admission was the price of a bond). Evaluation of those boats confirmed what was already known—that the Germans had the best diesels and range of operation. But there were some surprises. Where the atmosphere in U.S. submarines had not changed (literally!) from the earliest days—dank and damp—that in some of the U-boats was fresh and dry, because engine suction was taken from inside the hull rather than through an engine intake duct, thus constantly renewing the air inside the boat. The American S-boats required more than two minutes to dive; one of the smaller German UB-boats could make a crash dive in twenty-seven seconds.

The U-boats, however, offered some serious deficiencies: The compromises needed to put a reasonable operating efficiency in a small hull were obvious. There were only enough bunks for about two-thirds of the crew, there was no cold storage for food, the metacentric height was markedly less than that required on American boats, and the onboard machine shops were rudimentary at best.

The General Board—indeed, much of the Navy—ignored (or did not understand) the one clear lesson of the U-boat war: Merchant shipping was the logical, most valuable target for submarines. As one senior officer noted, "It is inconceivable that submarines of our service would ever be used against merchant ships as was done during the World War."

Therefore, the postwar design program was centered, as before, on the "fleet" submarine, to operate as a component of the battle fleet. The mission was to be scouting for, and taking out, enemy warships. This required good sea-keeping, long-range, *fast* submarines.

Those three key characteristics forced the boats to be large, and larger, and pushed engine requirements to the limits, perhaps beyond. If the General Board set a requirement for a minimum surface speed of fifteen knots, that boat must be equipped with sufficient engine power to make fifteen knots while also charging batteries, in a moderate sea with "normal" fouling (that which would be expected four months out of dock), at full war loading and in ready-to-dive trim, without putting more than an 80 percent load on the diesels. Easy to propose; difficult to achieve.

The U.S. had been sufficiently inspired by reports of the speedy "Ks" to launch design studies for its own large steam-powered boat. The caustic soda fireless boiler was revisited, but could not deliver the needed power. The Garrett-Nordenfeldt scheme—storing steam under pressure—also was considered, but the slow start-up and the excessive residual heat remained unassailable impediments. Heat remained a problem with "regular" steam, but the studies continued nonetheless. The heat might be ameliorated by inserting insulating spaces between the boiler room and the rest of the ship, but this would not solve the problem. As the study went on, other limitations became apparent: Getting under way would take too long, endurance would be limited, diving would be slow—estimated at twelve minutes to go from eighteen knots on the surface to eighteen knots submerged. All this—to gain three to five knots surface speed.

There were too many compromises, so the Navy accepted somewhat slower boats, achieving maximum power by putting two diesel engines on each shaft. This was not a good idea. One problem: It was very hard to synchronize the engines. Another problem: The awkward installation would on occasion set up an engineering anomaly known as "torsional vibration." If you've ever seen the newsreel footage of the Tacoma Narrows Bridge (given the nickname "Galloping Gertie") as it shook itself to pieces, you've seen torsional vibration in action.

These boats became the T-class, circa 1920. They were not successful. Where the predecessor class, the S-boats, could dive in sixty-four seconds, the T-boats took four to six minutes and engineering readiness was an abysmal 50 percent.

The next U.S. class of submarines would logically have been the "U" class, but with even greater logic, that letter was skipped over. The next nine boats were the V-class, three thousand tons with two 6-inch guns. This was the underseas cruiser that strategic planners wanted, but in which the submariners found too many limitations—again, too slow to dive.

In 1926, the Navy established a Submarine Officers Conference, bringing the people who used the boats in to talk with the people who set the requirements for them. But it was give-and-take between the operators—who had to live with limitations but could not provide the solutions—and the General Board—which had plenty of solutions but did not understand the limitations.

Eventually, under the influence of the operators, who could better evaluate the impact of compromises, the boats began to shrink a bit and habitability and endurance were improved. Air-conditioning was introduced. A high bow helped make for a (relatively) dry deck and conning station. However, the added buoyancy forward made for slow diving, so a "down express" tank was installed under the torpedo room, adding five tons of water to pull the bow under more quickly. The Navy shifted from riveted construction to welding, after dropping depth charges on test sections to prove that welding would hold up under the shock.

In a more elaborate series of tests in December 1940, the operational sub-

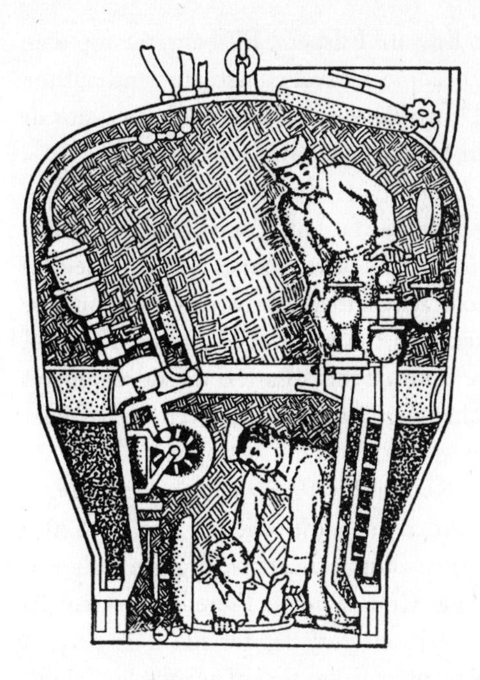

The McCann Rescue Chamber, developed after the December 1927 sinking of S-4; six crew members survived for a time, but could not be brought to the surface. The chamber pulled thirty-three men from Squalus, on the bottom at 240 feet, in May 1939.

marine *Tambor* was deliberately subjected to depth-charge attack. Initially, with the boat under way at periscope depth, charges were set off a distances of 510, 340, and 275 yards. The first was felt as a mild jolt; the second moved the boat sideways and broke a lightbulb; the third broke seven lightbulbs and caused some minor (but potentially serious) damage—two leaking valves.

For the rest of the series the boat was moored underwater, without crew; charges were set off at ranges from one hundred yards down to one hundred feet. The tests showed that depth charges were not as lethal as supposed: a three-hundred pound charge (until then, the standard) would have to go off within fourteen feet to break the pressure hull. As a result, improvements were ordered in shock-mounting of equipment and waterproofing of electric cabling. And the size of depth charge was doubled, to six hundred pounds.

Two American submarine disasters in the mid-1920s led to the development of a rescue chamber which proved of inestimable value in the next serious submarine accident, in 1939.

S-51 was rammed by a steamer off the Atlantic coast and sunk in September 1925. None of the crew survived; the challenge of raising the submarine from 130 feet of water fell to a team led by Captain Ernest J. King, assisted by salvage officer Lieutenant Edward Ellsberg. Ellsberg, the top man in the Academy Class of 1914, had become a naval constructor and author. His book about the salvage of *S-51, On the Bottom,* has long been a classic tale of innovative engineering and human drama.

Two years later, on December 17, 1927, *S-4* was rammed by the Coast Guard cutter *Paulding* while coming to the surface off Cape Cod, Massachusetts. There were survivors—for a time. Divers heard signals from six men (of the forty-man crew), but there was no practical means by which they could be rescued. The submarine also was raised by the team of King and Ellsberg. There were two "benefits" which grew out of this disaster: the development of the McCann submarine-rescue chamber, and the improvement of hazardous duty pay for submariners.

To that time, following Theodore Roosevelt's 1905 excursion in *Plunger,* enlisted crew members had been entitled to additional pay (up to $20 a month) for the discomforts of submarine duty; officers were given no additional compensation. After the sinking, Congressman Fiorello La Guardia spearheaded legislation that created hazardous duty pay for all (at 25 percent of base pay for officers; raised to 50 percent, equal to that of aviators, in 1942).

The rescue chamber was put to beneficial use within a dozen years when, on May 23, 1939, the new submarine *Squalus* sank while on sea trials. The engine room began flooding at fifty feet through an incompletely closed engine intake; Commanding Officer Lieutenant Oliver Naquin ordered that all watertight doors be closed and blew ballast in an attempt to surface; however, the boat settled to the bottom in 240 feet near the Isle of Shoals. Twenty-six men had been sealed into the flooded stern section; thirty-three were safe in the forward section—for the moment.

About five hours later, the submarine *Sculpin*—dispatched to investigate the apparently missing boat—found a marker buoy which had been released by the *Squalus* and hooked onto a telephone line. The line soon broke, but *Sculpin* was able to get enough information about the situation below. Within the next twenty-four hours, a flotilla of rescue vessels was assembled, especially including *Falcon,* bearing one of the McCann rescue chambers. Once the chamber had been mated to the submarine, divers rigged a fresh-air hose from the surface and passed food to the survivors, and then took all safely to the surface in four round trips. *Squalus* herself was raised in September and reconditioned and recommissioned as *Sailfish,* and served until decommissioned toward the end of 1945.

* * *

The British suffered a major disaster on June 1, 1939, when a junior officer unwittingly opened the interior door of a flooded torpedo tube aboard *Thestis.* The Royal Navy rushed through interlocks on torpedo doors to prevent similar accidents, and developed a "self-contained" escape system. The forward and after torpedo rooms were equipped with hose connections plugged into a compressed oxygen supply, one for each member of the crew; in the event of sinking, all crew members would assemble in one compartment or the other (or both), put on protective coveralls, and hook a mask into the piping. Then, the compartment would be flooded to about chest level, bringing interior pressure equal to outside pressure—whereupon an escape hatch could be opened. First, however, a heavy canvas chute, sealed at the hatch end, was pulled down below the interior water level; then, a valve at the hatch was opened to let water into the chute. The chute kept the now-compressed air in the compartment and provided an exit for one man at a time. No one rode an "air bubble" to the surface, because the rest of the crew would be drowned.

The first man out opened the hatch and clipped it open before leaving the submarine—a duty exacted for the privilege of being first. As each man moved to the chute, he would take one last deep breath from his mask, drop it, duck under the water, and swim up the chute, breathing out all the way to the surface (to compensate for the expansion of the air in his lungs as the pressure was reduced). The next man behind him would pick up that mask, make sure it was working, drop his old mask, and move into position for escape; each man behind him would likewise move up one position until all had left.

The Electric Boat Company had long enjoyed the position of being the major—virtually the only—supplier of submarine designs and accessory equipment to the U.S. Navy. However, Electric Boat continued to display a certain arrogance which did not set well with its primary customer. Relationships with Electric Boat had grown so chancy by 1915 that the Navy created a backup: an in-house design capability, and a government-operated construction facility at the Portsmouth Naval Shipyard, New Hampshire. For the record, the shipyard is actually in Kittery, Maine, but when the yard was established, the nearest post office was in Portsmouth.

Electric Boat Company also owned and operated the New London Ship and Engine Company, which manufactured the NELSECO diesel engines that powered most U.S. submarines. The engines were not reliable, and in a search for a better power plant the Navy tried Swiss-designed, U.S.-manufactured Busch-Selzer engines for the first boats of the V-class, then shifted to German-designed M.A.N. (Maschinenfabrik Augsberg-Nurnberg) diesels produced at the Brooklyn Navy Yard. They were not much better.

So the Navy held a competition. Five major U.S. engine companies were invited to develop—with Navy funding—small, reliable, yet powerful diesels; one goal was to reduce the weight per horsepower from sixty eight pounds to 27.5. If it was successful, everyone understood, the same engine could be turned into a railroad power plant, replacing expensive and inefficient steam locomotives. The winner: the General Motors Winton diesel. Fairbanks, Morse produced an engine on its own account which was also quite successful, and which shared the World War II submarine market with GM. By the end of the war, GM had cut the pounds per horsepower to eighteen.

The General Board was still convinced that the Navy needed two types of boats: the large, fast, long-range fleet boat, fully capable of working with and providing direct support to the battle fleet wherever required, and a smaller, coastal-type submarine—the Congressionally favored "harbor defense" boat. However, in the context of the times, these boats actually would be stationed in forward areas such as Guam, the Philippines, and Midway. Smaller boats could also be used for training, freeing the fleet boats for combat. The submarine officers wanted something in between; they didn't like the clumsy, slow-diving large boats, and saw the "coastal" boat as too much of an expedient, too much compromised by its limitations.

Germany had followed a similar plan in the Great War (coastal, midsize, and the U-cruiser), and as we shall see, had the same scheme in World War II. But Germany could effectively use the coastal submarine in its primary home-water operating area, the shallow and confined North Sea. The primary submarine operating area of the United States was the Pacific Ocean.

Nonetheless, the FY 1940 construction program (defined in 1938) included a pair of coastal-type boats of limited capability. Chief of Naval Operations Admiral Ernest J. King—one of the Navy's pioneering submarine officers—agreed with his compatriots that the small submarine was not only too small, but too limited, lacking the offensive capabilities—the sonar, the fire-control systems, the weapons-loading—of the fleet boat. It could not provide the habitability of the fleet boat, or the endurance. It could not dive much quicker, be handled any easier, nor really present any less of a target for a searching enemy. The only advantage: It (perhaps) could be built quicker. Therefore, King allowed, the small submarine could be built, as long as that program did not affect the increasingly vital delivery rate of fleet boats. In the event, the two were built, *Mackerel* and *Marlin*, and two only.

In fact, all the yards were too busy building fleet boats to allocate additional capacity to smaller siblings. The submarine construction program jumped from six a year to seventy-one for 1940, and 120 boats were launched the next year. Design had necessarily been standardized on that of the 212th U.S. submarine, *Gato.* Standardized, but not frozen. Changes were made as reasons were found until submarine construction was essentially ended in 1943. The major change, perhaps, was to increase the strength of the hull using an improved steel alloy and thicker plates beginning with SS285, *Balao*, allowing an increase in design operating depth from three hundred feet to four hundred feet. Thenceforth, *Gato* was known as the "thin-skinned" class and *Balao* as the "thick-skinned" class. *Gato* and *Balao* otherwise were almost identical; the original design had so much reserve buoyancy (both for safety and to allow for "growth") that lead ballast had to be added. The lead was traded for the heavier steel plate in *Balao.*

Gato and *Balao* displaced 1500 tons on the surface, were 312 feet long, and had a cruising radius of eleven thousand miles at ten knots and a submerged endurance of forty-eight hours at two knots. Armament: ten torpedo tubes: six forward and four aft. The torpedo load was twenty-four. Initial topside armament was a 3-inch deck gun and four machine guns; larger deck guns soon replaced the lightweight 3-inch. Crew: ten officers, seventy men. Modifications and improvements in machinery and equipment led to the virtually indistinguishable *Tench* class in 1943; twenty-five were built, most too late for the war.

One of the theoretical attributes of the smaller boat, that it could be built so much quicker, soon was overtaken by the reality of a wartime program. Before the war, it took on average thirty months to build a submarine; some planners estimated that this could be cut to twenty months under the forced draft of mobilization. Portsmouth Navy Yard pushed *Cisco,* SS-290, in 1942, in fifty-six days, keel-laying to launch, and delivered *Sea Poacher,* SS-406, in 173 days, keel-laying to commissioning.

The submarines that evolved, despite the fleet-support mission for which they had been designed, proved to be well suited to the eventual mission: destroying Japan's merchant commerce and logistics tail. Of course, the support-the-fleet-mentality had such an iron grip on the Navy that the prewar submariners never practiced attack on a merchant ship, nor practiced any surface attacks at night.

24 Interregnum: Germany and Japan

The hour of our salvation finally tolled. . . . Everyone will know that we former U-boat men welcomed it with particular joy when our government announced a few weeks later: "The first U-boat, U-1, entered service today. . . ."

—Günther Georg, Freiherr von Forstner, U-Boot Ahoi! *1938*

Karl Dönitz, first posted to U-boats in October 1916, earned a reputation (as reflected in his service record) as a "charming, dashing and plucky officer" with "above-average talent" who "enters into each duty with diligence and enthusiasm." By the end of 1917, he was slated for command, first of the Adriatic-based minelayer *UC-25,* then of *UB-68.*

His second command did not turn out well; while he was attacking a convoy on his first cruise, October 4, 1918, the diving system malfunctioned, and the boat stood on end with the bow down well below rated depth. Then, with tanks blown in a desperate move to avoid sinking, the boat shot to the surface stern-first, popping about one-third of her length into the air before plopping down on even keel—in the middle of the convoy. Under fire from all sides, Dönitz ordered the sea cocks to be opened and all hands to "abandon ship." All hands did—except the man who opened the sea cocks. He was trapped below when the boat sank in a reported eight seconds.

Dönitz was taken to England, where he quickly demonstrated some of that "above-average talent" by pretending to be crazy. He was sent to the Manchester Lunatic Asylum, where he gained somewhat easier treatment than other prisoners of war, and also gained a position at the front of the line when the time came for repatriation.

Then, his reputation put him in the group of 1,500 Naval officers permitted to postwar Germany by the Treaty of Versailles. He hoped to remain in submarines:

> I had been fascinated by that unique characteristic of the submarine service, which requires a submariner to stand on his own feet and sets him a task in the great spaces of the oceans, the fulfillment of which demands a stout heart and ready skill; I was fascinated by that unique spirit of comradeship engendered by destiny and hardship shared in the community of the U-boat's crew, where every man's well-being was in the hands of all and where every single man was an indispensible part of the whole. Every submariner, I am sure, has experienced in his heart the glow of the open sea and the task entrusted to him, has felt himself to be as rich as a king and would change places with no man.

That paragraph from Dönitz's 1958 memoir is pretty heady stuff. It's also pretty typical of the tone of U-boat propaganda of the Great War and after. "My crew readily took all hardships in their stride," wrote the commander of *U-53* in his 1926 memoir:

> [they] maintained in all dangers such placidity and in all storms such a sense of humour that I find them all the more heroic the greater the time-span that separates us from those events. Even if their names are scarcely mentioned, yet I believe that the deeds of my crew and the spirit that animated them should wrest them from oblivion. For it is meet that memory give us courage and hope for the future of our Volk and Fatherland.

"It would especially delight me," wrote another U-boat commander-turned-author in 1916, "to have awakened in many a youthful reader the joy and love for the fine and interesting service in the Imperial Navy, particularly so if our beloved U-Boat arm were brought closer in human terms to many readers."

For Dönitz's immediate future—as submarines were not permitted under the Treaty of Versailles—he soon became commanding officer of a torpedo boat: the next best thing, if you will, and the proving ground for submarine tactics for the "next war." What, after all, is a torpedo boat but a submarine on the surface? Later, he would command the cruiser *Emden.*

About the same time, Germany secretly acquired the Dutch shipbuilder Ingenieurs-Kantoor voor Scheepsbouw. The company designed submarines for international customers which were actually protypes for Germany. The first design was built in Spain and eventually sold to Turkey; sea trials for three submarines built for the Finnish Navy in 1930–31 were conducted by German crews.

Late in 1932, the Government approved—in secret—construction of the first sixteen new U-boats; in 1933, the Navy established an "ASW" school ("Unterseebootsabwehrschule" or UAS) staffed by submarine veterans. In 1934, they ordered parts for twenty-four boats to be smuggled into Germany for final assembly in 1935; secrecy was abandoned when Hitler publicly renounced the Treaty of Versailles on March 16th. Former U-boat commander Freiherr von Forstner would later write:

> The hour of our salvation finally tolled. An awakened Germany breathed in relief, when Führer and Reichs Chancellor announced on 16 March 1935 that after the final collapse of our former enemy's disarmament charade we had taken our own military sovereignty into our own hands. The announcement of the introduction of general military service triggered great jubilation throughout Germany. Everyone will know that we former U-boat men welcomed it with particular joy when our government announced a few weeks later: "the first U-boat, *U-1,* entered service today. . . ."

A submarine flotilla was established in September 1935, under the command of Captain Karl Dönitz. He soon had twelve U-boats in service with another sixteen on the way, but there were only two officers in the submarine force with combat experience: Dönitz and his Chief Engineer. They both knew that of the four hundred World War I U-boat commanders, only twenty-two accounted for

60 percent of the merchant sinkings. A clear signpost pointed toward the future: They began a training regimen as rigorous as any.

The submariners had to be, first of all, mariners. "The U-boat," Dönitz later wrote, "had to be equally at home on the surface and beneath it, for long periods at a time in the the widest possible expanses of the open seas and in all weathers. Our aim, then, had to be to ensure complete acclimatization of the crews to life aboard ship, complete familiarity with the sea under all conditions and absolute precision in navigation, and particularly astronomical navigation."

And, of course, tactics: how best to get into position for the attack; when to make the attack submerged or on the surface; how to take advantage of background light and the state of wind and sea to mask the submarine's silhouette; when to remain at periscope depth, preserving the ability to observe; when to dive, and go blind; when to dash away, when to creep away, when to stay motionless and silent. Every crew was required to make sixty-six surface attacks and sixty-six submerged attacks *before* ever being allowed to fire a torpedo. When the war started, it was little wonder that the German U-boat arm seemed not to have lost a beat since 1918.

Dönitz also had the luxury of thinking through the entire subject of command, control, and mission.

Should all control be local? Can a distant commander exercise close control? His search for answers led to the technical questions which drove tactical considerations—when there were answers. He needed to know—and test—such things as the range of radio at the various wavelengths and atmospheric conditions; the effect of antenna design and the height of the antenna on range and signal strength; the optimum size for a U-boat radio unit; various encoding and decoding systems and devices.

Then: Should boats sail singly, in pairs, or in groups? If in groups, how best to communicate with each other—on the surface and while submerged? What were the most effective formations—echelon, line ahead, line abreast, diamond—and at what intervals?

Out of these deliberations grew the fundamentals upon which the next U-boat war was to be based: *Tonnageschlact* (Tonnage War) and *Rudeltaktik* (Wolf Pack Attack).

In World War I, the U-boats only gradually took on the merchant fleet as the primary target. Had they done so from the start, and in strength, the outcome might have well been that predicted by Arthur Conan Doyle.

Tonnageschlact was a simple exercise in logic, a replay of the prescription of 1916: Allowing that the probable enemy was England—an island nation with few natural resources—Germany needed only to sink more supply ships than the enemy could replace, for a long enough period to strangle her.

Rudeltaktik would change the rules of engagement: seven or eight boats operating in concert; attack at night, on the surface (where the submarine's fifteen-knot surface speed was almost twice as fast as that of any convoy, and equal to that of most ASW escorts). Submerge to escape; surface again and speed ahead of the convoy for position for the next night's attack.

The Naval Staff was at first opposed, because the use of group tactics would require breaking radio silence, and the latest-model direction finders were a great improvement over those which proved so devastating to U-boats in the Great War. Döntiz's reaction was equivalent to a shrug. So what? The enemy would not be able to read the coded messages.

The chart having thus been drawn, what vehicles should be selected for the voyage? Dönitz was not interested in accommodating missions he regarded as peripheral: reconaissance, fleet-support operations, and surface gunnery. He was not interested in an antisubmarine role. In his judgment, the submarine did two things very well, launch torpedoes and sow minefields, and submarines should be designed around those missions.

Dönitz favored a boat based on the Finnish design of 1930; 626 tons, seventeen knots on the surface, eight knots submerged. The Naval Staff wanted larger boats, a 1930s version of the 1918 U-cruiser; two thousand tons, exceptional range, and able to carry its own in a surface gun battle.

In any event, another forty-six boats were laid down from 1936 to 1939: twelve were Type II (similar to the UB-UC of World War I), thirteen were the large Type IX, and twenty-one were Type VII, which Döntiz favored. This became the mainstay of the World War II U-boat fleet; 709 were completed by the end of the war (1,452 were planned), in six variations, VIIA through VIIF; the most numerous was the VIIC, 665 built.

Twenty-one years made a big difference; the Type VIIB of 1939 was similar in size to the later World War I models, but had vastly superior performance: more than double the range (to maximum 8,700 miles) with a functional endurance of seven or eight weeks without refueling; these boats could dive in one-third the time (twenty seconds rather than sixty seconds) to a maximum safe depth of 650 feet. Habitability was not much improved, however. There was an additional toilet, and the flush-depth had been increased to eighty feet, but the small compartment was usually given over to additional storage.

The other major aggressor of the 20th century gave useful naval support in the Great War, escorting Australian and New Zealand troops to the combat zone and providing a destroyer flotilla for antisubmarine duty in the Mediterranean. With the demise of the Imperial German Navy, the Imperial Japanese Navy became third in the world, behind Great Britain and the United States. In a world where the 12-inch gun had been king, the Japanese already had 16-inch-gun battleships, and more were planned.

The Japanese had dreams of expansion—not too thinly veiled—and were not accorded major-power status in the Washington Naval Conferences of 1921–22. The Japanese were embittered, and not deterred.

Emboldened by their surprising victory against the Russian colossus in 1905, the Japanese Navy began planning for an eventual showdown with the nation they viewed as their major and logical adversary—the United States. No other Pacific Rim nation was strong enough; no nation matched America's penetration into and interest in the Pacific. No other nation had incurred so much Japanese enmity through mistreatment of expatriot Japanese, as when the city of San Francisco blocked Oriental children from attending school with "Americans" in 1906. The around-the-world cruise of the American "Great White Fleet," 1907–1909, was meant to send a signal to Japan. It did. The message President Roosevelt meant to convey was, "Here's my 'big stick.' Don't mess around with the United States." The message they *received* was that they must be prepared to take on the American battle fleet.

Japanese planning for the great battle was much the same as that of the German Navy vis à vis the British. Japan assumed that the U.S. Navy would always be larger and stronger, at least in numbers, so the strategy was simple. Grind away at the American fleet as it moved westward across the Pacific; then, with the forces brought more or less equal, launch one grand, cataclysmic battle. A key element was to be the submarine; the planned targets were warships. The Japanese were not concerned about American submarines: "The number of submarines possessed by the United States is of no concern to the Japanese," noted Foreign Minister Sidehara, "inasmuch as Japan can never be attacked by American submarines."

Thus, the Japanese were chasing the same elusive "fleet boat" as every other navy, the high-speed, long-range component of the battle fleet.

Chasing—and catching. In the years between the wars, Japanese submarines were the largest and among the most capable in the world. Japan began with the examination of the surrendered U-boats—but went a step further: They imported as many as eight hundred German technicians, engineers, and Naval officers to help them learn how to design, build, and operate submarines. Several hundred were brought over on five-year contracts, some at salaries as high as $12,000 a year—plus bonus.

By the mid-1920s, their work was finished and they began going back home. A U.S. Navy intelligence report of 1928 noted, "As various submarines were completed . . . the Japanese staffs gradually took over the work, until finally a distinctly Japanese type of submarine was evolved. Consequently, during the past two or three years, the influence of the German designers with the Japanese lessened to such a degree that at present it is practically nil."

The Japanese developed three classes of submarine. The I-class was the fleet boat, produced in many versions but about the same size as the German U-cruisers and the American V-class. Several I-boats were designed as aircraft carriers, seaplane tenders, and cargo ships. The RO-class were coastal boats, similar in size but not as capable as the German Type VII. The HA-type were smaller, limited, intended for local defense of outlying island bases or special missions.

The most significant versions of the HA-class were the midget submarines, developed in secret in the 1930s. Transported to a combat zone aboard larger submarines, these were expected to swarm around and sink American warships. An early version, forty-six tons submerged, seventy-eight feet long, was armed with two torpedoes and achieved nineteen knots submerged in 1936. About fifty of this Type A version were built before the war. However, Type A could not recharge batteries and therefore was limited to eighty miles at six knots. An improved wartime version, Type B, added a small diesel for surface propulsion and battery-charging. Two seaplane-tender submarines were converted in 1941 to carry twelve midgets each.

An experimental HA-boat of 1938, number *71,* displaced 213 tons on the surface, was 140 feet long, and topped twenty-one knots submerged.

The most significant Japanese naval development, however, was a torpedo—known as Long Lance. The submarine version, Mk 95, had a nine-hundred-pound warhead, and was powered by a wakeless oxygen-fueled turbine and with a range of five miles at forty-nine knots. The nearest thing in the arsenal of the U.S. Navy had half the warhead and half the range—when it was working. See page 310.

Early in 1938, Germany began to move out—and over, into Austria. While England wasn't on the list for "annexation," the Naval Staff established a committee to assess the potential of, and recommend strategy for, war with Great Britain. The committee convened a study group, the study group suggested that the Royal Navy was too strong to be taken on head-on, and recommended a return to the strategy of the Great War: "Britain's vulnerability lies in her maritime communications. This postulates that all resources should be applied to mercantile warfare." To the committee, "all resources" meant commerce-raiding cruisers, not U-boats.

Based on the lessons learned in the last war—lessons being applied daily in the Dönitz training program—one would think that the committee would have recommended an expanded U-boat program. In fact, the committee did not even invite Dönitz to participate. The committee relied on the study group, which assumed that:

> . . . the English counter-measures against U-boats, in particular detection, have reached an especially high standard. U-boats' attacks on English forces will therefore not be too successful. So long as no unrestricted U-boat war can be allowed, "cruiser war" against merchant ships—if it is *only* conducted by U-boats—will have a limited effect. It comes down to the fact that the single U-boat by its nature does not come into question for "cruiser war" on the high seas, but must be employed in a more or less stationary role.

One voice was raised in support of U-boats. Admiral Hermann Böhm, soon to take command of the Fleet, gave "U-boats and mines" first priority in a war with Britain, followed by "a series of raids against the vessels of the British strategic blockade." In any case, on January 29, 1939, the committee announced the "Z Plan" for expansion of the Navy. This was for the "next" war, which the planners assumed to be ten years in the future. Hitler cut that to six years. "For my political aims," he told Admiral Raeder, "I shall not need the Fleet before 1946."

The Z Plan was so ambitious as to be impossible; heavily weighted, of course, and no pun intended, with capital ships. There were not enough shipyards or steel mills or trained workers in all of Germany to meet the ten-year plan, let alone do it in six years. But perhaps thanks to Admiral Böhm, the plan did include 249 submarines, 162 to be oceangoing, the remainder to be small coastal types.

While the committee had been setting the numbers—and Hitler had been setting the agenda—Dönitz was working on tactics. He assumed that merchant shipping would be organized into escorted convoys, and ran an anti-convoy war game over the winter of 1938–39. From this, he derived *his* numbers, and in April 1939, asked that the shipbuilding plan be modified to provide a fleet of three hundred oceangoing boats (preferably the Type VII). This would allow for one hundred on station, one hundred in transit and one hundred in training or under repair. Dönitz hoped that these boats could be finished by mid-1942.

But Hitler was going to need the fleet much earlier than 1946. Germany had marched into Czechoslovakia in March; in April, Hitler told his commanders to plan for an invasion of Poland on or about September 1st. By this time, there were fifty-seven U-boats in commission, of which only thirty-eight were the oceangoing Type VII and Type IX. And while the leaders of the Navy did not expect that Britain would go to war over Poland (and they had Hitler's personal assurance that war with Britain would be avoided), they would take no chances. Sixteen Type VII and Type IX U-boats were sent out to take stations in the Atlantic beginning August 19th; on August 25th, fourteen Type IIs moved into the North Sea.

At 9 a.m., Sunday September 3rd, the British issued an ultimatum: Get out of Poland and we're only giving you two hours to make up your mind. By 11:15, Germany not having responded, the British Admiralty sent the signal to all forces, "Total Germany," and the naval war began.

Handed a copy of the intercepted message during his daily staff meeting, Dönitz was visibly shaken. "My God!" he was heard to say. "So it's war with England again!"

25 1939–1941—Germany

Rescue no one and take no one aboard. Do not concern yourselves with the ship's boats. Weather conditions and the proximity of land are of no account. Care only for your own boat and strive to achieve the next success as soon as possible! We must be hard in this war.

—*U-Boat Command Standing Order 154*

World War II began with a loud and astounding echo. At nine p.m. on the first day, Lieutenant Julius Lemp, commanding the Type VIIA *U-30* on station 250 miles northwest of Ireland, without warning sank the liner *Athenia.* The ship was en route from Liverpool to Montreal with 1,100 passengers; 112 were killed, including twenty-eight Americans. The German High Command was stunned. Prize rules were in effect and Lemp had acted contrary to his instructions. No matter, of course—to the world, it was 1915 all over again.

Because the German Admiralty did not know for certain that *U-30* was operating in the area of the sinking, and Lemp was maintaining radio silence, they promptly claimed that no German warship had been in the area, that the probable cause of the sinking was a British mine or a British torpedo and the probable motivation was to inflame American public opinion.

Nonetheless, the ground rules were reiterated in succesive orders to the U-boats at sea: Attack only merchant ships which were clearly participating in direct military support (under armed escort, for example, or carrying troops). One message was even more direct: "By order of the Führer passenger ships until further notice will not be attacked even if in convoy."

Lemp limped into port under a disciplinary cloud. He and his crew were sworn to secrecy, the U-boat's log was modified to eliminate any reference to *Athenia,* and records at headquarters were likewise modified; the two torpedoes fired into *Athenia* were logged as having been expended in the sinking of two small ships.

Soon, the German Ministry of Propaganda and Public Enlightenment, a particulary effective arm of the National Socialist movement created by and under the direction of Joseph Goebbels, had invented a story about a time bomb put aboard at the direct orders of Winston Churchill; on October 19th—three weeks after Lemp had returned to port and told his story—the German Press Chief gave an "absolute assurance" that no U-boat had been anywhere near *Athenia.* In a radio broadcast on October 22nd and in a newspaper interview the following day, Minister Goebbels firmly asserted that Churchill had been proven guilty. On January 2, 1940, Goebbels ordered his staff "to continue running the *Athenia* propaganda . . . bearing in mind the fundamental principle of all propaganda, i.e. the repetition of effective arguments." The German public did not learn the true story until after the war.

September was a busy month for the Propaganda Ministry. On the one hand they were firmly disavowing a real sinking; on the other hand, they were claiming one that didn't happen. A German torpedo hit, but did not seriously damage, the British aircraft carrier *Ark Royal* on September 14, 1939; twelve days later an aerial attack did little more damage, although the Luftwaffe reported that the ship had been destroyed. The German propaganda machine tried to turn minor damage into a confirmed sinking, while the High Command successfully worked through the neutral press to coax information from the British about the true condition of *Ark Royal.*

By September 24th, the High Command seemed to be drifting toward the old rules—or lack of rules. "Armed force," the latest order read, "should be used against all merchant ships using their wireless when ordered to stop. They are subject to seizure or sinking without exception."

Poland fell to the German Army a few days later at a speed that astonished even Hitler; astonished, and emboldened. If it was going to be this easy, there were other worlds to conquer. September was a very good month for the U-boats, taking out forty-one ships; another eight fell victim to German-planted mines. The highlight of the month was the sinking of the British aircraft carrier *Courageous,* on September 17th, by *U-29.*

On October 2nd, orders went out to sink without warning any ship sailing without lights; the U-boats were instructed to enter a note in the log that the sinking was "due to possible confusion with a warship or auxiliary cruiser."

On October 14th, there was another glorious success for the U-boats. *U-47,* Kapitänleutnant Günther Prien commanding, had pulled another stunt from the last war into the present—incredibly, penetrating the British fleet anchorage at Scapa Flow and sinking the battleship *Royal Oak.* The British assumed that he must have shadowed a British ship past the defenses and into the harbor; in truth, the route of entry and exit had been determined on a prewar reconnaissance by Prien, a "tourist on holiday," supplemented by the study of recent aerial photographs.

As Tirpitz had used the "services" of an American journalist to bolster the morale of the German people, so too did the Propaganda Ministry ensure that the exploits of the U-boatmen were given international attention. William L. Shirer, a CBS radio journalist based in Berlin, was invited to a press conference four days after Prien's escapade. As recorded by Shirer and published in 1942 in *Berlin Diary:*

> Captain Prien, commander of the submarine, came tripping into our press conference at the Propaganda Ministry this afternoon, followed by his crew—boys of eighteen, nineteen and twenty. Prien is thirty, clean-cut, cocky, a fanatical Nazi, and obviously capable.

The last semblance of respect for international law was withdrawn with Dönitz's Standing Order No. 154, issued in November:

> Rescue no one and take no one aboard. Do not concern yourselves with the ship's boats. Weather conditions and the proximity of land are of no account. Care only for your own boat and strive to achieve the next success as soon as possible! We must be hard in this war.

For the time being, because of the approaching winter and the shortage of boats to cover the vast but reasonably well-defended area around Britain, and in yet another echo of the early years of the Great War, the focus of U-boat operations was shifted to the Mediterranean. The first two months of the war had gone well enough: sixty-eight ships sunk, with the loss of seven U-boats.

Shirer was invited to the U-boat support ship at Kiel for Christmas interviews, a Christmas(!) celebrated by a triumphant U-boat force. “Surprising,” Shirer wrote, “with what ingenuity these tough little sailors had fixed up their dark hole—for that it was—for Christmas.” There was a tree, adorned with lights, and “fantastic Christmas exhibits” of the sort usually put in department store display windows: figure skaters, motivated by “a magnetic contraption.” A diaorama of the coastline of England. A miniature naval battle. Rum and tea and “case after case of Munich beer” were served, and “towards midnight everyone became a bit sentimental.”

The main point was made: “Impressive, though, the splendid morale of these submarine crews, and more impressive still the absolute lack of Prussian caste discipline. Around our table the officers and men seemed to be on an equal footing and to like it.”

“Ingenuity.” “Tough little sailors,” but “sentimental.” And of course, that “impressive . . . splendid morale.”

Dönitz was a past master at ensuring high morale; not only had he been there, done that, and suffered just as these new crews were suffering through the hardships of combat patrol, he knew how to treat them once they returned home. Home? Well, he made it a home of sorts. No crews remained aboard the boats in port; they “lived” on the support ships or in barracks. No crew returned from patrol without a welcome, often by Dönitz himself—now promoted to rear admiral. Combat awards and decorations were passed out as soon as possible, often within a day. No committees, no red tape, no bureaucratic recommendation-endorsement-approval. No Congressional oversight. “I regard this practice of immediate awards,” Dönitz noted, “. . . as psychologically important.” He was well aware that if the award was not “immediate,” the recipient might not be around long enough to enjoy the boost.

While Dönitz did not have a large enough force to permit pack tactics in 1940, he nonetheless experimented with methods of control. One early idea—that an on-scene commander, standing off from the battle, could best exercise control—fell early victim to reality. If the commander was to stand off, he couldn't *see* the battle, and in addition kept a valuable resource—an additional submarine—out of the action. If the on-scene commander couldn't see the battle, why could not the battle be commanded from Berlin?

Why not indeed? Dönitz took over tactical control of the forces in the field, not necessarily from Berlin, but from wherever he established his headquarters. With judicious use of radio communications, headquarters could be kept up to date on the general situation, including sightings of the enemy, position of forces, weather conditions, availability of other assets, as well as any information gathered from intelligence sources, to assist in the initial vector and instructions. The *actual* attack, however, was at the discretion, skill, and initiative of each individual U-boat commander.

There were many deficiencies in British readiness for war, but they were at least ready with convoy. Late in August, the Trade Divison of the Admiralty assumed control of all British merchant shipping, whether in convoy or independent sailers. The first convoy actually sailed before the war began, Gibraltar to Cape Town, on September 2nd; within a week, organized convoys were leaving the British Isles for Freetown and Halifax. Initially, escorts were provided only to and from a point about two hundred miles to the west; the escorts would then steam in circles until they met the next inbound convoy.

There were fast convoys and slow convoys, but ships sailing over fifteen knots and below nine knots were on their own. Of course, a fast sailer always had the option of slowing down to join a convoy; the slow sailers did not enjoy an option. As had been the case in World War I, there were stragglers who couldn't keep up and "rompers"—independent-minded ships (well, ships with independent-minded masters) which romped ahead of the convoy, often to their peril.

At first, an effort was made to keep all inter-ship communications secure, using coded wireless or visual signals such as flashing light. To thus send any message was cumbersome and time-consuming; the message had first to be encoded, then sent, then decoded at the other end, all by hand. To thus send a tactical message was dumb, and soon enough speed was given precedence over security.

The air forces available for convoy escort and antisubmarine patrol were largely obsolete, of limited range and capacity. It is ironic, but a recently purchased American-designed-and-built airplane, the Lockheed Hudson, gained the RAF's first air victory (October 8, 1939, shooting down a Dornier flying boat) and became the first airplane in history to capture a submarine (August 27, 1941); see page 295.

The shortage of suitable aircraft limited the patrols to convoy close support. It may have boosted the morale of the merchant sailors to see airplanes circling overhead, but vast stretches of water—likely being used by shadowing U-boats—went unpatrolled.

Nonetheless, the early months of convoy were a great success. From September through December 1939, 5,756 ships sailed in convoy; four were lost to U-boats. One factor in this success, not known to the British at the time, was a shortage of U-boats which prevented Dönitz from launching his pack tactics. This would change, but convoy operators had about one free year to practice.

In that interim, factions of the British Naval Staff were hard at work to emasculate the convoy—by taking away the escorts. If "convoy" was a learned-lesson, the "offensive" fallacy remained offensive—and a fallacy. In June 1938, the Tactical Division of the Admiralty had proposed the establishment of nineteen hunter-killer groups, each with five ships and supporting aircraft. When the war began, there were 101 ships suitable for this—or, alternatively, convoy escort duty. Had all nineteen hunter-killer groups been established, six ships would have been left for escort duty. The proposal was quickly forgotten.

Well, *that* proposal was forgotten, but not the burning desire to go out hunting. Despite a staff recommendation that "for the present, every antisubmarine vessel with sufficiently good sea-keeping qualities should be employed with convoys rather than dispersed in hunting units," the Admiralty could not let go of the idea of an independent flotilla which could work like a cavalry division on the approaches. The fox, you understand, never comes looking for you—you must mount up and go on the hunt. Of course, while you're off leaping over the hedgerows, the clever fox is pillaging the chicken coop.

From the very start of the war, U-boat commanders began to complain about the performance—rather, the *non*-performance—of their torpedoes.

On September 14th, *U-39* fired three torpedoes at the aircraft carrier *Ark Royal;* all exploded about eighty yards short of the target. Escorting destroyers jumped on the U-boat, which was forced to the surface by depth charges, and the crew was captured just before the boat sank. It was the first U-boat lost in the war; news of the misfires reached Dönitz via a prisoner-of-war grapevine.

On October 25th, *U-48* had five torpedoes misfire; on October 30th, *U56* fired three torpedoes against the British flagship *Nelson*—none exploded. On October 31st, *U-25* had four failures. There were others. Dönitz's War Diary noted, "*At least* 30 per cent of torpedoes are duds. They do not detonate or they detonate in the wrong place." Senior staff officers attributed the problem to nervousness, inexperience, or incompetence on the part of the submariners.

It got worse. Germany invaded Norway on April 9th; Allied forces launched a counter-invasion on the 15th. The North Sea was filled with targets and thirty-one U-boats—including six from the training school—were sent off to battle. One U-boat fired torpedoes against "a wall of ships" with only one explosion, and that from hitting a rock. "I cannot be expected to fight with a dummy rifle," complained Germany's leading submarine ace, Gunther Prien. The net results of one period of this campaign: four attacks on a battleship, fourteen on cruisers, ten on destroyers, and ten on transports—with one transport sunk.

Germany entered World War II with two basic torpedoes—the G7a and the G7e—developed as part of the secret submarine program of the 1920s and early 1930s. "G" meant seventh model; 7 meant seven meters long; the first subtype was "a" (Whitehead) or "e" electric. These were differentiated only by the method of propulsion. Each could be fitted with a variety of detonators and control systems.

The basic Whitehead torpedo was powered by a small alcohol-fueled steam turbine; oxygen necessary for combustion was supplied from a flask of compressed air. The torpedo could be given one of three speed/distance settings: as fast as fourty-four knots with a range of six thousand yards. The slowest speed gave twice the range. The Whitehead was an efficient machine with one serious drawback: It left a trail of bubbles which often could be spotted at great distance, giving the target sufficient warning to turn toward the track (to present the smallest target) and giving escorts and aircraft an arrow pointing toward the submarine.

An electric drive used current either from an installed battery or a generator powered by a spinning propeller at the nose. The World War II-era electric drive was only about half as capable as the steam turbine if measured on the speed/distance curves—maximum speed thirty knots, range six to seven kilometers—but as it left no track, it was greatly preferred by submariners. Germany began development of the first electric torpedo just at the end of World War I, and refined the design in the 1920s under the cover of a German-owned company in Sweden.

The depth-keeping mechanism of the German torpedo measured the difference between a reference, atmospheric pressure in an onboard reservoir, and the pressure of the water outside. A simple, elegant concept—the horizontal fins would move up or down as necessary to position the torpedo to maintain the predetermined pressure differential. However, the seal around the reservoir was not airtight; if the ambient pressure within the submarine was higher than atmospheric pressure—and this was almost always the case because of leaks from all of the systems which ran on compressed air—it would seep into the reservoir when the torpedo was opened for routine maintenance. These torpedoes ran deep. This defect was especially pronounced in the Norwegian campaign because the boats remained submerged for up to twenty hours a day.

A useful German innovation was a launching device that did not at the same time launch a large telltale bubble of air into the water. The Germans used compressed air to drive a piston which drove the torpedo out of the tube; however, the spent air was released into the living spaces, which increased the ambient pressure, which affected the depth-control setting.

German skippers had a choice of AZ (contact) or MZ (Magnet Zunder) detonators. They discovered that the AZ worked only with a square-on, 90-degree contact and would not detonate on a slightly angled hit. The Germans solved this problem by copying a detonator recovered from a captured British submarine in May 1940.

The MZ had been developed just at the end of World War I, but had never been refined. In theory, the torpedo would be set to run deep—on purpose—so that it would pass under the target. The pistol would be set off by the magnetic field of an iron hull just at the closest point of approach—right underneath—with the upward force of the blast thus breaking the back of the ship. In 1917–1918, the MZ had been too sensitive to local magnetic anomalies (such as seabed deposits of iron ore) or even sunspots. In 1940, it was still too sensitive and would detonate at the slightest provocation.

The amazing thing about these problems is that many people in the German Navy had known about them, all of them, for several years and had taken no action—or at least, no action adequate to correct the deficiencies before the war began.

There were plenty of government offices with responsibility for torpedoes. There was the Torpedo Experimental Institute, and the Torpedo Test Institute, and the Inspector of Torpedoes—enough bureaucracy, surely, to ensure control and therefore success.

By early 1937, the Experimental Institute knew that there were depth-control errors of as much as twelve feet; the recommended solution was a "depth spring." It was tested two times, and approved, but because of the various problems with industrial production then endemic in Germany, none of the springs were ready before January 1939—after a large number of unsprung torpedos had been shipped to the fleet.

Also in 1937, the third-highest scoring ace of the Great War, Max Valentiner, dropped by to remind the High Command of the failures of magnetic pistols in 1917–18. He was told about the great improvements that had been made in the years since, assured that there were no grounds to doubt the "absolute reliability" of the detonator, and sent on his way.

At the same time, whether or not anyone was willing to concede the fact to Valentiner, there were enough torpedo failures in the then-ongoing Spanish Civil War—in which German Naval units were clandestine participants—to spur the establishment of yet another unit, the Torpedo Trial Command. It ran some trials. In a two-week test with a torpedo boat, "the torpedoes were declared thoroughly unreliable both mechanically and from the point of view of maintaining the required depth."

The Torpedo Test Unit ran a test of its own in October 1938, using a destroyer. The captain reported, "Most torpedoes ran erratically as to depth [with] variations up to four metres." The Torpedo Test Unit took no action; it didn't even notify anyone further up the chain of command.

It was not until the failures of 1940 that all of this came to light. A vice admiral was dismissed; the rear admiral in charge of the Torpedo Test Unit was found guilty of dereliction by court-martial, along with two of his subordinates. He was sentenced to six months in prison. In the context of the time (and of the often harsh summary justice of the Third Reich) that amounted to the merest slap on the wrist.

The conquest of France was next on Hitler's list, and as the conquest of France would not require any particular support from U-boats, they were far down on his priority list. The Z Plan, of course, had been abandoned, although U-boat production continued—but at levels far below what the U-boat Command had been seeking. It was taking between nineteen and thirty months to deliver a boat, with an additional three or four months required for outfitting and training. Field Marshal Herman Göring was the nation's production czar; he and he alone determined priorities and the allocation of limited resources. The U-boat program of the Z Plan called for twenty-nine boats a month; by March 1940, that target was dropped to twenty-five, and actual monthly deliveries for the first six months slipped to two. That rose to six for the last half of the year, then to thirteen for the first half of 1941, and twenty boats a month for the last six months.

On the other hand, in May 1940, the German Air Force sank three times as much shipping as the U-boats. This anomaly was caused as much by the problems with the torpedoes as anything else, but nonetheless was a clear signal that the modern airplane could indeed play a major role in the war against seaborne commerce. Dönitz wished that Göring had a better appreciation of teamwork—the aircraft made available for working *with* the U-boats were hard to get and not the most effective when they arrived.

France signed an armistice with Germany on June 22nd, and the advance U-boat support team was on the way to ports on the shore of the Bay of Biscay. No channel choke points at the Dover Strait, no more extended northern passage to endure. By August, three French bases were in operation and the U-boats could now push farther out into the Atlantic than the escorts and air patrol then covered.

Thus came the time, the eighteen months between July 1940 and December 1941, which was memorialized by the U-boat flotillas as "the happy time," a time when it seemed as if they could sink anything that floated any time they tried and the score was more limited by U-boat endurance and weapons loading than anything they did or did not do. It was the time of the "great" aces of the Second World War—Gunther Prien, Otto Kretschmer, and Joachim Schepke, all commanding Type VIIBs—and public acclaim like no other. Even Wilhelm Bauer, who had not found favor or support in his homeland during his lifetime, was raised to the Pantheon.

Bauer was shown battling bureaucracy and professional instransigence in the feature movie *Geheimakte WB1* ("Secret File WB [Wilhem Bauer] 1"), and honored in a "biographical novel" (with *Brandtaucher* given a less-incendiary name, "The Iron Seal") as the true progenitor of today's heroes:

> One and the same will, one and the same goal-oriented power stand in the damp, dark, hulking space of the Iron Seal of 1850 and in the machine-pounding technosaur of 1940 which the Dreadnought so fears: both are peering and waiting into the future. They are struggling for the morrow. And thus one and the same line across a whole century runs through both of them: through the crew that today sails off into the rising dawn of the Germans, and through Corporal Wilhelm Bauer, the first man who dove into the twilight.

The rest of the world was watching: the newsreels recording; the newspapers reporting. But just as in the last war, when Arthur Conan Doyle had so clearly prophesied the possibilities, so too were the fiction writers ahead of the reporters in this war:

> He reminded Vance of a turkey gobbler as he strutted back and forth. "I, Ernst Oberdorf, am the man to bring America into the war!" he boasted. "My sinking of the *Vyatka* has already enraged the United States. It is but a matter of time now."
>
> Vance went cold. The man was mad—stark, raving mad. It showed in the German's glassy eyes. So this was the brutal devil who had sunk the cruise ship *Vyatka* without any warning!
>
> "Twenty-five torpedoes," Oberdorf gloated. "Twenty-five American ships! There's an excursion steamer leaving Miami Sunday for an afternoon cruise. On it will be 2,000 orphans . . ."
>
> "No!" croaked Vance . . . "You couldn't . . ."
>
> "Torpedo it," the Kommandant said calmly, "then shell the lifeboats. *That* will bring America into the war."
>
> "Why?" Vance wanted to know. "Why on earth should the Germans want us in the war? I thought you were moving heaven and earth to keep us *out!*"

"Ach! You don't see? It is simple. While you go to fight in Europe, then Japan strikes a lightning thrust on your West Coast, while Russia pours troops into Alaska." He rubbed his hands gleefully. "Later we divide the spoils."

—Alan Anderson, "U-Boat Rendezvous,"
Thrilling Spy Stories, Summer Issue, 1940.

26 1940–1941—America First

. . . when you see a rattlesnake poised to strike, you do not wait until he has struck before you crush him. These Nazi submarines and raiders are the rattlesnakes of the Atlantic. They are a menace to the free pathways of the high seas. They are a challenge to our sovereignty. . . .

—*Franklin D. Roosevelt, radio address, September 11, 1941*

Some policies of the United States of 1996 are still moderated by memories of Vietnam; Americans of the Depression years were much closer to the bitter baptism of World War I and wanted nothing to do with European rivalries.

In 1935, when Italy overwhelmed Abyssinia (Ethiopia), when Fascists and Communists were squared off in Spain and Germany was beginning openly to rearm, the U.S. Congress made "hands off" the law of the land. No loans or credits or shipments of arms to belligerents; no more *Lusitanias* or *Arabics,* because U.S. citizens could not travel on the ships of belligerents. A proposed Amendment to the Constitution would have required a nationwide referendum before any declaration of war. Isolation was the name of the day; keep the European disease in Europe.

By November 1939, reality was beginning to intrude and the prohibition against weapons mongering was repealed in favor of cash and carry. Any nation with the cash could carry armaments away by whatever means was available to them. In 1940, this allowed Great Britain to carry away, for example, 500,000 rifles, eighty thousand machine guns and 130,000,000 rounds of small-arms ammunition.

But on May 15, 1940, even before the Happy Times began, before the French bases had widened the U-boat range and the wolf pack had found its bite, newly appointed Prime Minister Winston Churchill asked "former naval person" President Franklin D. Roosevelt for the "loan" of "forty or fifty of your older destroyers to bridge the gap between what we have now and the large new construction we put in hand at the beginning of the war. This time next year we shall have plenty."

Churchill was also casting a nervous glance over his shoulder: "But if in the interval Italy comes in against us with another hundred submarines we may be strained to the breaking point." The interval was very brief. Not quite a month later, Italy did indeed come into the war, bringing 105 submarines to the Mediterranean theater—countered by just over eighty British and French antisubmarine vessels (and the French could be counted on for only about a week).

If there was an Italian contribution, it was barely evident. A flotilla of twenty-seven Italian submarines joined the German forces at Bordeaux in the fall of 1940, where, as a period of orientation to Atlantic operations, they were put on a sort of picket duty. In slightly over a month and a half, the twenty-seven boats did not provide a single useful contact for the U-boats, and sank but one ship themselves.

Churchill made another appeal on June 11th, followed by an even stronger request on July 31st: "Mr. President, with great respect I must tell you that in the long history of the world this is a thing to do *now.*"

The isolationist sentiment in the Congress was as strong as any back in 1917, but politics is the art of compromise and persuasion. Within a month, an arrangement had been made to exchange fifty World War I American destroyers—on loan—for the lease of naval bases in Newfoundland, Bermuda, British Guiana, and the West Indies. (In 1995, the U.S. Base Realignment and Closure Committee turned Naval Air Station Bermuda, the last remaining vestige, back to local control.)

The first eight of the lend-lease "flush deck and four pipe" destroyers were turned over to the British in Halifax, Nova Scotia, four days after the deal had been ratified by the Senate. As ships go, these jerry-built, twenty-year-old destroyers were not in the best of condition and they were not well adapted to the conditions in the North Atlantic, but they could make thirty five knots on a good day, and after they were brought up to "contemporary" antisubmarine standards with sufficient modifications in armament and equipment, they did what they had been asked to do and the best that could be expected: They filled in.

On August 17th, 1940, Hitler formally declared a total blockade of the British Isles—as if the world was not already aware. On September 6th, the first true "wolf pack" went into operation. The Germans intercepted a radio signal which designated the point of rendezvous for a convoy and escort. Four U-boats were sent off in search, found the convoy, and chased along with it for four days, making surface attacks at night, sinking five of the fifty-three ships in the convoy. Four of the ships were sunk by a single U-boat—Prien's *U-47*—but that does not invalidate the tactic. In World War I the simple fact of "convoy" kept the U-boats at bay, but in the early months of *this* war the shortage of escorts, the lack of escort coordination (they do not appear to even have talked together, let alone trained together, before meeting up in mid-ocean), the absence of any tactical plan—all contributed to a continuing success for the wolf packs.

From July through September, the French-based U-boats averaged about fifty ships and 252,926 tons a month. By October, they were picking off more than two ships a day—the best month of the first year of the war.

One example suffices. A convoy was spotted on October 16th by one U-boat, and six more were quickly vectored in by headquarters. The three escorts were on station more than six miles apart, leaving a wide path for U-boat penetration. After the first sinking, one of the escorts detached herself from the convoy and went off searching; she did not find the U-boat, nor did she find the convoy again. Two more escorts arrived—along with an additional U-boat.

By dawn on October 19th, it had all come together—or fallen apart. Of the thirty-five ships in the convoy, seventeen had been sunk. Three of the U-boats left the scene and headed home, having fired all of their torpedoes. Five others remained in the vicinity to be rewarded by the sight of another convoy the very next day; they sank fourteen of forty-nine ships, and later the same night, yet *another* convoy hove into view. There were seven more victims by dawn, a total of thirty-eight in three days, without any corresponding U-boat losses. Total tonnage for the month: 352,407. This with not more than nine or ten U-boats operational.

One key element in the early success of the U-boat fleet was the insistence on training, training and more training. A new boat was not given a green crew and a quick shakedown and sent off to dubious battle; a new boat and crew might first work together through seven months of non-combat operations in the Baltic.

However, the number of boats now available fell off. Those which had been on the line in October moved back home for a brief rest and to rearm; some boats were being assigned to training duties for crews for the newly constructed boats which soon would be coming on-line. From November through January, there were only three boats on station. As Dönitz was to note, "The war against Great Britain, the mighty sea-power and our principal adversary, was being waged by 120 to 240 men."

It may not yet have been apparent to the Germans, but the most valuable addition to the ASW war was just then coming into play: ASV I ("Air to Surface Vessel") radar. Not many units were available at first, enough to equip only twelve out of 226 antisubmarine patrol aircraft, but before long an improved radar was on hand, and by December 1940, surface-running submarines could be picked up at seven miles.

At the end of 1940, the big-ship champions in the German Navy were no less dense than their British counterparts. Despite the clear evidence of U-boat and Luftwaffe success against merchant shipping, a staff study called for more battleships:

> The Naval Staff holds that the salient lesson of the war to date should be the recognition that the remarkable development and performance of the Luftwaffe, and the achievements of the U-boat and mine-laying campaigns, have done nothing to undermine the importance of the capital ship. . . . The main protagonist in the war against the enemy's ocean communications is the battleship itself.

Of course, at the time, Germany *had* no operational battleships; every battleship on the list either was not yet finished or was undergoing repairs; the only major warship ready for sea was a cruiser. It was acknowledged, therefore, that the U-boat would have the duty until new battleships could join the fleet. When some of the big ships were operational—and participated in several notable wide-ranging cruises—they (combined with merchant ships fitted out as surface raiders) still only managed to sink about half as many ships as the handful of U-boats then in service. From November 1940 through March 1941, surface ships sank ninety ships totaling 472,381 tons; the U-boats took out 170 ships, 925,788 tons.

On February 9, 1941, the ideal was realized—one time. In the only combined air, surface ship, and U-boat operation in the Atlantic, *U-37* spotted a convoy and sank three ships; air strikes called in by the U-boat sank another five; the *Admiral Hipper,* also summoned by *U-37,* arrived late, but nonetheless sank a straggler.

In March, the Allies hit their own "ideal"—sinking or capturing, in a quick series of actions, the leading U-boat aces of the war. Gunter Prien, in *U-47,* was sunk by the depth charges of a convoy escort; a ten-second-long red glow was seen, deep in the water, followed by pieces of submarine bobbing to the surface. Schepke was killed in the sinking of *U-100*—among the first U-boats to be spotted on the surface by newly installed radar. Otto Kretschmer, in *U-99,* was captured when his boat sank. Two less experienced commanders were victims of the same series of actions between March 7th and March 23rd. U-boat Command was certain that the British had introduced some new, secret weapon—but except for the contribution of radar to one kill, there was nothing new, and there was no common denominator in these losses. Just, perhaps, bad luck.

German preoccupation with surface raiders should have ended on May 27th, when the battleship *Bismarck* was sunk in open battle by the British Navy. However, German hopes were pinned on *Scharnhorst, Gneisenau* and *Prinz Eugen*—which were pinned down in the harbor at Brest and kept under relatively heavy air attack for about nine months. In the broad scheme of things, the attacks were not very effective. In one eight-week period, Bomber Command flew 1,161 sorties against the heavily camouflaged ships at Brest—with four hits. However, after each wave of bombers, the shipyard workers would work to repair whatever damage had been done to the yards and the warships—workers who otherwise would have been assigned to U-boat repair and maintenance.

In his frustration, Dönitz sent an impassioned plea to Commander in Chief Grand Admiral Erich Raeder:

> In view of the urgent need of maintenance personnel for the U-boat arm, U-boat command is of the opinion that the whole question of repair of battleships and cruisers . . . should be re-examined . . . We are in conflict with the two strongest maritime powers in the world, who dominate the Atlantic, the decisive theatre of the war at sea. The thrusts made by our surface vessels into this theatre were operations of the greatest boldness. But now, principally as a result of the help being given to Britain by the USA, the time for such exploits is over, and the results which might be achieved do not justify the risks involved. . . . the invevitable and only logical conclusion must be that they are not longer of major importance to the prosecution of the war as a whole. That being so, maintenance personnel, which is most urgently needed for the vital U-boat arm, should no longer be wasted on repairing battleships and cruisers.

The memorandum had no effect. Repairs continued. Finally, on February 11, 1942, the three ships with heavy escort and Luftwaffe cover made a dash up the Channel. Within two weeks, each was heavily damaged—one by mines, one by a torpedo, one by air attack—but repairs were now shifted to another set of shipyard workers, and the men of Brest were at last "released" for U-boat maintenance.

The role of intelligence is often like divining the future from symbolic tea leaves. The British maintained two command centers—one in the Admiralty and one in Liverpool—with duplicate plotting boards, on which were displayed the location of every convoy, escort, air patrol, rescue ship, and reported U-boat. The information was updated at least every four hours.

The information, of course, came from a variety of sources but among the most useful were the U-boats themselves, which too freely exchanged intelligence of their own by radio. The transmissions usually were brief, vectors to help other U-boats intercept a sighted convoy. However, each transmission would result in a direction-finder line of position (or a fix if several D/F stations picked up the signal).

Until another transmission, or an attack, or a sighting by surface ship or aircraft, that U-boat would for all practical purposes disappear, destination unknown. But certain things could be assumed. The length of time a U-boat could remain on station was reasonably finite (unless, of course, all weapons had been expended in quick action); individual boats often could be identified by radio signature or sightings or whatever and their movements, toward the end of a patrol, predicted—heading for home. Or, suppose that such-and-such a U-boat sent in a lengthy dispatch; lengthy dispatches were unusual, and usually meant that she was either reporting her kills and headed for home, or reporting damage and headed home. In any of these circumstances, surface traffic could be routed away from her probable track.

The United States moved closer to war on April 11, 1941, when the boundary of the designated "U.S. Security Zone" was moved from 60 degrees west longitude to 26 degrees west. That line was devised when President Roosevelt and his Secretary of War, Henry L. Stimson, took out an atlas, picked a point midway between the bulges of Africa and Brazil, and converted that to longitude. In a stroke, about 80 percent of the Atlantic Ocean was made part of the Western Hemisphere. The line, in truth, was more than 2,300 miles from New York, but only 740 miles from Lisbon. Within this zone, U.S. Navy ships would track—but not directly attack—German ships and U-boats. In the same month, the U.S. opened the "loaner" base at Bermuda and closed down some German weather stations in Greenland.

The first U.S. casualty of the U-boat war was the freighter *Robin Moor,* en route to South Africa, sunk by *U-69* on May 21st.

The commingling of American, Canadian, and British escort ships posed a dilemma for the Germans. On June 21st, under orders from Hitler to avoid any incident which might incite the United States, Dönitz signaled his forces "til further orders" to restrict all attacks "to cruisers, battleships and aircraft carriers and then only when identified beyond doubt as hostile."

This order was preciptated by two unrelated events. The first was *U-203*'s misguided attempt to attack the American battleship *Texas* on June 20th—without result and even without notice by the intended victim. The second was Hitler's planned attack on the Soviet Union, which began June 22nd.

"Operation Barbarossa" was to prove the major German blunder of the war, absorbing resources, consuming energies, and—for the U-boat Command—realigning priorities away from the Battle of the Atlantic to the great Eurasian land mass. U-boats were diverted—temporarily, as they found no targets—to the Baltic and the Arctic, and sinkings in the Atlantic fell from June's third-highest monthly total of the war to date to less than one-third that score in July and about one-fourth for August.

However, by August the restrictions against attacking escorts had not only been lifted, escorts were now given priority as targets. But at this very time, in another major blunder of his war and in an effort to protect the supply lines to his forces in North Africa, Hitler ordered a shift of U-boats from the Atlantic to the Mediterranean. Other than to attack those few British warships which were on the prowl, it was not clear how a submarine could "protect" surface ships; the principal threat was from Allied submarine and air attacks. Hitler ignored the protestations of both Raeder and Dönitz on this point. The first of the transferred U-boats achieved some success—against warships, with the sinking of the carrier *Ark Royal,* a battleship, and a cruiser; then, under the (mistaken) impression that Italy was threatened by a rumored invasion of North Africa by British and Free French forces, the Naval High Command ordered the U-boat Command to consider the Mediterranean as the main theater of operations. Dönitz was directed to move *all* operational U-boats to the Mediterranean and the Atlantic approaches to Gibraltar.

John Terraine offers what is perhaps the best single explanation for the shift from probable victory to certain defeat in German fortunes:

> This situation well illustrates that has been called "the extraordinary amateurishness of the German war effort." "Germanic efficiency" withered under the Nazi touch: Göring, as economic overlord or as head of the *Luftwaffe,* was a disruptive disaster. Hitler, as head of the Armed Forces, was the same: ignorant, dogmatic, suspicious, operating not by principles but by "intuition," he in fact reduced military policy to gambler's throws, expedients and quackeries of which the diversion of the U-boats to the Mediterranean is a perfect example.

A caution here. While denigrating the Nazi leadership, we must resist the temptation to endow Dönitz with any more insight or exceptional powers than his peers. History does show that, had his early prescription for concentration on merchant traffic been followed, the war might have been won in the Atlantic before American interference could have had any effect. History, however, is often the accumulation of accidental occurrences into what seems like some foreordained conclusion. Dönitz was running *his* war with a staff of half-a-dozen young assistants, none of whom had the education, training, or experience to provide serious analysis of the always changing problems of U-boat operations—and Dönitz spent too much time playing operational commander, reading and sending messages to the U-boats, and not enough time as the force commander in chief. He should have been working for more supplies, more aircraft, more submarines—not directing day-to-day tactics.

Under an agreement reached between Roosevelt and Churchill in August, the U.S. Navy would assume full, not shared, escort responsibilities for fast convoys in the Western Atlantic as far east as Iceland; the Canadians would escort the slow convoys; the British would be free to concentrate their forces in home waters.

Before those arrangements had taken effect, *U-652* fired two torpedoes at the American destroyer *Greer* inside the "U.S. Security Zone." They missed, but this September 4th incident gave Roosevelt the opportunity to denounce the attack, in his radio address of September 11th, as "piracy legally and morally":

> Hitler knows that he must get control of the seas. He must first destroy the bridge of ships which we are building across the Atlantic and over which we shall continue to roll the implements of war to help destroy him, to destroy all his works in the end. He must wipe out our patrol on sea and in the air if he is to do it. . . . when you see a rattlesnake poised to strike, you do not wait until he has struck before you crush him. These Nazi submarines and raiders are the rattlesnakes of the Atlantic. They are a menace to the free pathways of the high seas. They are a challenge to our sovereignty. . . .
>
> From now on, if German or Italian vessels of war enter the waters, the protection of which is necessary for American defense, they do so at their peril.

Public opinion in the U.S. was running about 70 percent in favor of neutrality. In his radio chat, the President did not mention that *Greer* had assisted a British patrol plane in tracking and attacking the submarine and that, once the first torpedo had been fired, *Greer* herself counterattacked with depth charges.

Not long after, an American destroyer was hit by a torpedo (*Kearney* survived, but eleven sailors were killed); two weeks after that, the destroyer *Reuben James* became the first American warship casualty. Only forty-five men of a crew of 160 survived. The folk song said it all:

> Tell me, what were their names,
> Tell me, what were their names,
> Did you have a friend
> On the good *Reuben James*?

The U.S. Congress voted to moderate the Neutrality Act. In the meantime, trouble was brewing in another part of the world entirely. On October 16th, a relatively benign Japanese government was toppled by a fiercely aggressive military group. Roosevelt remarked, "I simply have not got enough Navy to go round."

27 1941–1942—America, at Last

. . . merchant shipping will be sunk without warning with the intention of killing as many of the crew as possible. Once it gets around that most of the seamen are lost in the sinkings, the Americans will have great difficulty in enlisting new people.

—Adolf Hitler, conversation with the Japanese Ambassador, January 1942

Whereas in World War I the presence of "escort" had been sufficient to ensure convoy safety, the increasingly aggressive pack tactics of World War II brought results, often startling success, for the Germans. In one running battle along the coast of Greenland in September 1941, sixteen ships, 24 percent of the total in the convoy, were sunk—despite the eventual presence of as many as eleven escorts. "Eventual" is the key word; the convoy started with only four, the others joining along the way but after battle was well joined.

Another convoy in the same month lost seven of the twelve escorted merchantmen. A third convoy in the same month had nine ships lost out of twenty-five. The lesson for the Allies: The mere *presence* of escorts no longer counted for much, and weak escorts were, well, weak. Heavier ships were better than light ships—more room for ASW equipment and weapons, more room for fuel and ammunition, more room for crew. The larger the crew, the more efficient the watch-on, watch-off operation, the more people on hand for necessary but non-combat chores such as cooking.

All crews needed more training. Signal crews who barely had learned the Morse code in a classroom were suddenly called to read dimly flashing lights amidst sleet and spray, all the while standing on a deck heaving up and down forty or fifty feet once, twice a minute.

At this stage of the war, aircraft were valuable—vital—for feeding information into the intelligence grid, and for the "scarecrow" effect which forced the U-boats to keep their heads down and consume vital battery power. As an offensive *weapon,* the aircraft were not effective. In the first twenty-six months of the war, British Coastal Command could claim only three kills shared with surface craft, one U-boat sunk by aircraft alone—and the rather freakish capture in August 1941 of *U-570.* The submarine was brought to the surface by bombing, and surrendered. To an airplane. *U-570* had a green crew, the ideal training regimen now having slipped in a world where the need for operational boats was critical. Not only was she captured (physically, after a surface ship arrived to take over), but *U-570* was "transferred" to the Royal Navy and commissioned as *Graph,* where she served until wrecked off the west coast of Scotland in March 1944.

Notwithstanding the low kill-rate of the last war, the value of maritime patrol aircraft had been amply demonstrated and was no less valuable in this one—but the bureaucrats who controlled the distribution of assets seem not to have noticed. The Coastal Command would have liked more of the new long-range aircraft which were coming into service. However, the service the planes were coming into was the Air Force, the Bomber Command, which had its own ideas about the beneficial employment of scarce assets. In August, the Deputy Chief of the Air Staff, Air Marshal Sir Arthur Harris (known to history, not necessarily as a compliment, as "Bomber Harris"), unwittingly added a new capability to the submarine's portfolio when he applauded the refusal of the Air Ministry to transfer new long-range bombers to coastal patrol; in that event, he asserted, "twenty U-boats and a few Focke-Wulf in the Atlantic would have provided the efficient antiaircraft defence of all Germany."

Let Britain bleed to death. The bombers must be free to bomb! Suffice to say that the postwar Strategic Bombing Survey showed conclusively that

Bomber Harris and his American counterparts more or less bombed out; they killed a lot of civilians ("unrestricted warfare" is not limited to the sea!), but only had marginal impact on German industrial capability. Long-range aircraft, assigned to antisubmarine patrol as early as possible and in meaningful numbers, might have—*would* have—ended the U-boat threat two years before it was finally throttled.

The Royal Navy exercised initiative—clever, but not effective—by rigging some merchant ships to launch a shore-type fighter for a dismal four-hour mission at sea. When the fuel ran out, the pilot was expected to ditch his plane close to the mother ship. The plane would sink; the pilot would be hoisted aboard and readied for another mission. The pilots, we may assume, were not highly motivated. This expedient lasted but a few months.

A more useful innovation was the small escort or "jeep" carrier, which first went on patrol in September 1941 with six Grumman fighters: F-4-F Wildcats in the USN, Marlets for the British. These were tough little birds, and immediately earned their keep by taking on marauding German long-range Condor aircraft. One Condor was shot down on the first cruise.

In the meantime (1940–1941), the Germans had constructed virtually bomb-proof submarine shelters in the Atlantic ports—with reinforced concrete roofs sixteen feet thick. These were proof against all attacks. Ironically, the British had watched as the pens were being built and could have interfered—had the Bomber Command not been wrapped up in mission priorities. The Bomber Command was opposed to any diversion from what it viewed as its "proper" role: attacks on ball-bearing plants and runs against civilian morale. (As we now know, other sources for ball-bearings were easily found, and civilian morale was strengthened, not weakened, by the bombing.)

In the fall of 1941, the convoys seemed to disappear from the ocean. They had to be *somewhere,* but they weren't being spotted by the U-boats. Why not? Perhaps the convoys were being tipped off to the U-boat locations. How? Well, HF/DF intercepts had proved to be of some value in the last war, but back then, operations were relatively close inshore. In *this* war, the Germans assumed, operations were too spread out, too distant, for HF/DF tracking to be accurate. This "invisibility" had to be temporary. Reasoning thus, the Germans urged "caution," but continued to exercise command and control by radio from headquarters. It was a big mistake.

From June 1941 until February 1942, the Allies not only could track U-boats in real time using HF/DF, but were in some measure able to read the message traffic itself and thus track them to their destinations. This was one of the big secrets of the war, not revealed until many years later—indeed, not until after Karl Dönitz had died. The Allies had developed a mechanism, a machine, for breaking the principal German operational code, Enigma.

And for the whole of World War II, the Germans—certainly the U-boat Command—did not realize that many of their secrets were wide open. *Even though* there were questions, unanswered, about failed operations. *Even though* they had themselves been able to break into Allied encrypted traffic. *Even though* the basic system had been available on the open market, for encryption of commercial messages, since the mid-1930s.

Even so—Engima should have been unbreakable; it was a most elegantly simple device with an elegantly complex method of operation. The machine looks like a large, clumsy, old-fashioned typewriter in which the type-heads have been replaced by three alphabetic rotors. By following the current "key list," the operator prepared the machine for encoding (or decoding) by switching the internal wiring by which each key on the typewriter keyboard became a different letter, then selecting the correct three rotors for the day from the kit of eight and repositioning the outer ring on each to match the key list setting, and finally, putting each rotor in the machine with a different starting point. When any key was pressed, the signal was routed through each rotor, and changed each time, before a letter was printed on a strip of paper.

An unbreakable system—just look at the odds: eight wheels from which three would be used, 336 possible combinations; twenty-six ring settings for each of three rotors, 17,576 possible combinations; plug connections (1,547 possible); and rotor settings—that is, on which letter of the alphabet did the rotor start (another 17,576 possible combinations). As many as 160 trillion *possible* combinations to produce any one letter. An unbreakable system. Until it was broken.

Working with the commercial device as a pattern, with the assistance of two Polish engineers who recreated a rudimentary Enigma machine from memory, and with a few captured rotors and some typical key lists, a group of British mathematicians, chess players, and eccentrics established a code-breaking project, code name Ultra. They created a machine which would test every possible combination of settings, one after another after another, looking for that combination which would produce a key test word. When successful, the machine would stop, the setting would be recorded, and messages transmitted around the same time would be fed in for decoding.

At one point, the team was aided by the capture of a complete machine, taken off just before the sinking of *U-110*. The commander was Julius Lemp—the officer who sank the first ship of the war. There was a report that Lemp allowed himself to go down with his boat, committing suicide when he realized that he was responsible for the loss of Engima. Lacking evidence, we'll assume that Lemp drowned along with other members of his crew, and didn't have much chance to think about Enigma, one way or the other.

Suspicious—but not convinced—that there may be *some* problem with the codes, the German Navy added a fourth rotor (and renamed the system Triton) in February 1942. This brought on an extended, but not eternal, codebreaking blackout. Persistent diligence once again paid off.

The Germans used other codes and codings; some were broken early in the game and remained relatively unchanged throughout the war: the Tetis cipher, used by U-boats in the training squadron, and Hydra, used in home waters (including Norwegian and French areas). For its part, Germany had occasional success in breaking the British Naval code, reading as much as 50 percent of the traffic during the Norwegian campaign, but in the early days of the war, even though the Germans had captured a copy of the Merchant Navy Code, merchant ships were pretty good at maintaining effective radio silence. But the British Navy, just like the German, tended to hang on to a "successful" code too long.

There is belief, in some quarters, that the Allies won the war because of Ultra. A caution: Just because you know the enemy's plans doesn't mean you'll win the war; you may be able to set some traps or avoid some confrontations, but you won't win a battle, much less a war, unless your military forces have adequate training, equipment, and motivation.

The Japanese attacked the American Naval base at Pearl Harbor on December 7, 1941, with the intention of striking a near-fatal blow on the American fleet—accelerated attrition, if you will. They may also have hoped to force the U.S. to stay out of the way of their Far Eastern conquest. They did not succeed on either count.

The battle fleet was damaged, seriously, but in good time all ships were back in service except for two obsolete battleships—*Arizona*, sunk at her moorings, and *Oklahoma*, which sank while under tow back to the West Coast for repairs. Another "victim" of the attack was the already out-of-commission *Utah.* The carriers were all at sea; the submarines were overlooked—as were the machine shops and some 4,500,000 barrels of fuel oil stored nearby.

Perhaps the two major effects of the attack were: first, to coalesce American public opinion as never before, and second, to force the U.S. Navy to abandon the ingrained, worldwide, time-honored fascination with battleships and shift the burden to the new-generation warships: the aircraft carriers and the submarines.

Japanese submarines participated in the attack, but without effect. Twenty-five I-class boats had been put on station around the islands. They did not see any American warships, although one Japanese boat was herself spotted and sunk by airplanes from *Enterprise.* Five HA-midgets, carried to the area aboard I-boats, attempted to penetrate the harbor before the air attacks began, but achieved nothing but their own destruction. One became the first casualty of this war. She was sunk by gunfire and a depth-charge attack from the venerable World War I destroyer *Ward*—the ship with which the Mare Island Naval Shipyard had set a speed record in 1918, sixteen-and-a-half days, keel-laying to launch.

Ward sent a flash dispatch, "We have attacked, fired upon, and dropped depth charges upon a submarine operating in defensive sea area." The message reached senior commanders at Pearl Harbor—who asked that *Ward* be contacted for verification. False alarms had been the order of the day.

The United States was thus thrust into a war for which it was far from ready, despite a recent and rapid buildup of forces and equipment. However, Roosevelt and Churchill had made some plans against this day, most notably that, in the event of war, the first priority for the Allies would be the defeat of Germany. Unless the German threat was removed, it didn't make any difference what happened in the Far East.

In September 1941, Dönitz had prepared to greet the eventual American entry into the war with a *Paukenshlag,* or "roll of drums." He would immediately dispatch a flock of U-boats to the East Coast of the United States, where he was certain they would find rich and unprotected pickings. But when he asked for authority to shift some assets to American waters, he was turned down.

Hitler had a wary eye on Norway, expecting an Allied invasion, and put twenty U-boats in Norwegian waters; he refused to release the thirty-two U-boats already assigned to the Mediterranean. Twenty-nine boats were in repair and maintenance and eleven were in transit to or from the other Atlantic operating areas. So, once again, perhaps the most critical element of the war came up short just at a time when it might have been at its most effective.

Dönitz sent what he had: seven Type VIIs to Newfoundland, where the weather limited their success, and five larger boats, Type IX and XC, to U.S. waters—where they found the whole East Coast lit up like Times Square on New Year's Eve: no blackouts; all navigational aids fully aiding; all ships sailing with normal lights. Among other delights, it was tourist season in Miami and the war was three thousand miles away. Because of the northward-flowing Gulf Stream a few miles offshore, ships heading south stayed well inshore. They presented a magnificent black silhouette against a miles-long stretch of the glowing Florida skyline.

The U-boats loved it, no doubt, sitting out the day submerged, coming to the surface at night and picking off the fat happy targets trooping by, with often as many as ten in sight at the same time. There were twenty-three sinkings off the East Coast in nineteen days in January; thirty-one in February, forty-eight in March—all in American coastal waters. March was an exceptional month worldwide: 273 ships, 834,164 tons. The coastal communities did not go under blackout until April 18th. Hitler's birthday.

Some writers have been highly critical of the slow start made by the leaders of the U.S. Navy, and especially of the delay in adopting convoy along the East Coast and in Caribbean waters. They should widen their aim to take in a lot more people—or try better to understand what America was like in 1941. The United States *was not* prepared for war; isolationist sentiments had continued at such a high pitch that the Selective Service Act of 1940 was extended in 1941 by a *one*-vote margin in the House of Representatives. The U.S. Army learned close-order drill with broomsticks because there weren't enough rifles to go around—and those which were in the inventory were obsolete Model 1903 Springfields. The total strength of the U.S. Armed Forces in 1940—while Europe was erupting in flames—was about the same as the level maintained by the U.S. Navy alone in most every *peacetime* year since the Korean War. We should resist the temptation to judge the state of readiness of 1941 by the standards of the Cold War years.

The resources immediately available for convoy escort were few to none. Admiral King resisted organizing convoys with inadequate escort; he believed that was an invitation to attack. As a first step, beginning in April, the U.S. Navy set up a sort of "bucket brigade" tactic: The convoys would be handed off from one group of escorts to another for daytime steaming, then put into guarded harbors or anchorages at night when the U-boat menace was greatest.

To most quickly create an escort and patrol force, the U.S. revisited World War I and mobilized a fleet of fishing boats and yachts. These were soon enough augmented by sixty-seven escorts which had been built in slightly more than a month—American industry to the rescue!—and a couple of hundred minesweepers which were pressed into temporary escort duties. By May, enough escorts were on hand to allow the establishment of coastal convoys.

Army Chief of Staff General George C. Marshall sent Admiral King an anguished memo on June 19th; "anguished" as in "I am fearful that another month or so of [losses by submarines] will so cripple our means of transport that we will be unable to bring sufficient men and planes to bear against the enemy in critical theatres to exercise a determining influence on the war." Admiral King offered reassurance ("the situation is not hopeless"), and acknowledged that he had now embraced convoy: "Escort is not just *one* way of handling the submarine menace; it is the *only* way that gives any promise of success. The so-called patrol and hunting operations have time and again proved futile."

Japanese forces swept across Asia, taking Singapore, the Philippines, and Burma before six months had passed; the Germans took a version of *blitzkrieg* to North Africa. Expansion of the war expanded the theaters in which the Allies were forced to operate, and even though all agreed to put the near-term focus on the defeat of Germany, the Pacific couldn't be abandoned to the Japanese.

Thus, the U.S. Navy shifted focus to the Pacific, and U.S. escorts began to disappear from the Atlantic (to the chagrin of the British and the consternation of the Canadians, who had been thrust into a big war with a minimum naval presence).

About this time, the airplane came into its own as a U-boat killer. It had been a hard birth. The airdropped bombs which had been developed between the wars (and put into "service" without testing) were useless. They would detonate on contact with the water, or take an erratic direction once in the water, or not detonate at all and certainly not do much—if any—damage unless scoring a direct hit.

Then, after a flurry of development (some of which had been in progress for years), radar, the radar altimeter, the aircraft searchlight, and improved depth charges all arrived.

Guides for their proper operations and employment were developed by newly formed Operations Research teams, which brought statistical analysis to the war: What were the best settings for airdropped depth charges? (Initially fifty feet, later changed to twenty-five feet.) How close together should they be spaced? (Initially sixty feet, later increased to one hundred feet.) Once a U-boat had disappeared from sight, how far ahead along the track should the charges be placed? If the U-boat had disappeared for more than thirty seconds, don't bother. "Only those attacks made on a still visible U-boat or within 15 seconds of disappearance resulted in damage or destruction." In other words, there was about forty-five seconds from the time the U-boat spotted the airplane to get the weapon in the water. Beyond that, attack was a waste of time and weapons.

Some of this was guesswork, some was based on interviews with air and ship crew members, some on interviews with captured U-boat crews, some on intercepted and decoded radio traffic. Before long, aircraft were accounting for 50 percent of all U-boat sinkings.

The Operations Research teams also determined that convoy losses were proportional to the number of U-boats and/or the number and quality of escorts, but appeared to be independent of the size of the convoy. Ergo, they recommended that the size of the convoys be increased, thereby reducing the numbers of convoys sailing and reducing the losses. However, practical considerations kept size of convoys well below infinity.

On January 3, 1942, Hitler told the Japanese Ambassador that he was not too concerned with American entry into the war; the Americans might be able to build ships quickly, but would not be able to find the crews to take them to sea:

> For that reason, merchant shipping will be sunk without warning with the intention of killing as many of the crew as possible. Once it gets around that most of the seamen are lost in the sinkings, the Americans will have great difficulty in enlisting new people. The training of

> seagoing personnel takes a long time. We are fighting for our existence and cannot therefore take a humanitarian viewpoint. For this reason I must give the order that since foreign seamen cannot be taken prisoner, and in most cases this is not possible on the open sea, the U-boats are to surface after torpedoing and shoot up the lifeboats.

There is no record that any orders to "shoot up the lifeboats" were ever issued, but the philosophy was clear. Sink ships without mercy, anytime, anywhere: "The enemy's shipping constitutes one single, great entity," as Dönitz put in his War Diary, 15 April 1942. "It is therefore immaterial where a ship is sunk. Once it has been destroyed it has to be replaced by a new ship; and that's that." The key was to sink them better. Dönitz met with Hitler on May 14th to explain the advantages of an improved magnetic pistol; it would "accelerate the sinking of torpedoed ships" with "the great advantage that in consequence of the very rapid sinking of the torpedoed ship the crew will no longer be able to be rescued. This greater loss of ships' crews will doubtless aggravate the manning difficulties for the great American building programme."

This was drifting perilously close to the rocks and shoals of international outrage, if not law. Soon enough, Dönitz would issue specific orders to the U-boats which helped to bring him before the Nuremberg War Crimes Tribunal and almost put his neck in the noose.

On September 12, 1942, *U-156* torpedoed and sank the large, armed troopship *Laconia* just off the west coast of Africa. (*Laconia* was the second ship to bear that name; ironically, the first had been torpedoed and sunk in World War I.) The attack was proper, under the "rules." However, it turned out that in addition to some eight hundred Allied servicemen and eighty women and children, *Laconia* was carrying 1,800 captured Italians. This became apparent to U-boat Commander Werner Hartenstein when he heard cries for help in Italian. He started picking up survivors and at the same time requested instructions from headquarters. Dönitz gave tepid approval, ordered *U-506, U-507,* and the Italian submarine *Capellini* to assist, and called on the Vichy French at Dakar to send ships.

U-156 took on 193 survivors and Hartenstein sent a clear-text message in English apprising any and all listeners of the situation and promising that unless he was attacked, he would not interfere with rescue efforts by any other nation. The Allies regarded the message as a hoax, to lure ships into ambush, and no rescuers were sent.

Hartenstein continued his own rescue work, organizing rafts and lifeboats for two more days; *U-506* and *U-507* arrived on the 15th, the Italian submarine on the 16th, and took on many of the survivors. *U-156,* towing a string of lifeboats and with a large jury-rigged Red Cross flag draped over the forward gun, started toward safe haven—and was attacked by an American B-24 operating out of a new, secret air base which had been carved into the lava on Ascension Island.

Communications between the British area commander and the garrison on Ascension Island were, well, inexact. The Americans were notified on September 15th that *Laconia* had *just* been sunk—three days after the actual event—but were not told of the rescue effort, or of Hartenstein's call for assistance and safe passage. The base supported limited-endurance B-25 bombers; when sent out to the limits of their range to search for *Laconia* survivors, they found nothing. One long-range B-24D, which had been diverted to Ascension for emergency repairs, was requisitioned to help in the search.

At 0930 on September 16th, B-24 pilot Lieutenant James Harden spotted a submarine which was then taking four lifeboats under tow. Harden circled overhead, puzzled by the scene below, and radioed back to base for instructions. He challenged the submarine for identification, with uncertain response: No one in his crew could read the signal reply.

The dilemma posed to officials back on the island was clear, if not easily resolved. There were no Allied submarines in the area; therefore the spotted boat was most certainly German. If it was attacked, some *Laconia* survivors would be put at risk. If it was allowed to escape—apparent rescue attempt or no—the submarine would remain a clear threat to Allied shipping in the area, and many more Allied lives would be in jeopardy.

After timely deliberation (the B-24 could not remain on station much longer) the pilot was told, "Sink sub." He dropped some depth charges and reported, "The sub rolled over and was last seen bottom up. Crew had abandoned sub and taken to surrounding boats." As it turned out, the pilot—on his first combat mission—was mistaken. One depth charge had fallen in among the lifeboats, one of which capsized and became the "rolled over" sub. *U-156* was damaged, but not sunk.

The affair made Dönitz very nervous, and shortly after midnight on the 17th he sent an order to *U-506* and *U-507*:

> Boats will at all times be kept in instant readiness to dive and must retain at all times full powers of underwater action. You will therefore transfer to the lifeboats any survivors you have aboard. Only Italians may be retained aboard your boats.

Later in the day, he added:

> Do not hoist Red Cross flag, since (i) that is not a recognized international procedure; (ii) it will not in any case, and most certainly not as far as the British are concerned, afford any protection.

That same day, the same B-24 attacked *U-507* (carrying more than one hundred *Laconia* survivors) without significant effect. That evening, *U-507* transferred her passengers to a Vichy French sloop and passed along the approximate position of two groups of lifeboats. All told, more than one thousand *Laconia* survivors were rescued. The B-24 itself did not long survive that mission, crashing a month later en route to rejoin the squadron from which it had separated in September. The crew was not injured.

Dönitz had another message to send—the most damning of the war. To all commanding officers:

> All attempts to rescue members of ships sunk, therefore also fishing out swimmers and putting them into lifeboats, righting capsized lifeboats, handing out provisions and water, have to cease. Rescue contradicts the most fundamental demands of war for the annihilation of enemy ships and crews. . . . Orders for bringing back Captains and Chief Engineers remain in force. . . . Only save shipwrecked survivors if statements are of importance for the boat. . . . Be hard. Think of the fact that the enemy in his bombing attacks on German towns has no regard for women and children.

We must note that, at just the same time, the German propagandists were showing quite a different image of the U-boat war. In a clear reprise of the first war's *Die Magische Gurtle*, the feature film *U-Boote Westwärts!* ("U-Boats

Westwards!")—produced with the full support and cooperation of Admiral Dönitz—was reassuring viewers of the selfless honor and integrity of the day's gallant U-boat commanders. They follow prize rules; surprise attacks are only directed against ships in convoy; death is not death, but something transcendent. "It is sweet and honourable," a dying young officer affirms with his last breath, "to die for the Fatherland."

28 1941–1945—The Pacific War

Any nation that attempts commerce destruction by submarines will tend toward certain of the same practices that the Germans arrived at; how far it will go depends on its racial characteristics and, very likely, by how hard it is pressed.

—*Captain Thomas C. Hart,* Director of Submarines, 1920

On December 31, 1941, Admiral Chester Nimitz took command of the U.S. Pacific Fleet—on the deck of the submarine *Grayling.* It was a moment of mixed symbols: One, the pioneering submariner had now risen to the most important operational job in his navy; two, the traditional battle fleet was *hors de combat* and only those naval stepchildren of the 20th century, the submarine and the aircraft carrier, were left. *Grayling* was chosen because the carriers were at sea.

In what was a startling break with long-standing American policy, submariners had been authorized to attack any and all targets the day the war began. No restrictions. No "prize rules." The nation went to war in 1917 in large part because of the unrestricted U-boat war; now what?

The official *Instructions for the Navy of the United States Governing Maritime and Aerial Warfare,* issued shortly before the war, held the clear statement:

> In their action with regard to merchant ships, submarines must conform to the rules of International Law to which surface vessels are subject.
>
> In particular, except in the case of persistent refusal to stop on being duly summoned, or of active resistance to visit or search, a warship, whether surface vessel or submarine, may not sink or render incapable of navigation a merchant vessel without having first placed passengers, crew and ship's papers in a place of safety.

However, during the Naval War College 1941 version of the annual war game against Japan (annual, at least since the cruise of the Great White Fleet, 1907–09) the players, taking note of the manner in which submarines were then being used in the European war, recommended that, in the event of war, the waters of the Far East should be declared as "war zones which all merchant ships would enter at their peril."

The General Board took the high road. It reminded the Secretary of the Navy of long-standing U.S. policy and asserted that "war zones" had no standing in international law.

The General Board was living in a world fast slipping by. The rest of the Navy was making plans to take just such a step, and the officer most directly affected was the commander of the Asiatic Fleet, Admiral Thomas C. Hart.

As a captain, Hart had been the American submariner sent to coordinate European operations in 1917, and in 1920—then detailed to Washington as "Director of Submarines"—told a Naval War College audience:

> I shall pass over the unhumane features of German submarine warfare because their ways were characteristic of the race. Any nation that attempts commerce destruction by submarines will tend toward certain of the same practices that the Germans arrived at; how far it will go depends on its racial characteristics and, very likely, by how hard it is pressed.

Another submariner was about to be hoist on his own petard. He now was "pressed" hard enough to want the authority, and in November 1941 was privately advised that it would likely be forthcoming.

When Chief of Naval Operations Admiral Harold Stark called the President to report the attack of December 7th, Roosevelt authorized immediate implementation of the then-standing war plan for the Pacific. Admiral Stark made certain the President knew that unrestricted submarine warfare had become part of the plan; by dinnertime, the Pacific forces were ordered to "Execute Unrestricted Air and Submarine Warfare Against Japan."

As a practical matter, those Americans who were concerned about this bold reversal, in the face of so many years of anti-German vituperation, allowed themselves to define *all* Japanese shipping as being in the service of the military and in support of the war effort, thus not "merchant vessels." Specious at best; hypocrisy at worst; reality in the end.

Having the authority was quite a different matter from having the capability. It would be several years before the submariners began to be even reasonably successful. On December 7, 1941, the U.S. Navy had 111 submarines in commission—sixty in the Atlantic, fifty-one split between Pearl Harbor and Manila (soon moved to Australia) in the Pacific. As with any other mature submarine fleet, they comprised a mixed bag of age and capability; the older World War I S-boats were used largely for training, but some were on distant station and others were quickly pressed into combat service. The class which was to be the mainstay of the war—*Gato* and successors—had not yet entered service. There was no common doctrine or strategy from one fleet to the next, which caused some problems—but not as many as the limited prewar training and the shortage and frequent malfunction of torpedoes. Malfunction? Non-function. Sound familiar? Reread chapter 25, or wait a few paragraphs.

The Pacific submariners had their first opportunity to score the first day of the war. Twenty-eight submarines of the Asiatic Fleet were in defensive posture around the Philippines. More submarines than the entire German U-boat fleet at the beginning of World War I; indeed, more submarines than had ever been assembled for one battle at the same time. They might as well have been in San Diego.

A Japanese invasion force which included seventy-six loaded transports and supply ships approached Lingayen Gulf on December 21st. It was opposed by seven American submarines—which between them sank one freighter and a minelayer. For the whole of the losing three-week Philippine campaign, U.S. boats fired ninety-six torpedoes in forty-five attacks (on average, somewhat less than two attacks per submarine). Total score: three Japanese ships. All of the American submarines but one (damaged in an air attack and scuttled) were safe and sound and shifted to Australia.

Prewar training had centered on the "fleet" mission: Find the enemy's battle fleet, seek out and destroy warships. Since warships, by their very nature, are prepared to resist destruction, and battle fleets—by their very nature—were by this time designed to resist penetration by submarines, it's not surprising that "caution" was at the heart of American submarine doctrine. "It is bad practice and is contrary to submarine doctrine," noted the Report of Gunnery Exercises 1940–41, "to conduct an attack at periscope depth when aircraft are known to be in the vicinity. . . ." Keep your periscope down in calm seas and any time within five hundred miles of an enemy airfield; stay deep; fire on sonar target data. Of course, sonar data proved to be relatively useless.

The physical conditions in the submarine training areas pushed caution even more. The waters were often so clear and the sonar conditions so fine that the surface forces always had the advantage. As a result, the average submarine commander at the beginning of the war was cautious to a fault, and unprepared for the aggressive, daring tactics which the war would require. In fact, in the first year of the war, about 30 percent of submarine commanders were relieved for poor performance.

The problems with torpedoes were more significant. A timid commander could be relieved, but unless there was an assured supply of weapons which worked, it made no difference how bold his replacement was.

America entered the war with two basic torpedo models:—the 21-inch Mark X, with a TNT warhead of 497 pounds and a thirty-six-knot range of 3,500 yards, and the newer dual-speed Mark XIV with a Torpex warhead of 650 pounds, a dual magnetic/contact pistol, and a thirty-six-knot range of nine thousand yards, or 4,500 yards at forty-six knots. However—another example of how America 1941 backed into the war—torpedo production was limited to sixty a month; for all of 1942, even under wartime pressures, total production was 2,382. The submarine commanders, already too cautious, were cautioned not to waste their precious ammunition; for the year, they shot 2,010.

But it soon developed that their precious ammunition was just as bad, and shared many of the same problems, as the previous year's German models. They ran too deep; the magnetic exploder was erratic; the contact exploder did not work at all angles of contact.

Isaac Newton warned us: Bodies at rest tend to remain at rest. So too with entrenched bureaucracies. It was precisely the German experience. The problem was most certainly not with the equipment; it must therefore be with the unskilled or self-serving operators. When complaints from the fleet reached the Navy's Bureau of Ordnance, it was so confident (or arrogant) that it would not even run tests.

So the operators in the field took that chore upon themselves. For a test on depth settings, held in June 1942, they fired torpedoes into nets suspended from booms; the holes poked through showed that the torpedoes ran on average eleven feet deeper than set. As they discovered, the Mark XIV had never been tested with a live warhead; too expensive. Just as the British had to learn in World War I and the Germans had to learn earlier in World War II, water-filled exercise warheads at neutral buoyancy were not of the same weight as those with dense explosive warheads.

Next, when the submariners set their torpedos to run shallow, they still did not work. They wouldn't explode.

Unlike the German fish which had two separate exploders (AZ and MZ), only one of which could be fitted at a time, the MK XIV had a dual contact/magnetic pistol. The Naval Torpedo Station at Newport had developed the Mark VI exploder in such secrecy that it had been tested only once—for fear that the secret would get out—and that in Narragansett Bay, where the magnetic conditions were quite different from those in the operating areas of the Pacific Ocean. In addition, it soon was apparent that the Mark VI would generate its own magnetic field while trying to maintain the set depth. When that fact was uncovered, in one of those classic issues of command and control, Pacific Fleet Commander Nimitz told his boats to get rid of the Mark VI magnetic trigger, but the submarine force commander in Australia—who had been one of the originators of the Mark VI, back before the war—refused to give it up until ordered to do so almost a year later.

The U.S. submarine score for 1942 was 180 ships, 725,000 tons (about equal to a monthly U-boat total). The Japanese replaced 635,000 tons in the same period. It was going to be a long war.

Once the problems with the magnetic trigger had been solved—by removing the trigger—the submariners discovered that the contact trigger did not work very well either. On one patrol in the fall of 1943, *Halibut* expended twenty-three torpedoes without magnetic trigger. Only one exploded. Many of those were visually tracked to the targets, and three were known to hit but to no effect. Well, *almost* no effect. At least one ship was sunk when torpedoes simply punched a hole in her rusting hull plates.

Another set of tests. First, torpedoes were fired against a cliff wall until there was a dud, and the failed warhead was recovered by divers. (*That* would have been a challenging mission!) Examination showed that the firing pin had not traveled far enough to pop the fulminate-of-mercury firing cap. Then, using a ninety-foot-high crane, dummy warheads fitted with live exploders were dropped onto a steel plate, and the problem was uncovered. The firing pin would jam against the guide rails on a head-on, perfect shot. The pistol worked just fine with a less-accurate, glancing blow. This was exactly the reverse of the German experience! Before all the problems had been acknowledged—and solved, two years into the war—almost four thousand torpedoes had been fired against the enemy.

The abysmal American performance in the first two years did nothing to offset entrenched attitudes among her enemies. The Japanese shared the German opinion of American submariners: They were too enamored of air-conditioned luxury to ever be a threat. The Japanese did not even establish a dedicated antisubmarine "Escort Command" until November 1943—and even then, it had a very low priority in the assignment of effective warships.

Nor did they see any reason to change their basic submarine doctrine and strategy, and here they differed from the Germans. The battle fleet remained the threat and the target; merchant commerce might be attacked if the submarines didn't have anything better to do. In an ocean war, where the only support for Allied forces would come by sea, the Japanese were transfixed by a handful of obsolete British battleships in Southeast Asia—which posed no threat—and never developed a useful plan for the employment of their submarine forces.

This was not immediately apparent to American commanders, especially when Japanese submarines soon began to appear off the West Coast of the United States. Over a four-month period, December through March, nine I-boats went in (unsuccessful) search of anticipated reinforcments headed for Hawaii.

The material impact of these operations was negligible. No warships were attacked (only one seems even to have been spotted); perhaps five merchant ships were sunk, another five damaged. The true impact was psychological, as Americans on the West Coast were suddenly brought face-to-face (so to speak) with an enemy at the shoreline. On February 23, 1942, *I-17* spent twenty minutes lobbing shells—with minor effect—at an oil derrick and pier near Santa Monica. The Japanese Naval Information Department asserted that the "bombardment" had "unnerved the entire Pacific coast." They were right. "Americans know," Radio Tokyo proclaimed on March 3rd, "that the submarine shelling of the Pacific coast was a warning to the nation that the paradise created by George Washington is on the verge of destruction."

One small victory for the American submarine force grew out of this excursion: their first Japanese warship. Three of the I-boats returning from patrol off California exchanged radio messages so often that their progress was tracked, predicted—and *I-173* was intercepted and sunk by *Gudgeon.*

Another weak pass on the mainland came in June when five more I-boats were sent to the Pacific Coast; they were relieved on station a few weeks later by another seven, which remained until the middle of July. In total, these twelve submarines sank only three cargo ships, although they repeated the psychological blow of Santa Barbara by shelling a radio station in Vancouver on June 7th and Astoria, Oregon on June 20th.

There was one last run on the U.S. mainland. In a most tepid retaliation against the April 1942 carrier-launched B-25 raid on Tokyo (led by Brigadier General Jimmy Doolittle), the Japanese sent the seaplane-carrier *I-25* to attack the Pacific Coast. The airplanes made two runs against the forests of Oregon, September 9th and September 29th, and dropped a total of four 170-pound incediary bombs—only one of which worked, starting a small fire. *I-25* also sank two tankers and a cargo ship—and a Russian submarine, in a case of mistaken identity. Russia and Japan were not at war (in fact, Russia did not declare war on Japan until two days after the first atomic bomb annihilated Hiroshima), but Russia seems not to have known the truth about the loss of *L-16.*

The lackluster West Coast performance notwithstanding, the Japanese submarine force was encouraged by some remarkable victories in the early months of the war. Twenty-five submarines participated in the invasion of the Dutch East Indies, sank perhaps forty cargo ships, and damaged six more; only two submarines were lost. Submarines on picket duty reported the locations of the British battleship *Prince of Wales* and the cruiser *Repulse*—both of which were sunk the next day by air attacks.

At the Battle of Midway at the end of May, the Japanese *I-168* sank the already crippled carrier *Yorktown* and a destroyer alongside; tit for tat, the American *Nautilus* gave the coup de grâce to the crippled Japanese carrier *Soryu.* (Another American submarine was indirectly responsible for the sinking of a cruiser. Two Japanese cruisers collided while maneuvering to avoid contact with the American submarine *Tambor;* aircraft later sank the crippled *Mikuma* and sufficiently damaged *Mogami* that she was out of action for a year.)

On May 30, 1942, the World War I British battleship *Ramilles* was put out of action for a year by a midget submarine, which one hour later sank a tanker. It was the highest accomplishment of any Japanese midget—almost the *only* accomplishment of any—during the war, although a day later another midget did launch two torpedos at the Spanish American War-era American cruiser *Chicago* in Sydney harbor. They missed, but one sank a ferry boat.

And then on September 13th came perhaps the single most spectacular, albeit unplanned, event of any submarine war: *I-19,* commanded by Takaichi Kinashi, sank the American aircraft carrier *Wasp* with three torpedoes of a spread of six; the other three torpedoes continued for *twelve miles* into another task group, where one caused fatal damage to the destroyer *O'Brien* and another sent the battleship *North Carolina* to the shipyard for two months. The sixth torpedo steamed on into the unknown.

Through much of the war, the Indian Ocean and adjacent Australian waters were a happy hunting ground for the Japanese submarines—when they were allowed to go hunting. There were merchant targets aplenty and few Allied warships, fewer still with any ASW skills. In six months beginning in the middle of May 1942, Japanese submarines sank at least forty-two cargo ships and damaged eleven. Only one was lost: *I-28,* torpedoed by the American submarine *Tautog.*

This unqualified success against commercial shipping encouraged some Naval officers to plan full-scale "unrestricted submarine warfare" against the United States; however, before this radical idea could be given serious consideration, the tide of battle had turned. Submarines were pulled from their secondary mission of commerce raiding and put into their primary mission of fleet support, and then and for the rest of the war, the submarine force was whipsawed to meet immediate tactical needs—patrol this area, protect that landing force, supply some distant island garrison—while strategic possibilities went untouched.

Realism began to intrude with the Allied invasions of Guadalcanal and Tulagi, beginning August 7, 1942. Twenty Japanese submarines were put under the control of the fleet commander, who shifted them around in tactical formations like units on a drill field and with as little effect. They started as a patrol line one hundred miles to the north, then were shifted two hundred miles in another direction, then—chasing Allied formations sighted by aircraft—moved three hundred miles to yet another area. From August through October, the submarine force scored one minor victory: damage to the carrier *Saratoga,* forced to withdraw to Pearl Harbor for repairs. Eight midgets were launched against convoys; one damaged a transport, all were lost. One warship was sunk for the whole of November, *Juneau,* by *I-26* the morning of November 13th.

On November 16th, as the situation deteriorated—more and more troops surrounded, trapped without supplies—submarines were pulled out of fleet support and put in support of the Army. They carried in more than 1,100 tons of cargo. They took troops into the combat zone; they brought them back out again. There was no apparent plan and little coordination. Eventually, more than two thousand troops were safely pulled out by submarines.

By February, the Battle for Guadalcanal had ended and the Army garrison at Lae, in eastern New Guinea, was under siege. The submarines were pressed into another major supply effort, forty-eight sorties moving one thousand soldiers and 1,400 tons of supplies.

Submarines were once again brought in under central tactical control to oppose Allied landings on Tarawa in November 1943, and the controllers appear not to have learned anything from the Guadalcanal experience. The boats were constantly shifted from one picket line to another; six of nine assigned submarines were sunk, against a score of one Allied ship.

Again, during the invasion of Saipan in June 1944, fourteen of twenty-one deployed submarines were lost. No Allied ships were even damaged by submarines. Eleven Japanese submarines took part in the defense of Okinawa; eight were sunk.

These lopsided scores reflected both Japanese tunnel vision and Allied ASW skill. The Japanese boats had been designed for attack, with little thought having been given to defense. They lacked effective electronic sensors, were too slow and too sluggish when submerged, and many could not operate below three hundred feet. When deployed on the picket line, they aligned themselves with military precision, each thus giving searching ships a marker for the others. They were no match for trained and aggressive ASW forces; over a twelve-day period in May 1944, one hedgehog-equipped destroyer escort alone, *England,* sank six Japanese submarines. An appreciative Admiral King sent the message "There'll always be an *England* in the United States Navy."

Well . . . it was a wonderful play on the words of the song, but an unenforceable promise. The last *England* was decommissioned a few years ago, and in the reality of today's shrinking fleet—where ship's names are chosen more for politics than tradition—there are no immediate plans for a replacement.

Operational experience fed the wartime evolution of the American submarine. The *Gato* boats, which had entered service in December 1941, had a relatively massive superstructure; on a bright, clear, star-filled night the boats stood out like the targets of the U-cruisers against the Miami skyline. Accordingly, the fairwaters of existing boats were cut down radically, leaving only the ribs needed to support the periscopes and twin open decks—a machine gun platform aft, a small open bridge forward.

As already noted, operating depth was increased to four hundred feet (with collapse depth estimated at nine hundred feet) in the *Balao* class, commissioned February 1943; changes in the *Tench* class included the conversion of an external main ballast tank to fuel storage, adding twenty thousand gallons and four thousand miles at ten knots. To preserve the weight/volume balance, the weight of consumed fuel had to be replaced with seawater flowing in behind; however, since salt water is about 18 percent heavier than diesel fuel, this tank would never be full; therefore, it had to be strengthened to withstand water pressure. The standard surface displacement of *Tench* was, therefore, about forty-five tons greater than *Gato/Balao.*

There were some moves toward the development of cargo-carrying submarines; in 1942, Simon Lake was still pursuing his favorite scheme, and shared his thoughts with the Congress. President Roosevelt was intrigued with the idea, and the Navy explored the feasibility of refitting some prewar boats no longer qualified for combat. That plan was rejected—the boats were notorious "leakers" of fuel, which made them too easily discovered by the enemy, and perhaps more to the point, cargo capacity would be too limited. There was not enough need to warrant embarking on a totally new design.

By the middle of 1943, the American submarine program was cut to seven per month; there was not much need for new boats. A year later, fifty nine contracts were canceled, and no new contracts were issued for the rest of the war. The production of diesel engines was shifted to landing craft.

But improvements continued. Until 1944, reduction gears transferred the power of the high-speed motors to the propeller shaft; the gears were a major source of noise. BuShips developed a slow, variable-speed direct-drive motor. The first set was installed on *Sea Owl, SS 405,* in service July 1944; conversion of older boats continued until well after the war.

The average diving time for the submarine of 1941 was fifty seconds; with a larger down-express tank, more and larger limber holes in the superstructure (to let trapped air escape more quickly), and improved arrangements for tanks valves, that had been reduced to forty seconds (and some boats could submerge in thirty seconds) by 1945.

Not bedeviled by air patrols and free to operate extensively on the surface, U.S. boats had more opportunities to use their deck guns than any of the Axis submarines. They went to war with a single 3-inch/.50-caliber gun aft of the conning tower; this proved to be an awkward postion, so the guns were moved forward, beginning early in 1942. However, a 3-inch/.50 is not a very potent gun; by mid-1943, 5-inch/.25s were being installed on some boats, and became the standard for all new construction in 1944. By the end of the war, submarines were being given two 5-inch/.25s and a Mark 6 surface-ship gunfire computer system. The commerce raider in the classic form—with no commerce left to raid.

Most wartime Japanese submarine developments were expedients, not improvements. There was a 3,512-ton cargo-carrier for the relief of beleaguered garrisons. It was of little consequence. There were eleven submarine landing ships (each of which could carry two amphibious landing craft, one hundred soldiers, and eighty-five tons of cargo). The Japanese tried a submarine-towed underwater cargo container; it was too difficult to manage. The Japanese *Army* built twenty-eight cargo submarines on its own, without Naval assistance; no one seems to know what happened to them.

There were three high-speed submarines, similar in size and capability to the late-war German Type XXI (page 329) and able to manage submerged bursts to nineteen knots. They did not make it into operational service. There were some other anti-ASW efforts, also matching German developments—notably, the use of anechoic hull coatings of synthetic rubber to absorb, rather than reflect, sonar impulses.

The most impressive development—if you will allow that sheer size is impressive—was the *I-400* class. Japan had comissioned its first aircraft-carrying submarine in 1932: the 2,243-ton, 320-foot *I-5* equipped with one reconaissance floatplane. Over the next twelve years, another twenty-eight submarine carriers were built in several versions, each an improvement over the other. Evolution culminated in the 5,223-ton *I-400* class, each of which carried three dive-bomber seaplanes and was designed for attacks against the Panama Canal and the West Coast of the United States. Eighteen were planned in 1942; only two, *I-400* and *I-402,* were finished by war's end. They were teamed with two of the smaller submarine aircraft carriers, *I-13* and *I-14,* each with two airplanes, and sent off to do battle in July 1945. *I-13* was sunk on July 16th; the war ended before the other units saw useful action.

In another echo of the last war, submarines were pressed into service to carry critical cargo to a blockaded Germany. She needed materials such as tungsten, tin, and rubber which Japanese-held areas had in abundance; in exchange, the Japanese wanted advanced technology. Blockade-running surface ships found the transits between Germany and the Far East to be increasingly difficult; in June 1942, *I-30* was dispatched on a trial cargo run to Germany. Her early August arrival in France was personally attended by Grand Admiral Raeder; her return voyage was uneventful . . . until she hit a mine outside of Singapore, and sank. Most of the crew was rescued; most of the high-priority cargo was lost.

On April 23, 1943, *I-29* made an Indian Ocean rendezvous with *U-180.* The German submarine was carrying technical data on rockets and jet engines and an Indian revolutionary leader who wanted to join forces with the Japanese; the Japanese submarine carried three of the latest-model torpedoes, two boxes of gold for the Japanese embassy in Berlin, and two engineers who were being sent to study U-boat construction. The cargoes were safely swapped during the ten-hour meeting and transported on without incident.

Four round-trip missions were undertaken in 1943 and 1944; *I-8* had the only complete success, taking a Japanese submarine crew for training in Germany and carrying critical cargo and technical advisors both ways. The crew was to bring a U-boat back to Japan—one of two Hitler transferred as his "personal" gifts to Emperor Hirohito. The first, *U-511,* left Germany in April 1943 with a German crew and a Japanese vice admiral as a passenger, then sailed from Penang to Kure with a Japanese crew, arriving August 7th. *U-1224,* with the newly trained Japanese crew, sailed from Kiel on March 30, 1944, and was sunk by an American destroyer on May 13th.

I-29's voyage was an almost-success; she made it to France and back as far as Singapore, with radar, twenty Engima machines, and plans for bombsights, superchargers, and the Messerschmidt Me 262 turbo-jet. Some of this material was taken on to Japan from Singapore by air; the rest went down when an intelligence intercept sent three America submarines to set a trap.

If these efforts accomplished nothing else, they served as rather remarkable examples of the endurance of the Japanese boats and their crews: *I-8,* sixty-one days from Penang to Brest and seventy-eight days from Brest to Kure; *I-29,* a round trip to and from Singapore lasting from December 16, 1943 to July 14, 1944.

The Allied ASW effort had ended the Battle of the Atlantic, although the Germans continued a desultory effort. The submarine battle of the Pacific ended because the targets disappeared. American efforts at breaking the Japanese Naval code rivaled the Ultra effort in Europe; to the later confusion of historians, the Pacific-based Americans came to call their program the "Japanese Ultra." In a mirror image of the European theater, the Americans in the Pacific could plot merchant convoys in advance—no need for long open-ocean hunting expeditions. The Americans tried their own version of the wolf pack, the "Coordinated Submarine Attack Group," although it wasn't much needed.

By the middle of 1944, perhaps 140 American submarines were operating in the Pacific; effectiveness had increased to such a level by the end of the year that for that year alone the Japanese lost more than six hundred ships and 2.7 million tons. This was almost half the total of the prewar merchant fleet and more than the cumulative total for 1941, 1942, and 1943. By the end of 1944, the few straggling merchant ships stayed close inshore; the warships steamed at top speed, heavily guarded. American submarines spent much of their time picking off barges and fishing boats. They spent their most profitable time on picket duty—to rescue downed aviators; four submarines were put in place for each B-29 raid on Japan. For 1943, sixty-four days on rescue station, forty-three pilots rescued; 1944, 469 days on station, 117 rescues; 1945, 2,739 days on station, 380 rescues.

By the middle of 1944, the Japanese had no more than twenty-six fully operational submarines, and desperation drove them to desperate measures. The most spectacular—and most effective—were the *kamikaze* ("divine wind") suicide aircraft, which had great impact on the ships of the Allied forces. But the Japanese could not build airplanes fast enough for the *kamikaze* missions—more than four thousand were launched—and they put great hopes in the parallel development of an underwater version, the *kaiten* ("revolution") suicide torpedo. This was the large, twenty-four-inch Type 93 surface-ship version of the Long Lance, modified to provide a control compartment for the pilot—locked in for a one-way voyage which was destined to last not more than five hours, no matter what. With speeds up to forty knots, *kaiten* were considerably faster than any submarine—indeed, faster than any surface ship; control was difficult, training accidents were common, but enough of the problems were solved that the units became operational in November.

The *kaiten* were carried into battle aboard I-boats, and the first attacks were against ships anchored in Ulithi Atoll in November 1944. After another mission in January, the Japanese claimed four aircraft carriers, three battleships, ten tranports, and one tanker at a net cost of twenty-two *kaiten.* The actual score, against the loss of two *I-boats* and thirty-two *kaiten:* the tanker *Mississinewa* and possibly a small landing ship sunk, and two transports damaged. Several attempts at Iwo Jima in February and March produced no result, nor did efforts off Okinawa in April. However, more and more submarines were fitted to carry *kaiten,* which made a number of open-ocean attacks toward the end of the war. The record is ambiguous. One destroyer escort may have been fatally damaged, but actual results are unknown.

One ominous note: the number of suicide weapons which Japan had assembled against the anticipated invasion of the home islands. In addition to as many as 180 *kaiten,* the Japanese were prepared with about the same number of two-man midget submarines (with another five hundred five-man "coastal defense" versions under construction) and perhaps two thousand explosive-laden motorboats. They had also trained a large group of divers, who would be stationed off the landing beaches and provided with hand-held spar torpedoes.

29 1942–1945—The Battle of the Atlantic

. . . if [the bomber offensive] is reduced to lesser proportions by further diversions of large numbers of bomber aircraft for seagoing defensive duties, it will fail in its object. . . . This in my opinion would be a far greater disaster than the sinking of a few extra merchant ships each week.

—Air Field Marshal Sir Arthur Harris, March 1943

Dönitz became Commander in Chief of the German Navy on January 30, 1943, relieving Admiral Raeder, who had resigned that post over a disagreement with Hitler and finished the war as the Navy's Inspector General. The disagreement: What to do with the heavy surface ships? They weren't going anywhere, and Hitler wanted their guns de-mounted and installed in shore defenses. Surprisingly, after a month in his new job, Dönitz took up Raeder's position: To scrap the big ships would send a signal to the British that they had "won" the surface war—great for the British morale, a disaster for the German. In addition, the scrapping operation would require more scarce shipyard workers and facilities than keeping the ships in limited active service. Hitler backed down.

The new four-rotor Triton code, now in operation, had not been broken by the Allies; the U-boats were again running into convoys all over the ocean. They made their targets in May and June; by the beginning of August, more than 350 submarines were in the water (although many were not ready for service) and wolf packs of up to twelve boats were sweeping the convoy routes, sometimes three at a time.

To extend U-boat range, Dönitz had established seagoing refueling stations: Type XIV, a boat of 1700 tons with the capacity for seven hundred tons of fuel and extra supplies rather than armaments, designed for mid-ocean rendezvous. Dubbed *milchküe* ("milk cows"), one of these boats could keep a dozen Type VIIs at sea for another month, or five Type IXs for two months. The first of the *milchküe* was on station in July 1942.

There were two serious problems with the mid-ocean service stations. One, they gave U-boat commanders a dangerous temptation to wait until the last minute to replenish fuel; this left some boats in trouble when storms or enemy patrol craft delayed rendezvous. Two, a dozen U-boats clustered together were a lot easier to spot than Mahan's needle in the haystack and made a very tempting target. By the next summer, the U.S. Navy had assembled "hunter-killer" groups centered on small "jeep" carriers, with extended search range. Of the nine *milchküe* in service in June 1943, seven had been sunk by August.

However, from July 1942, wolf packs could be directed to follow and attack any westbound convoy, refuel, then pick on an eastbound convoy. Thus began the true Battle of the Atlantic. It was to continue for eleven months and score some 712 merchant victims. Total Allied losses for the war to that point, from all causes: fourteen million tons. Total replacements: less than half. New U-boats were joining the fleet at a rate of almost one a day.

It looked as if Dönitz had won the war.

He had an unwitting ally in Bomber Harris, who argued against diverting any more bombers to the U-boat campaign, taking as his text the vital support needed by the Russians:

> It cannot be pointed out too strongly that in the Bomber Offensive lies the only hope of giving really substantial help to Russia this year . . . and that it if is reduced to lesser proportions by further diversions of large

> numbers of bomber aircraft for seagoing defensive duties, it will fail in its object and the failure may well extend to the whole of the Russian campaign. This in my opinion would be a far greater disaster than the sinking of a few extra merchant ships each week.

He was also opposed to equipping maritime patrol planes with radar; he didn't think the Coastal Command pilots would ever get the hang of it. "Our experience, which is considerable," he wrote, "is that even expert crews find it no easy matter to attack with accuracy even a city by means of [radar]. I am therefore rather sceptical of the prospects of inexperienced crews with ASV. Indeed, I feel that the provision of aircraft equipped with this apparatus will mark the beginning rather than the end of the difficulties involved in sinking U-boats."

At this time, British Coastal Command had thirty-two B-24s. Total wartime production of B-24s was 19,203. Fewer than a hundred, perhaps, ever reached U-boat patrol status. And yet—these were the planes which eventually broke the back of the U-boat effort. The bombers were modified for antisubmarine patrol by taking out extra weight, removing the belly gunner's turret, replacing self-sealing gas tanks with normal tanks, and removing most protective armor. This allowed them to carry two thousand gallons of fuel and eight 250-pound depth charges. As a result, eighteen-hour patrols were routine, and some were stretched to twenty.

Radar was constantly being improved, but so were German radar detectors. The *Metox* could pick up a radar signal out to thirty miles—about twice the effective range of the radar itself—making the surface once again safe for nighttime cruising. The aerial for *Metox* was a simple cross-shaped wooden frame, which became known as the *Biscayakreuz*—the "Biscay Cross"—and which had been installed in all boats by October 1942. Well, "installed" is an elegant description for a repeated manual process. Upon surfacing, a crew member would carry the Biscay Cross topside and hang it as high as possible, the connecting wires running back down into the conning tower; before submerging, the Cross had to be taken down and carried below. The conning tower hatch could not otherwise be closed.

Metox was soon retired when a British POW "warned" the Germans that the sets gave off sufficient radiation permit easy detection. The Germans developed a replacement—*Naxos-U*—which they discovered in tests of their own could itself be detected out to thirty miles. Ironically, that fact does not seem to have been known by the British, but German efforts to correct the problems delayed introduction of the systems until late in the year.

The most lethal Allied ASW weapon, the hedgehog, was put into service in 1943. A primitive version had been introduced, but not perfected, in World War I. The hedgehog may be compared with a multiple mortar launcher, mounted forward of the bridge of an escort ship, carrying twenty-four sixty-five-pound projectiles arranged to cover a hundred-foot circle 230 yards in front of the launcher.

The hedgehog was a contact-detonating weapon, but if a U-boat was within the hundred-foot circle, it would almost certainly be hit by at least one of the weapons. When the hedgehog was under development, and especially when it first began showing up on ships (and was therefore visible to enemy spies and eyes), it was always referred to as an "anti-dive-bomber" defense.

March 1943 saw one hundred ships lost to the U-boats. In just one extended action forty-one U-boats hit two convoys over four days, sank twenty-two ships (146,000 tons), and lost only one submarine. It must have have seemed another high point in the U-boat war (almost as high as the summer of 1942), but it was really the watershed. To the true point of the era: German planners had seriously underestimated the ability of the Allies to replace ship losses. In May 1942, they determined that the Tonnage War would be won with a monthly target of 700,000 tons; perhaps 150,000 for the Air Force and surface Navy, the rest left to the undersea boats. They predicted that, for 1943, new merchant shipping launched by Britain and the U.S. together would be less than eight million tons. The U.S. alone launched almost double that figure.

One major contribution came from industrialist Henry J. Kaiser, who devised an assembly-line method for building welded, rather than riveted, merchant ships. The first such "Liberty" ship was launched in September 1941; by April 1943, a shipyard could produce a Liberty every forty-two days—and eighteen shipyards, with 171 slipways, were turning them loose at a rate of more than 140 a month. By the end of the war, 2,710 had been built and slightly more than two hundred had been sunk.

In fact, new tonnage was to pass all wartime losses by July 1943. To win the Tonnage War, the U-boats would have to sink 1,400,000 tons a month, not 700,000, and this at a time when new ASW technology and forces had the U-boats, literally, on the run.

By April, thanks in large part to the great increase in air and surface convoy protection—which more and more forced the submarines to remain submerged—the rate of sinkings had been cut in half. By May, with merchant sinkings still on the decline, U-boat losses jumped to forty-one, 25 percent of current operational strength!

Indeed, in the first week in May 1943, a convoy sailed right into the middle of a pack of thirty-one boats—and only a few miles ahead, another group of eleven waited to join in battle. It was a battle indeed, but not the slaughter the U-boat command expected. The final tally: twelve merchant ships lost, but also nine U-boats sunk (two from a collision), five badly damaged, no escort even heavily damaged. There were more U-boats on the attack than there were ships in the convoy.

Another convoy in same month (May 13–14) met thirty-six U-boats in a pack; the U-boats were not able to get through the convoy screen and only three stragglers were sunk—along with the loss of three U-boats.

The Allied airborne homing torpedo had a successful debut on May 13, 1943, sinking *U-266,* with an encore the next day, *U-657.* Known as "Fido" or "Wandering Annie," this weapon would home in on the sound of propeller cavitation. It was a slow-speed and relatively shallow-running torpedo—slow speed to prevent confusion from the sound of its own propeller, and shallow running because cavitation diminishes with depth.

In yet another convoy action, on May 19–20, there were thirty-three U-boats with not a single ship sunk—if you discount the five U-boats lost. Lost with them was Dönitz's twenty-one-year-old son Peter, on his first patrol. When given that news, Dönitz showed no visible emotion, but within a few hours, the following message was sent from U-boat Command to all units:

> If there is anyone who thinks that fighting convoys is no longer possible, he is a weakling and no real U-boat commander. The Battle of the Atlantic gets harder but it is the decisive campaign of the war. Be aware of your high responsibility and be clear that you must answer for your actions.

The next day, two convoys were attacked by twenty-one boats—no sinkings, two boats lost. Dönitz conceded that, with them, the Battle of the Atlantic had been lost. On May 24, 1943, in the face of diminished success and mounting losses, he recalled most of the submarine force.

However, even in the face of certain defeat, the German Navy was unwilling to abandon the field and thus free Allied assets to other operations; despite all, there was yet some hope, with radical new weapons about to join the fleet if only the fleet could hold on. Dönitz put the question to his senior commanders. There was never a question about the answer.

Late in 1942, the German Navy had embarked on a crash program to build high-speed U-boats—boats which could run faster submerged than most escorts could run on the surface, boats that would confound all existing ASW doctrine and weapons. They would be powered by a turbine driven by the decomposition of highly concentrated hydrogen peroxide. This wasn't some crazy Nazi pipe dream, but grew out of some very real experiments which had been conducted by Dr. Helmuth Walter in the 1930s.

The Walter system could result in the world's first *true* submarine—one which could operate submerged for some period of time, not limited by battery capacity and free of the need for atmospheric oxygen. The first test boat, *V.80,* was launched in 1940 and reached a speed of twenty-eight knots submerged.

When hydrogen peroxide (H_2O_2) breaks down, it releases water (H_2O) and oxygen (O). The Walter system used a stabilized form of concentrated (above 95 percent) H_2O_2 known as Perhydrol (the hydrogen peroxide in your medicine cabinet is a 3-percent solution). The water came off at 1700°F—superheated steam to drive a turbine; some of the oxygen was used to support combustion of conventional fuel, and the exhaust gases were added to the power stream. Surplus oxygen replenished the atmosphere in the crew compartments.

In 1940, Hitler thought his war was won and production of a series of Walter boats was put in limbo, although design work continued. Some problems surfaced: Fuel consumption was enormous—twenty-five times as much as for a diesel engine. According to one source, it took $200,000 worth of Perhydrol for one trial run of 6.5 hours.

A three-hundred-ton *V.300* (re-named *U-791*) was launched in 1943, but never completed; after juggling space for the turbine, the Perhydrol storage, and a supplementary diesel, there wasn't enough room left in that hull for crew and munitions. Eventually four 250-ton boats, designated Type XVIIA, were built (1943–44) and put into service. U-792 and U-793 had hull design Wa201, with a submerged speed of twenty-five knots; U-794 and U-795 had hull design Wk202, and were slightly smaller and one knot slower. Each boat tested a different arrangement of the power plant, and each had a small supplementary diesel for increased range (to 1,800 miles). Armament was limited to two torpedo tubes, with one reload for each.

A larger variant, the Type XVIIB of three hundred tons, offered greater fuel oil storage for double the range on the auxiliary diesel; twelve were ordered, three put in service—*U-1405, U-1406,* and *U-1407.* These were scuttled at the end of the war; some components of *U-1406* were brought to the United States for evaluation.

Another variation, proposed because of the high cost and uncertain supply of Perhydrol, eliminated the turbine and ran on a closed-cycle diesel engine with oxygen provided from storage tanks in what was known as the "Krieslauf" cycle. This boat was designated Type XVIIK; one was ordered and launched as *U-798,* but was never completed.

Late in 1942, when the U-boat fleet began running into problems, Walter started work on a 1,600-ton Walter-cycle Type XVIII—with Hitler's enthusiastic support—and two prototypes were ordered. This model had a figure-eight hull cross-section, with the top half the larger for the turbine, and living, control, and weapons spaces, the smaller lower half for Perhydrol storage tanks. Within two months reality caught up with wishful thinking—the reality of production problems, of fuel supply problems, of high cost—and the project was shelved.

Other new and promising equipment was put into play just at this time, primarily the breathing tube called the "schnorkel" or snorkel. This permits a slightly submerged submarine to take in fresh air for running the diesels, and to remove engine exhaust. In theory, a submarine with a snorkel could run submerged almost indefinitely. In practice, at least in the beginning, there were a few problems. The tube might break if the boat's speed exceeded eight knots. The intake was fitted with a ball float, to close the tube whenever a wave passed over. When the tube closed—with the engines still running—engine intake air was sucked out of the submarine itself, often with sufficient force to pop the eardrums of crewmen (although the number of such incidents is probably much less than the wartime rumors).

Snorkel-like devices have been used on submarines from the earliest days—but the "modern" snorkel was developed by the Dutch in the 1930s, and some snorkel-equipped boats were taken over by both the British and the Germans when Holland fell to blitzkrieg in 1940. Neither the British nor the Germans saw any value in the units, and had them removed.

However, in the dark days of 1943 when air patrols made daytime surface running almost suicidal, Dr. Walter had remembered inspecting the snorkel, and it was resurrected. First installation was in July 1943; within a year, thirty boats had been converted.

In the meantime, the Type XVIII hull design was modified to carry two lightweight Type VIIC four-thousand-horsepower diesels and a 4,200-horsepower electric motor in the upper half, with an extra-large bank of batteries in the lower hull, replacing the Perhydrol. Two new classes were developed on this model: the 1,600-ton Type XXI and the 230-ton coastal version Type XXIII, both of which became known as "Elektro-boats."

The Type XXI had only half the range of the conventional 1,600-ton Type IXD—but could manage bursts of seventeen knots underwater, (compared with seven knots for the Type IXD), could snorkel at twelve knots (compared with six knots), could dive three hundred feet deeper (to almost one thousand feet), and could stay totally submerged at economical creep speed for eleven days. Habitability was improved to include a food-storage freezer, a watermaker, air-conditioning, and a garbage-disposal system. Equipped with fifty-mile-range hydrophones and a sophisticated fire-control system, Type XXI could launch an attack from a depth of 150 feet.

The Type XXIII offered a similar comparison to the small prewar Type IIA—twice the submerged speed, five times the underwater endurance. However, the combat effectiveness of the Type XXIII was severely limited: only two torpedo tubes, no internal reloads.

All other submarine construction was phased out in favor of the Type XXI and Type XXIII. In a radical change in method, both to speed up construction and to minimize interruptions from Allied bombing, these boats were built in finished sections, each about twenty-five by twenty-seven feet with all equipment installed. There were eight sections for a Type XXI, four for a Type XXIII, constructed in widely scattered parts of the country, then moved on Germany's extensive canal system to one of three assembly yards: Hamburg, Bremen, or Danzig.

The idealized production schedule called for thirty-three Type XXI boats a month, but there were too many errors in the rushed detail drawings and too many material shortages. Then too, Allied Bomber Command came through with an unexpected success in basically destroying Hamburg. More than fifty thousand civilians were killed and the Blohm & Voss Shipyard, one of the assembly yards, was damaged enough to curtail, but not halt, production.

Undaunted—well, at least not defeated—the Germans rushed through the construction of a reinforced concrete assembly building, one thousand feet long with a sixty-foot overheard clearance. U-boats could not only be assembled, but also given final underwater testing in a dredged-out section at one end—still under the roof.

Type XXIII went into service first, heading out on patrol by February 1945. By the end of the war in early May, six Type XXIIIs were in service, with another fifty-three in the water and nine hundred under construction or on order. The first Type XXI entered service in March 1945; another thirty were in the trials-and-training cycle. The first boat to go on an operational patrol, *U-2511*, left Hamburg on April 30th; when she returned home to surrender, 121 Type XXI were in the water and another one thousand were under construction or on order.

Each improvement in Allied ASW made it more difficult for the U-boats to get into the optimum position for attack, and once there, to take careful aim. The need, therefore, was for torpedoes which could be fired with less care in aiming (or almost no aiming at all) and find the target by themselves.

Very early in the game, submariners discovered that they could hear their targets before they could see them; with some practice, a hydrophone operator could determine the bearing of the target (and even to some degree, the range) with amazing accuracy. Germany turned that discovery into an acoustic torpedo. Fitted with a pair of sensors on each side of the nose, this weapon could steer itself toward the loudest source of noise in the ocean, expected to be the propellers of the target ship. The torpedo would start out in the general direction of the target, adjusting course to this side or that to keep the noise dead ahead. The first adjustments would be rather broad, but then they'd become increasingly narrow, until the last few yards would become a dead-on line to the screws. Also, since warships had more powerful engines and were therefore noisier than merchant ships, an acoustic torpedo could be in theory an ideal anti-escort weapon.

In actual practice, when the T4 ("Falcon") version of the torpedo was first put in service early in 1942, the fast-moving escorts turned out not to be a very good target; they could—albeit unwittingly—simply sail ahead of and therefore away from a slow torpedo. So the Germans improved the weapon, issuing the T5 ("Wren") model. This had 25 percent greater speed and a finer acoustic sensitivity, specially tuned to the frequency generated by an escort running at the normal working speed of fifteen knots. T5 entered service in August 1943.

In September, the Germans made another try for the North Atlantic convoys with a pack of twenty-one boats now augmented by one of the newer radar detectors and the T5. They went forth full of hope. They did battle. They reported the destruction of twelve destroyers, all by the new torpedo, and the sinking also of nine merchantmen with the loss of two U-boats.

Dönitz was delighted; Hitler was overjoyed, calling the U-boat the bright spot in a dark time and calling for more of the same. The actual score, however, was six merchantmen and three escorts sunk, two damaged, three U-boats lost.

Bad enough; but the Allies, now alerted to the new torpedo, quickly brought forth some anti-acoustic-torpedo defenses which had been waiting in the wings. Through a misinterpretation of early intelligence reports about another advanced-model torpedo, the FaT (see below)—the British thought that a "seeking" torpedo must be using an acoustic sensor—the Allies had been working on countermeasures for the acoustic torpedo since 1941. American-developed mechanical noisemakers—called "foxers"—were provided within a month of first deployment, to be streamed astern by the escorts. Designed to create a noise much louder than that of the ship, the foxers did decoy an acoustic torpedo, but with serious penalities: noisier than the escort, they could be heard at a greater distance by any theretofore unware U-boat; they were so noisy that they masked the escort's Sonar; since they were mechanical (the noise was generated by the passage of water through the mechanism), the only way they could be turned "off" was to haul them in, a slow and cumbersome process.

Captured U-boat crewmen soon enough revealed the frequency to which the T5 sensor had been tuned, and all convoy escorts were ordered to increase and decrease speed ("High Safe Speed" or "Low Safe Speed") radically and frequently when the presence of a U-boat was suspected.

The Germans countered with a retuned sensor, better able to discriminate between a ship and a foxer; the Allies redesigned the foxer. Then the war ended.

The Germans had great hopes for the T5—from April 1944, the combat patrol plan called for half of ready tubes to be loaded with T5—but the Allied countermeasures appear to have been very effective. Postwar accounting estimated that 640 T5s had been fired in combat, with only thirty-nine hitting a target.

There has always been one major problem with noise-seeking torpedoes: The loudest noise in the vicinity might well be that of the submarine itself. German doctrine called either for a crash dive to sixty meters, or an immediate shift down into silent running.

Another approach was the *Flächenabsuchender*—literal translation, "shallow-searching"—*Torpedo* (FaT). This played on the fact that the ships in most convoys were fairly tightly bunched together. Once the torpedo had turned to its initial course to intercept the convoy, it began a back-and-forth searching, much like a bloodhound crossing and recrossing a trail. A straight leg of eight hundred or 1,600 meters, followed by a 180-degree turn to the right (or left, as preset) to another straight leg, then a 180-degree turn to the left, another straight leg—the sequence repeated until the torpedo ran into something, or ran out of fuel.

The FaT entered service in November 1943, with a very favorable result: One U-boat reported four hits against one convoy, two of which came after more than eight minutes running. There were two operational cautions. One, the steam-driven FaT left a bubble wake and thus could only be used at night; two, a FaT could run into another U-boat as easily as a convoy ship, so the attacking boat was required to broadcast a short "FaT warning" to alert other members of the pack to dive or move away smartly. An electric-drive FaT was soon introduced, although it was not as useful because it had a much shorter range.

A greatly improved version, which came in both steam and electric models, appeared early in 1944: the *Lagenunabhängiger Torpedo* (LuT)—the "bearing independent" torpedo. A FaT could only make one initial course change before beginning the back-and-forth tracking; this meant the attacker almost always had to be positioned ahead of and square to the right or left of the convoy track. The LuT could make a second "initial" course change, and thus could be fired from almost any bearing on the convoy.

The Germans began installing radar on U-boats in December 1943, and developed radar decoys—notably a foil-bedecked balloon anchored to a float. Allied radar operators quickly learned that targets which moved only with the current or the wind were not likely to be U-boats. To combat sonar, the U-boats deployed the *Pillenwerfer,* Submarine Bubble Target. This was a small perforated metal canister filled with a gas-generating chemical; a sonar signal would bounce off the cloud of bubbles thus released, sending a false echo back to the operator. It was not very effective.

In the meantime, the U-boat war continued the slide into irrelevance. For the months September and October 1943, 2,468 ships sailed in sixty-four North Atlantic convoys—only nine were lost, against thirty-five U-boats sunk. For November and December, 2,218 ships sailed in seventy-two convoys; no losses, twenty-seven U-boats sunk.

The U-boats would have one more grand chance to influence the course of history. The Allied invasion of France was coming, although no one knew when or exactly where. Dönitz vowed to be there whenever and wherever, and the Allies expected his boats to join their party. How many boats? No one knew, not even the Germans. The estimates were all over the ocean (so to speak). According the British intelligence estimates in February, the Allies might expect to see 175 to two hundred U-boats. By the end of May, this had been adjusted downward to a estimate of approximately forty boats.

In March, Dönitz had begun to assemble his anti-invasion force—which turned out to be approximately forty boats! He also had begun preparing them for the ultimate sacrifice. Attack any target, no matter how seemingly insignificant, regardless of risk; do not let navigational hazards or possible minefields stand in the way; any loss inflicted on the enemy before landing is a fair trade, even if the U-boat is lost in the process.

Although the invasion had long been expected, the High Command was nonetheless caught off guard when the invasion force began pounding the beach defenses of Normandy on June 6th. Only thirty-five U-boats could be sent out—to take on a force of almost seven thousand ships and landing craft, including 1,213 warships, and some 350 maritime patrol aircraft.

The U-boat score among such rich prospects: June through August total, five escort ships, twelve merchantmen, and four landing craft; another seven ships damaged. For the same period, in the same operation, the Germans lost forty-two U-boats, six of which were scuttled as the Allies moved on the French coastal bases.

In June 1944, an ASW Group led by the jeep carrier *Guadalcanal* became the first American force to capture an enemy warship on the high seas since the War of 1812. *U-505* had been brought to the surface by depth charges; quick action by a boarding party saved the boat from sinking (the crew had opened the sea cocks, but not completely). *U-505* was used in some War Bond rallies, and eventually put on display at the Museum of Science and Industry in Chicago.

Just as Japan, Germany was by this time pursuing weapons of desperation. One was the *Seehund* two-man, two-torpedo midget submarine: thirty-nine feet, fifteen tons, powered by a six-cylinder, eighty-five-horsepower diesel truck engine and a seventy-horsepower electric motor. *Seehund* could dive to 165 feet and run on the surface for 120 miles at eight knots, or 250 miles at five knots; submerged, twenty miles at five knots, sixty miles at three knots. From September 1944 to April 1945, 286 were made ready for service. Near the end, a closed-cycle version was under development; the Soviets captured a building yard and acquired eighteen finished and thirty-eight partially finished boats. The Soviets have never admitted to midget submarines, but in 1980 the Swedes complained of encroachment by Soviet midgets.

Two of Germany's weapons of desperation were not as suicidal. They included a pair of unmanned flying bombs, directed at targets in England. The V-1 was a pulse-jet airplane, launched from easily rigged portable ramps. When the fuel ran out, the bomb stopped flying. The V-2 was a ballistic missile, which could carry a two-thousand-pound warhead with some accuracy for 180 miles—in five minutes. V-2 launch sites were complex, and when they were put in jeopardy by Allied advances, the Germans tried to develop a submarine launch system. Large 120-foot-long submersible barges had been developed to resupply German units in Norway. The barges were at slightly positive buoyancy when at rest, and preset diving planes would drive them under the surface when pulled by a towing a U-boat. The plan was to mount one missile on each of three barges in a string, tow them into the Baltic, surface, erect the missiles by sinking one end of the barge, and launch.

It was never tested, although the conquering Allies seemed to accord the barges the same mythical status they gave any German weapon, experimental or proposed. Thus, when the Soviets captured much of the equipment, they assumed it was a proven system. The Germans had built some three hundred launching barges—which never worked.

As their home ports were lost to Allied advances, the remaining French-based U-boats were shifted to Norway. They hoped to take full advantage of the now-operational snorkel, and a few boats were always out on patrol for the rest of the war. They did not have much effect. The U-boat war had ended, just as the submarine war had in the Pacific. One more try at a wolf pack in March was a failure. Six Type VIIs sent to American coastal waters, and four did not return. There was still hope, *real* hope, with almost two thousand fast new U-boats in the works, if only the nation could hang on. The eventual deployments of Type XXI and XXIII were lost in the chaos of collapse, when the Allied barbarians were at the gates and U-boat crews were pulled ashore as guard units.

Hitler committed suicide on April 30, 1945, having designated Dönitz as his successor. The admiral assumed duties of Chief of State on May 1st, issuing useless fight-to-the-death orders to all armed forces, followed three days later by orders to cease fire; the U-boats were recalled at the same time.

Dönitz had vowed not to allow a repeat of the final shameful event of World War I, the proud High Seas fleet herded into Scapa Flow like so many dumb sheep. He had planned to order all ships scuttled on his signal "Operation Rainbow." However, he bought a few days to shift German soldiers and refugees away from Russian forces in the east by promising the western Allies that naval resources would be surrendered intact, and did not execute "Operation Rainbow." But individual units had not been waiting for orders, and some 215 U-boats had been scuttled by May 7th, the last full day of the European war.

Ignoring (or not receiving) the May 4th cease-fire order, the Type XXIII *U-2336* sank two small British steamers at midday on the 7th, and the Type VIIC—in a fitting coda for the most useful submarine of the war—scored the last U-boat victim, sinking a Norwegian minesweeper at 1952 the same day.

* * *

The 1700-ton Type XB minelayer *U-234*—Germany's largest submarine—was at sea when the war ended, and surrendered in mid-ocean to an American destroyer escort on May 12th. Her original destination had been Japan, and among the passengers was the German Air Attaché to Tokyo, General Ulrich Kessler. Among the cargo were three of the newest-model electric torpedoes, two complete Me 262 jet fighters (disassembled in crates, but with complete technical data), and 550 kilograms of uranium$_{235}$ packed in lead containers. The reason the uranium was being sent to Japan has never been determined—or, at least, revealed.

30 1945–1971—Final Accounting, and the Future

U_{235} is capable of liberating energy at such an unbelievable rate that one pound of it [is] the equivalent of 5,000,000 pounds of coal. . . . A five pound lump of only 10 to 50 percent purity would be sufficient to drive ocean liners and submarines back and forth across the seven seas without refueling for months.
—*William Laurence,* Saturday Evening Post, *September 7, 1940.*

The last American surface ship of the war to be sunk by a submarine was the cruiser *Indianapolis,* returning unescorted from delivering components of the first atomic bomb to Tinian Island. Commander Mochitsura Hashimoto's *I-58* fired six torpedoes around midnight of July 29th and had three hits; Hashimoto thought he had sunk a battleship. He denied postwar assumptions that he had also used *kaiten* in the attack, asserting that his normal Type 95 torpedoes were quite adequate.

Hashimoto's message report was intercepted and decrypted—and discounted. Since *Indianapolis* was on a classified mission and maintaining radio silence, the staff at fleet headquarters was not aware that the cruiser had been passing through those waters. Further, because no one was adequately tracking *Indianapolis,* she was not reported "missing" for several days. There were 316 survivors, of a crew of 1,199.

Indianapolis is the best-known submarine victim of the war, for two reasons unrelated to the war but related to each other. One, a large number of survivors were attacked in the water by sharks, and two, a retelling of that tale was a highlight of the 1975 feature film *Jaws.*

For the record: The largest ship ever sunk by a submarine was the brand-new aircraft carrier *Shinano,* 71,890 tons, on November 28, 1944, just outside of Tokyo Bay by *Archerfish. Torsk* sank the last cargo ship torpedoed by an America sub on August 13th; the last Japanese warships sunk were *I-373,* torpedoed early in the morning of August 14th by the submarine *Spikefish,* and two frigates sunk within ten minutes of each other by *Torsk* later in the day. Frigate No. 47 apparently won the prize as the last warship to be sunk in World War II. The war ended the next day.

At war's end, the entire surviving Japanese submarine force comprised fifteen operational I-boats and seven in the training command, five *RO*-class, four ex-German U-boats, and two ex-Italian cargo submarines. American submarines sank 4.75 million tons: 1,300 merchant ships, one battleship, eight carriers, eleven cruisers, and 180 smaller warships—including thirty-two submarines. Japanese submarines sank one-fifth as many ships.

In return, the Americans lost fifty-two submarines; 22 percent of the submarine personnel who went on wartime patrol did not return. It was the highest casualty rate of any branch of the service. Germany lost an astonishing 63 percent—630 of every thousand men who served with the submarine force did not survive the war.

Postwar analysis of 821 U-boat sinkings credited the largest number, 303, to shore-based aircraft; seventy-nine were lost to naval aircraft, six to a combination of shore-based and naval aircraft, and forty-five to a combination of surface vessels and some sort of aircraft. Thus, aircraft were involved in more than half (433) of the sinkings, a remarkable increase over World War I's *one.* Surface ships took out 252, mines thirty-four, accidents forty-five, other submarines twenty-five; in 1945, the British submarine *Adventurer* conducted the only attack in both wars where both victor and victim were submerged. (An American submarine, *Tautog,* sank the submerged Japanese submarine *RO-30,* but was herself on the surface at the time.) Unknown causes took fifteen, two were interned, fourteen scuttled by their crews (not including the

"Operation Rainbow" tally) following battle damage or in advance of Allied occupation. *U-78* achieved the distinction of being the only U-boat sunk by a shore battery in either war, on April 16th, 1945; the guns were manned by Russians.

The Soviets entered the war with the world's largest submarine fleet, 218 boats; they added fifty-four during the war and lost 109. Many were tied down to inactive service with the Pacific Fleet, but the Soviets claimed to have sunk 160 merchant ships (402,437 tons) in the Baltic and Black Sea. The Germans rarely bothered with convoy in waters patrolled by Soviet submarines, because Soviet submarines were rarely on patrol. When they did venture forth to seek battle, their attacks were often too tentative to be of much effect, and they encountered the torpedo problems which seemed common to all navies except the Japanese: The torpedos ran too deep, the exploders frequently failed to explode.

However, a Soviet submarine caused the single greatest maritime disaster in world history—the sinking of the German liner *Wilhelm Gustloff* by the *S-13* under Commander Alexander Marinesko. On January 30, 1945, *Wilhelm Gustloff* was part of a massive effort to shift German soldiers away from the advancing Red Army and into German territory; loading out at the port of Gotenhafen (now Gdynia, Poland), the ship had counted 6,050 troops and civilian refugees across the brow when perhaps another two thousand rushed aboard in the last moments before sailing. She was sunk soon after, and fewer than one thousand people were rescued.

Admirals Raeder and Dönitz were arrested in May 1945 and brought before the War Crimes Tribunal held in Nuremberg—the only city in Germany with an undamaged court and prison large enough to handle the demands of a trial involving twenty-four individuals and six organizations. Raeder and Dönitz were each charged on three counts: one, as part of a conspiracy which planned to conduct "aggressive war"; two, conducting aggressive war, and three, war crimes.

Raeder had been a member of the inner circle from before the war; Dönitz was then only a captain and was hardly a key player in planning to take the world to war. He seems to have been included in the indictments for the general reason that he had briefly acted as Chief of State: Dönitz himself believed he was a stand-in for Hitler.

The British Admiralty felt that the German Navy generally had behaved itself during the war and that Dönitz should not be charged with war crimes.

To the Admiralty's recommendation that the charges be withdrawn, the Foreign Office responded: "Typical Admiralty whitewashing of the German Navy."

For his defense counsel, Dönitz chose a German Navy lawyer—who wore his uniform in court, the better to emphasize that this client was a professional naval officer serving his country, not a criminal.

Once Dönitz was charged, the prosecutors began looking for specific crimes. They focused on the *Laconia* order of September 17, 1942 (page 305), which discouraged the rescue of survivors because "it runs counter to the elementary demands of warfare for the destruction of enemy ships and crews." The prosecutors charged that with this—and in pep talks to U-boat crews—Dönitz had "encouraged" the killing of crews. However, that "encouragement" could not be linked with any actual instance where survivors were gunned down in the water.

In fact, just as with World War I, there seems to have been only one such proven incident. Following his sinking of the *Peleos* in March 1944, Lieutenant Heinz Eck ordered his crew to machine-gun and throw hand grenades at survivors hanging onto life rafts and bits of wreckage. Eck was on his first combat mission; his motive, he testified at his own trial before a British military court in October 1945, was to hide the sinking from patrolling aircraft (and thus conceal his own presence in the area).

Two of his officers, also charged, claimed in defense that they were following the orders of their superior; the court invoked the precedent of *Llandovery Castle* and denied the claim. Eck and the others were found guilty and were executed by firing squad on November 30th.

In testimony in his own defense, Dönitz differentiated between killing crewmen on their ship, who were able to fight the U-boat, and "shipwrecked" crew members "who, after the sinking of their ship, are not able to fight any longer . . . Firing upon these men is a matter concerned with the ethics of war and should be rejected under any and all circumstances."

The court was tracking crimes, not ethics, and seems to have taken his testimony at face value. However, the admiral drew a strange line between killing and humanity; as he told Hitler in May 1942, a properly functioning magnetic pistol would "have the great advantage that the crew will not be able to save themselves on account of the quick sinking of the ship."

The best witness in Dönitz's defense turned out to be Admiral Chester Nimitz. He did not appear in person, but answered a series of interrogatories:

2.Q. Did the U.S.A. in her sea warfare against Japan announce certain waters to be areas of operation, blockade, danger, restriction, warning or the like?

2.A. Yes. For the purpose of command of operations against Japan the Pacific Ocean areas were declared a theater of operations.

3.Q. If yes, was it customary in such areas for submarines to attack merchantmen without warning . . . ?

3.A. Yes, with the exception of hospital ships and other vessels under "safe conduct" voyages for humanitarian purposes.

4.Q. Were you under orders to do so?

4.A. The Chief of Naval Operations on 7 December 1941 ordered unrestricted submarine warfare against Japan. . . .

Admiral Nimitz also affirmed that "U.S. submarines did not rescue enemy survivors if undue additional hazard to the submarine resulted or the submarine would thereby be prevented from accomplishing its further mission."

Dönitz was found "not guilty" on the charge of planning aggressive war, and "not guilty" on the war crimes charge. In particular, he was excused of "breaches of the international law of submarine warfare" largely because of "the answer to interrogations by Admiral Nimitz . . ."

But Dönitz was held "guilty" on charge two, that he waged aggressive war—a finding which surprised observers at the time and which some later felt went beyond the charge and specification. "It is clear," the Tribunal explained, "that his U-boats . . . were fully prepared to wage war." Under that construct, *any* military command which is ready for war when war starts would be guilty. Dönitz was given ten years in prison.

Raeder was found guilty of "planning to wage aggressive war" and "waging agressive war," and also on a war-crimes charge in which he was indirectly involved in the summary execution of two captured British commandos. His sentence was life imprisonment.

Their world had disappeared, and with it, civility: Raeder charged that Dönitz was "conceited;" Dönitz returned the compliment by calling Raeder a "jealous old fogey."

* * *

With the end of the Second World War, new tensions began to build and Soviet naval strategy began to evolve from the support-the-army mission to that of a blue-water navy. They quickly regained their prewar status as operators of the world's largest submarine fleet; between 1950 and 1958 they built about 235 "Whiskey" class submarines, using the Type XXI as template, one of a number of new classes put into production. The weapon of the weaker nation was about to be balanced in a Cold War against the naval forces of the world's strongest maritime nation—the United States. For the rest of the century, Soviet actions were directly to influence American strategy and weapon development.

And for the rest of the century, the center of submarine development shifted back to the United States, where the dream was finally to be realized: the true submarine, free from the atmosphere, free from the limits of batteries, free from the need for frequent replenishment of fuel. Jules Verne's *Nautilus* took its power from seawater; the *Nautilus* which steamed out of Groton, Connecticut, in September 1954 took its energy from an even more basic source: the atom.

Atomic fission was first demonstrated in Germany in 1938, and reports that "splitting the atom" could release tremendous levels of energy excited the scientic world. An application to naval power plants was immediately apparent; Dr. Ross Gunn of the Naval Research Laboratory appears to have the first to suggest (March 1939) that "fission chambers" using the U-235 isotope could be used to power submarines. He was given a $1,500 allocation for further study, and his first report a few months later concluded that an atomic power plant could operate without oxygen, thus freeing the submarine from any surface connection: "a tremendous military advantage [that] would enormously increase the range and military effectiveness of a submarine."

Gunn saw a number of problems, not the least of which was separating U-235 from uranium ore. The Manhattan Project, which developed the atom bomb, started in October 1939—and in January 1941, Gunn and fellow-researcher Philip Abelson developed a process for separating U-235 which was used in the work on the atom bomb.

Science writer William Laurence—call him a latter-day William Bourne—described the German experiments for readers of the *Saturday Evening Post* in September 1940. U-235, he wrote, was "capable of liberating energy at such an unbelievable rate that one pound of it has the equivalent of 5,000,000 pounds of coal . . ." He also noted that "A five pound lump of only

10 to 50 percent purity would be sufficient to drive ocean liners and submarines back and forth across the seven seas without refueling for months."

As early as February 1946, the General Board recommended "taking earliest possible advantage of developments of atomic energy for application to submarine main propulsion."

And in March 1946, Abelson wrote a paper, "Atomic Energy Submarines," in which he optimistically predicted that "with a proper program, only about two years would be required to put into operation an atomic-powered submarine mechanically capable of operating at 26 knots to 30 knots submerged for many years without surfacing or refueling. In five to ten years a submarine with probably twice that submerged speed could be developed." Abelson envisioned a nuclear power plant married to a Walter hull, and he added, "This fast submerged submarine will serve as an ideal carrier and launcher of rocketed atomic bombs."

Vice Admiral Charles Lockwood, commander of Pacific Fleet submarines during the war, later recalled a briefing by Abelson:

> If I live to be a hundred, I shall never forget that meeting . . . [the] walls lined with blackboards which, in turn, were covered by diagrams, blueprints, figures, and equations which Phil used to illustrate various points as he read from his document, the first ever submitted anywhere on nuclear-powered subs. It sounded like something out of Jules Verne's *Twenty Thousand Leagues Under the Sea.*

However, no one in the Navy had any illusions that nuclear power would be so quickly acquired. Despite the obvious advantages, no agency at that time was pursuing nuclear power for ships; no one in the Navy had any experience with nuclear energy, and no one in the government seemed interested in sharing scarce, vital nuclear raw materials in the face of an obviously growing Soviet instransigence. Weapons came first; power, someday.

On March 14, 1946, Secretary of the Navy James Forrestal signed a letter to Secretary of War Robert Patterson (who at that time had administrative control of the Manhattan Project), expressing an interest in developing atomic power for ship propulsion. Patterson's man on the scene, General Leslie Groves, affirmed that "the most attractive use for atomic power was the military one of submarine propulsion." The two men recommended that Forrestal detail some Navy personnel to Oak Ridge, Tennessee—home of the atomic bomb. The Navy sent five officers and three civilian engineers; the senior officer was Captain Hyman Rickover, of whom more later.

The 1946 Submarine Officers Conference had a gloomy assessment of the current state of affairs and a hope for the future. "Present anti-submarine techniques and new developments in submarine design have rendered our present fleet submarines obsolete," they reported, "to a greater degree than any other [warship] type." They called for the "development of a true submarine" using nuclear power, and in the interim asked that lessons learned from the war (and from the Germans) be applied to existing boats as soon as possible.

As at the end of World War I, the victors picked through the spoils and took home bits and pieces of the newer U-boats. The U.S. Navy received two of the Type XXI—*U-2513* and *U-3008;* the latter soon became a spares locker to keep the former in service. Both were eventually used as targets in weapons development programs, but not before they had been thoroughly studied. (The U.S. Navy also took over the three surviving Japanese "aircraft carriers" for study, and two experimental high-speed boats, *I-201* and *I-203,* which proved to be too difficult to operate; they were sunk in torpedo tests in March 1946.)

The resulting upgrade, dubbed "Greater Underwater Propulsive Power" (GUPPY), was applied to fifty-two fleet boats. Snorkels were installed; the sail and superstructures were streamlined, guns removed, battery-power greatly increased. For all the improvements, it was a bit like playing catch-up ball with a team that had already forfeited; the boats had borrowed features from, but could not duplicate, the Type XXI.

Nonetheless, the Guppies were a revelation to ASW forces. Underwater speed was about doubled, to the range of fifteen to 17.5 knots (depending on the boat), and snorkel-enhanced endurance seemed almost infinite. The first Guppy deployed to the Far East, *Pickerel,* made the return trip, Hong Kong to Pearl Harbor, on snorkel. This 1950 record of twenty-one days, 5,194 miles submerged held until 1958, when the nuclear powered *Seawolf* ran submerged—no snorkel—for sixty-one days.

There was a downside to the conversions: Because space in the hull was finite and limited, the increase in battery power was achieved by increasing the number of lead plates in the batteries, which exposed more surface to electolytic action. This could only be done by making the plates thinner—to allow room for more plates—with the result that batteries which theretofore had lasted six years now had to be replaced every two years.

Another nineteen older boats were given some streamlining and a snorkel but not the batteries. The first of the fully postwar boats, *Tang* and *Trigger*—which more closely resembled the Type XXI—entered service in 1952. *Tang* set an American depth record, 713 feet, on April 30th.

During this immediate postwar period, researchers in the science of "sound" began to show phenomenal progress. First, they verified that sound travels at a constant speed and in a straight line in the ocean's isothermal layer (the layer where temperature is relatively constant); the depth of that layer varies from ocean to ocean, and with the seasons—in the Atlantic Ocean, typically three hundred feet in the summer, six hundred feet in the winter. Then they discovered that below the isothermal zone sound velocity varies with depth, and rather than travel a straight line, the path is refracted just as light passing through layers of glass. This explained why a submarine in the right position could actually hide beneath the isothermal layer—the "pings" from a searching sonar are bent off and away.

Another discovery: Sound passing through deep water at certain angles will actually refocus at special "convergence zones," typically every thirty-five miles from the sound source (in the Atlantic Ocean). When a tracking source and a target are in optimum positions, active sonar and passive listening range are both greatly extended. In 1947 (with the listening submarine stopped and crew members almost holding their breaths) a destroyer was detected in the third convergence zone at 105 miles, and depth charge explosions at six hundred miles.

Yet another: Sound bounces off the bottom and off the surface, going out and coming back, which under the right conditions can also greatly extend sonar range—perhaps to fifty miles for a unit with a normal, straight-path range of ten miles. Improvements in sonar range also came from the use of lower frequencies, although lower frequencies required bigger installations.

And still another discovery, perhaps the one of greatest significance: Researchers found a deep ocean channel over which sound is carried for hundreds, thousands of miles—tens of thousands under the right conditions. That discovery encouraged the development of the long-range fixed sound-surveillance system SOSUS, which began in 1956 as a series of one-thousand-foot hydrophone arrays planted in the deep ocean. The original installations could detect a snorkeling submarine at five hundred miles.

A key element of SOSUS was Lofar, a technique of electronic spectrum analysis so precise that a target submarine could not only be detected, but also be identified by individual characteristics. "Lofar" has sometimes been called "low-frequency analysis and ranging"—a logical error. But Lofar is a made-up word without significance, chosen merely because of its similarity to radar and sonar.

In the late 1950s, the success of SOSUS led to the development of portable arrays—hydrophones mounted on wire networks towed several miles behind a surface ship or submarine. Towed arrays are very effective, especially if they are long enough and properly "tuned" so that the noise of the towing ship does not interfere with the listening. Towed arrays are obviously cumbersome, and various methods have been tried to simplify the mechanical process of launching and retrieving an array. One method was to use a harbor craft to clip an array onto a submarine leaving on patrol and unclip it on return. More sophisticated methods have been developed; for example, winding the towline around a drum set inside a main ballast tank.

Also during this period, several unique classes of submarines were put into the inventory: the antisubmarine submarine SSK (a K-boat, only here K stood for "killer"), a seaplane tender, and some amphibious assault boats.

In 1948, the Navy envisioned a fleet of 245 relatively inexpensive SSKs. They would not need great range, because they would be operated from forward bases, such as Scotland. They would be so basic that equipment would be replaced, rather than repaired—even the power plant could be swapped out by a tender. Silencing might best be achieved by simply anchoring the boat underwater as it waited for its prey. K-1 through K-3, the only three actually built, came in at 196 feet, 740 tons. Sonar performance was greater than that of any surface ship, which opened the possibility of SSKs acting as manned, submerged sonars. Some fleet boats not suitable for Guppy conversion were turned into SSKs; one later nuclear SSKN, *Tullibee,* was effective, but superseded by advancing technology. The SSK designation was dropped in 1959; thenceforth all attack submarines were considered to be killers.

Guavina was converted as a seaplane tender, to support the P6M Seamaster minelaying seaplane. This 600-mph jet made its first flight in July 1955

and its last in December, when it blew up in midair. Tl
Jet-powered seaplanes were never successful.

The Marines wanted assault submarines which cou
operations with standard landing craft; that proved in
marines were converted as commando delivery systems by adding recycled missile hangars (see Chapter Thirty-Two) and that mission survives, although the Navy's "commandos" are called the SEALS. *Perch* landed a British commando unit in Korea (after *Pickerel* made a periscope evaluation of potential landing sites) to cut Korea's only railroad line, and participated in Vietnam operations. Several older nuclear boats were converted to SEAL support beginning about 1982, and two older ballistic missile boats, *Sam Houston* and *John Marshall,* were given a new version of the hangar, now called a dry deck shelter (DDS), to hold swimmer-delivery vehicles and carry sixty-seven SEALs; they were retired in 1992, and are being replaced by other boats.

The U.S. Navy embarked on a High Submerged Speed Submarine Program to pursue the two major lines of development suggested by Abelson: propulsion and hull forms. Since nuclear power was still in the future—no matter how optimistic some officials had been—the Bureau of Ships proposed to study the next best thing: the Walter cycle and other variations of a closed-cycle system.

Closed-cycle, by itself, was nothing new. It was merely a method of running engines while submerged, without access to atmospheric oxygen. Holland and others had given it a trial; but just as with the early versions of the snorkel, the ideas were either not suited to the technology of the day—or were simply forgotten, to be reinvented at a later time.

Three Walter boats had survived the war—more or less. The United States got *U-1406,* but it was scuttled, heavily damaged by fire and flooding and not again suitable for service. However, the vital element—the 2,500-hp Walter power plant—was removed, and set up at the Naval Engineering Experimental Station at Annapolis along with a 7,500-hp unit intended for one of the larger, unbuilt Walters.

In addition to the Walter cycle, the Navy's formal evaluation was to include a closed-cycle diesel (an outgrowth of the German wartime Krieslauf engine), a semiclosed-cycle gas turbine which used a heat-exchanger to transfer energy to the turbine, a steam plant which would operate on a high-pressure closed-cycle when submerged, and a free-piston engine (the comstion chamber is between a pair of opposed pistons; most of the expanding gas flows out to run a turbine—the remainder pushes the pistons back for another compression cycle).

The Walter system combined steam and diesel combustion to run a turbine. A major inhibitor remained the cost of peroxide—$20,000 to $30,000 per hour of full-speed operation, compared with $100 to $200 an hour for existing diesel engines.

Oxygen was considered as an alternative oxidant. The cost was about 6 percent of the cost of peroxide, but liquid oxygen had to be stored in pressure tanks, which would require special trim tanks and plumbing to compensate for the consumption of oxygen. In this, peroxide had an advantage: It is liquid at atmospheric pressure and—as it was in the Walter boats—could be stored in plastic bags between the inner and outer hulls. Trim would be automatic; seawater would replace the volume lost as the fuel was consumed.

Hardware trials for each of the variants were planned for 1950–51, but the schedules kept slipping. The test program was too complex; the Navy was trying to sort out too many systems at the same time, and kept bumping into itself. Also, unlike the diesel-engine competition of the 1930s, there was no clear commercial application for any of these submarine propulsion systems. American industry did not jump in with great enthusiasm. The "interim" program was overtaken by the unexpectedly fast development of the nuclear power plant, and was canceled in October 1953. Most of the technology eventually found a home in surface ships; one closed-cycle plant did enter underwater service—aboard the midget submarine X-1, described on page 352.

The British explored—briefly—the free-piston engine, the gas turbine, and the Krieslauf, and worked on several versions of the Walter cycle, one of which used liquid oxygen rather than peroxide. The ostensible reason for the shift was the uncertain supply of concentrated peroxide; the true reason may have been the uncertain safety of the peroxide engines. In his postwar memoir, former U-boat commander Peter Cremer noted, ". . . when the Royal Navy tried to set a captured XXVI [sic] Walter U-boat in motion, the hydrogen peroxide propulsion exploded."

Cremer's designation of a Type "XXVI" illustrates a continuing confusion in later U-boat nomenclature. Some authorities label all Walter boats with the double-X, others hold the position that the Walters of 1943–44 were single-X, falling between the 1941–42 Type XIV "Milk Cow" and the 1944–45 XXI oceangoing Elektro. In any event, Cremer was referring to *U-1405*, which was a Type XVIIB; there probably was no XVI (or XXVI) as such; seven XVII Walter boats were finished in various size and configuration, as noted above; the XVIII was not built, and the hull design became the XXI.

Why the confusion? A better question would be, why not? While Germany was falling apart, almost two thousand boats with a double-X designation were in the pipeline; U-boats were being launched even while Allied troops were entering the coastal cities. The four completed Walter XVIIAs and the three XVIIBs were of a superior technology; it made sense to some people to give them a superior number.

The Soviets had no success with their Walter-cycle submarine; they did build some Kreislauf-powered boats, the Quebec class. Beginning in the 1980s and to the present day, several closed-cycle systems are being offered for sale by Russia—and Sweden and Germany as well—for contemporary diesel boats. These include the Krieslauf, fuel cells, and the Stirling engine.

When it came to hull forms, the Navy found that it was literally sailing into uncharted waters. The Walter boats had demonstrated high speeds, but there also had been reports of problems with control—nothing major, but nerve-racking. The American crews which sailed the Japanese high-speed boats took them as far as Hawaii, and wanted nothing more to do with them for the same reason.

American submariners long had been dealing with one-speed-related instability: The broad, flat decks of the fleet boats became control surfaces on a down-angle, thrusting the bow below that ordered by the man at the controls, and even at eight knots, threatened a too-rapid descent below crush depth. High speed would compound the problem: a boat with a 30-degree down-angle moving at thirty knots would dive five hundred feet in twenty seconds.

As it turned out, the Navy did not know very much else about high-speed hydrodynamics. The David Taylor Model Basin in Carderock, Maryland, ran a series of tests using abstact hull forms, later modifying the shapes and adding such appendages as the bridge, control surfaces, and device housings. One series of tests used the airship as a model form—totally symmetrical with no parallel planes—and established that the ideal length to diameter ratio was 6.8 to 1. (The fleet boats were 11.5 to 1.) They determined that a single screw was more efficient that twin screws. They were repeating, perhaps unwittingly, much of the data Holland already had developed almost half a century earlier.

Tests in the model basin raised questions about, but did not offer solutions to, the problems of high-speed control, which clearly had to be addressed before the Navy began investing in some very expensive hardware. Thus, the Navy created the experimental test-bed submarine *Albacore* (AGSS 569).

The designers were given a free hand, unconstrained by operational considerations or the need for a weapons suite. In fact, *Albacore* had no snorkel, no weapons, and a minimal diesel plant for running on the surface—but was given a 7,500-hp electric motor (almost twice the output of that on the Type XXI). She lived up to expectations: In her first configuration out of five (1953 through 1971), *Albacore* reached a submerged speed of twenty-six knots. She also demonstrated serious instabilities. The General Board of 1906 had argued that a submarine would become unstable at submerged speeds greater than six knots. They had the right concept, the wrong threshold.

For example, normal-sized control surfaces proved to be *too* effective at high speeds—in a tight turn, the boat would execute an uncommanded snap-roll, whereupon both the sail and the rudder began to act as a diving plane. As a result, any turn led to a dive—the sharper the turn, the steeper the dive. Aircraft-type trim-tabs, which provided minimal control surface, helped correct that problem.

In Phase II (1956), the control surfaces were moved *forward* of the propeller, a position which Holland had used in the initial configuration of *Holland VI* and had changed, under pressure from his colleagues, when "erratic steering" was cited for the failure of the first Navy trial. Holland had the right idea after all.

In her fourth configuration, completed in 1965, a second 7,500-hp motor was installed to drive a counter-rotating propeller in line with the first. According to *Jane's Fighting Ships, 1967–68, Albacore* then set an underwater world speed record of thirty-three knots. By 1972, her original diesel engines were shot, and replacements of the long out-of-production model were not available. New engines would have required enlargement of the hull; this would have changed hull dynamics; the cost was prohibitive. *Albacore* was retired, and is now on display as a museum near Portsmouth, New Hampshire.

Albacore introduced the concept of a single pilot—borrowed from the Germans—to American submarines. Theretofore, separate operators controlled the rudder, the bow diving planes, and the stern diving planes; at higher speeds, coordination was tricky. Control proved tricky for the single pilot in *Albacore* as well; the similarities to an airship were significant enough that her pilots first trained on blimps.

Many other navies adopted the single-pilot configuration, adding computer control to assist in fast-changing maneuvers. American submarines after *Albacore* kept one planesman in addition to the helmsman. The Americans were not in favor of taking control away from a human being and giving it to a computer.

A recent event suggests that this was not necessarily reactionary. Freidman (*U.S. Submarines Since 1945*) notes an almost-fatal event during trials of the British submarine *Upholder*, which has a computer-controlled electric motor. The computer was programmed to trip the circuit if the motor became overloaded—an electronic circuit-breaker, if you will—but without a "slow-blo" feature. The commanding officer ordered emergency back full from ahead full, the motor overloaded, the computer did what it was supposed to do, and the boat was perilously close to sinking. The computer programmer did not understand the dynamics of an electric drive: The propeller continued to move forward for a time after the field in the motor had been reversed. A time-delay was incorporated.

The *Albacore* hull form was transferred to the *Barbel* class of late diesel boats, the *Skipjack* nuclear class, and the experimental midget *X-1*.

The Navy became interested in small submarines, similar to World War II British X-craft and her German and Japanese cousins, in 1949. These were true submarines, not torpedoes with a driver's seat. They were designed to place a mine against or under a moored warship—Bushnell and Fulton would have approved—or to plant a mine in a channel. A British midget had been used against *Tirpitz* in Norway, and another sank the Japanese cruiser *Takao* in Singapore. Midgets offered certain quantifiable advantages: too small to present a viable radar or sonar target, too quiet to be detected by hydrophones, not enough mass to put a blip on a harbor-defense magnetic detector.

Perhaps of more immediate import, the Soviets were known to have captured the most advanced of the German midget designs and were rumored to be advancing the start of the art. The U.S. Navy at the least wanted a midget type to play "aggressor" to test harbor defenses.

The Navy sketched out two models. Type I was an "attack" version, fifty-six feet long with a surface displacement of seventy-three tons—just slightly larger than *Holland No.1* (fifty-four feet, sixty-four tons). A small diesel-electric power plant and a snorkel comprised the propulsion system; armament would have been homing torpedoes. Price: about $1 million. The Type I project had been abandoned by April 1950.

However, a Type II "harbor attack" version became the forty-nine-foot, twenty-nine-ton, four-man *X-1*. This boat was a more direct descendant of the British X-craft, and the concept was being revisited because of the advent of nuclear weapons. Where the World War II version could carry enough explosive to take out a single warship, a midget-planted nuclear weapon could take out the whole harbor. The power plant for X-1 was a dual-cycle diesel engine, operating in normal fashion on the surface, but burning a combination of diesel oil and hydrogen peroxide while submerged. Batteries were used for slow speed, long-endurance running. The power plant exploded on May 20, 1957, and was replaced with a fully conventional diesel-electric plant. *X-1* was decommissioned in 1973 and put ashore as a memorial at Annapolis.

The U.S. Navy has continued with efforts to develop a midget; one of the more interesting projects, proposed by a team at the Naval Ordnance Test Station, China Lake (California) was a two-man interceptor submarine called *Moray* in 1960. *Moray* would loiter in a high-threat area, diving as deep as six thousand feet to wait for a target—then dash off at forty knots to close the range and fire a battery of eight underwater rockets. The original scheme was propelled by a closed-cycle steam engine fueled by diesel oil and hydrogen peroxide; gaseous oxygen was later substituted. *Moray* itself was never built, but the closed-cycle combustion chamber was tested. An electrically powered test version of *Moray,* dubbed *TV-1A,* was completed in 1962 and reached depths of two thousand feet.

Should you think that a six-thousand-foot goal was over-reaching, know that the Deep Submergence Rescue Vehicle (page 365) is rated to six thousand feet, and the Lockheed Aircraft Corporation built a deep-ocean research boat in the mid 1960s, *Deep Quest,* with an eight-thousand-foot capability.

A footnote to history: After the war, Professor Walter emigrated to the United States, where he assisted in the evaluation of his system, and later took a position with the Worthington Pump Company of New Jersey. In 1965, Roy Crane, author of the nationally syndicated King Features comic strip *Buz Sawyer,* called on Dr. Walter for advice; Crane was obsessive about accuracy, and anything he put in the strip had to be both plausible and technically possible. Walter helped develop an episode which featured the demented commander of a twenty-year-old Type XVII Walter boat—the "only one to escape at the end of the war"—long hidden away in a Venezuelan coastal submarine pen.

The fictitious Captain von Spitz had been waiting all those years for a chance to carry on Hitler's dream of world conquest, and had created an underwater-launched missile powered by hydrogen peroxide. From Cuban waters, he would launch his missile on Miami and—he believed—provoke a cataclysmic nuclear exchange between the United States and the Soviet Union. He and his crew would be standing by to pick up the pieces. Commander Buz Sawyer, USNR, of course, stepped in and saved the day—saved the world.

Dr. Helmuth Walter, the genius who was an advisor to Adolf Hitler and who had a solution which might have won the war for Germany, was now advisor to an American comic strip.

It was his last involvement with submarines.

31 1950–1968—*Nautilus* Reborn

One of the most wonderful things that happened in our Nautilus *program was that everybody knew it was going to fail—so they left us completely alone so we were able to do the job.*
—*Hyman Rickover,* Reader's Digest, *July 1958*

More than any other man, Hyman Rickover brought forth not only the nuclear-powered submarine, but commercial nuclear power as well. For his reward, he was often ignored, disdained, and penalized by the corporate Navy—which even made so bold at times as to deny Rickover the credit, while attempting to give that honor to others. The Rickover story has been told, from all points of view, in a number of fairly recent books, and the controversies surrounding the later career of the man known with derision as the "kindly gray-haired old gentleman" are outside the scope of *this* book. However, as they involve his initial work in bringing nuclear power to the submarine, they merit our attention.

Hyman Rickover was born in Poland just after the turn of the century and came to the United States at the age of six, the family eventually settling in Chicago. In 1916, working as a Western Union delivery boy, Hyman had a brush with fame when a newspaper photographer snapped a photo as he delivered a telegram to President Warren Harding at the Republican National

Convention. Unlike some young men of our time who were photographed with Presidents, there is no evidence that this event was later to shape his life; he seems not to have harbored any political ambition.

However, of regular ambition he had plenty. His local congressman—a friend of his uncle—offered him a chance to get an otherwise virtually unobtainable college education with an appointment to the Naval Academy; Rickover graduated 107th out of 540 in the Class of 1922. It was a respectable ranking, reflecting both Rickover's fairly good academic performance and his fairly abysmal athletic and military performance.

Over the next fifteen years, he served in a variety of engineering positions—on a destroyer, two battleships, a submarine. Naval Postgraduate School (which included a year at Columbia University) earned him a master's degree in electrical engineering. He demonstrated a range of interest wider than that of the average Naval officer; in an essay published in the September 1935 issue of the Naval Institute *Proceedings,* Lieutenant Rickover speculated that the "changed conditions of modern warfare" mandated a restudy of international law, citing not only the advent of unrestricted submarine warfare, but also the declaration of war or danger zones for minefields, the devastation of the British blockade, and aerial bombardment. On his own time and for his own interest, he also produced the first English translation of Admiral Hermann Bauer's *Das Unterseeboot;* Bauer had spent most of World War I as the commander of all German submarines.

In 1937, Lieutenant Commander Rickover became commanding officer of the China-based minesweeper *Finch.* He was the third commanding officer within a year; there would soon be a fourth, as he was accepted for appointment as an Engineering Duty Only (EDO) officer. This was Rickover's last sea duty and his only command.

It was in the engineering community that he soon made his mark, especially during the war. It was a mark that would have remained if nuclear energy had never been unleashed. His primary assignment during the war years was as head of the electrical section of the Bureau of Ships. When Rickover reported aboard, the section had twenty-three people; by war's end, it had grown to 343 people. Yes, everything grew during the war, but *this* growth was matched with accomplishment. Among other contributions:

—The section developed electrical cable which prevented water from leaking from a flooded compartment to its neighbor.

—They developed electrical insulation that would resist high-temperatures and not give off noxious fumes.
—They rationalized the cataloging of electrical equipment; Rickover found one item which was listed under twenty-five different part-numbers.
—They developed the casualty power system, whereby cables could be plugged into fittings throughout a ship to bypass battle damage.

Rickover's last job before he got involved with the nuclear program was helping put ships into mothballs—the "Reserve Fleet"—against any possible future need. When he got his orders to Oak Ridge, he grabbed everything he could find on the subject of atomic physics (there wasn't much). When he arrived at Oak Ridge, he exercised his authority as the senior Naval officer on-scene (although not given orders as such) and took control of the Navy contingent, forcing study hours and discussion. Most of the others were to become key members of the Rickover team—once he was able to assemble a team.

There would be a slight delay. When the Oak Ridge assignment was completed, Rickover expected to be named to head the Navy nuclear program, but was bitterly disappointed when several other Naval officers, none of whom had been at Oak Ridge and none of which seemed to have any special qualifications, were given nuclear-power jobs. Rickover had been cut out of the loop, and was slated for orders back to Oak Ridge to help declassify documents. He managed to avert that disaster and wrangled assignment as Special Assistant for Nuclear Matters to the Chief of the Bureau of Ships, and what may have appeared to others as an impotent desk job became, for Rickover, the heart of the nuclear program. He spearheaded the preparation of a classified memorandum for Chief of Naval Operations Nimitz, describing the military advantages "of a *true* submarine, that is, one that can operate submerged for very long periods of time and is able to make high submerged speeds." By this document, Nimitz would recommend to Secretary of the Navy John L. Sullivan "that the Navy initiate action with a view to prompt development, design and construction of a nuclear-powered submarine." The memo also noted the potential for arming the submarine with nuclear guided missiles. This marked the true beginning of the Navy's nuclear power program.

Rickover also established himself on the staff of the Atomic Energy Commission; this gave him a voice outside of the Navy which proved helpful in resolving Navy-AEC issues in his favor. He also courted the Congress with skill and purpose. He knew indeed that Congress bought ships, not the Navy.

Rickover brought the skills of the engineer to the program, rather than the instincts of the scientist. A scientist would build and test a series of systems before settling on perhaps two, then narrow to one, build a prototype, and finally go into production after several years. The engineer—especially a Naval engineer, trained to make quick decisions—picked a likely contender, and built it. Rickover carried the concept almost to the extreme: To accelerate production, he had construction of the hull started well before the prototype power plant was even tested.

Abelson had suggested a liquid sodium potassium alloy as the heat-transfer medium; Alvin M. Weinberg at the Argonne National Laboratory of the University of Chicago proposed pressurized water. Rickover decided to try both—starting with pressurized water for the first submarine, and then following up with a second reactor for the next submarine using the more challenging liquid sodium. Challenging in that, should the liquid sodium ever come in contact with water or steam—quantities of which would be all around the power plant—the reaction could be violent in the extreme.

The prototype Submarine Tactical Reactor (STR) was built into an actual section of submarine hull (constructed at Electric Boat and hauled across county to the AEC test site at Arco, Idaho) to ensure fit and compatibility; when the reactor went on-line, Rickover staged a mock full-power submerged trans-Atlantic run: 2,500 miles, ninety-six hours, at an average "speed" of 14.4 knots. To that time, the longest that any submarine had ever run on full power was twenty hours—on the surface.

Rickover might logically have expected a promotion—to rear admiral—in recognition of his accomplishments. His record was before the 1951 selection board, but he was passed over, not selected. Privately, his staff complained to Navy officials and was told, essentially, that promotion was never a "reward" for past service but a recognition of future potential, and that no one could be promoted unless there was a need for someone with his skills. Rickover was in a captain's billet; there were other qualified captains who could fill that billet; there were no flag officer billets in nuclear engineering.

They were not told, but presumably understood, that Rickover was not popular in the engineering community. He was abrasive, demanding, antisocial, not a "team player." A year later, Rickover was passed over a second time, and under law soon would be forced to retire when he completed thirty years active service. Rickover's associates began an open campaign in the Congress to get him promoted, and some in the Navy began a hidden campaign to discredit Rickover. An article in *Collier's* magazine (December 20, 1952), extolling the virtues of the "atomic" submarine as "America's Dreadful New Weapon" and ghost-written for Chief of the Bureau of Ships Homer N. Wallin, did not even mention Rickover.

Clay Blair, a staffer for *Time-Life* and a Rickover partisan, put together a draft ten-thousand-word article about the Navy's nuclear program which he duly submitted for "security and policy review." The draft was not intended so much for publication as to smoke out the Navy opposition to Rickover. It did just that.

The Navy's "Security and Policy Review"—a commander in the Office of Information—wrote an internal memorandum, not intended for public consumption, allowing that "This Time-Life article as a whole cannot said to be false, since perhaps 90 percent of the raw factual data appears to be accurate. . . ." But the commander certainly understood the position of the Chief of the Bureau of Ships: ". . . to combat the effects of the article . . . a story [should be] placed in a magazine friendly to the Navy and highly competitive to Time, to appear before the Time article does. . . . this would have to be handled with extreme delicacy, preferably by the Chief of Information personally."

And so it was, in a memo signed by the Chief of Information, Rear Admiral Lewis Parks. For his reward, it was all to come back and bite him in the very public forum of Drew Pearson's widely syndicated column:

> President Eisenhower picked up a newspaper shortly before the launching of the atomic submarine *Nautilus* last week, and almost spilled his breakfast coffee. What he read was a news report that the *Nautilus* was not battle-worthy, was merely a test, and in effect was not an important naval vessel at all. The President was furious. . . . The allegation that the *Nautilus* wasn't battle-worthy; that her torpedo tubes were added only as an after-thought; and that her delicate equipment would not work at high speeds, was prepared first by Cmdr. Slade Cutter. Later it was put in a memorandum signed by Parks.

If nothing else, that sort of behavior reinforced the eventual shift of the position of Chief of Information from that of an unrestricted line officer passing through a staff billet, whose primary responsibility was to "protect the Navy," to that of a restricted line specialist in public affairs—who presumably knows that the Navy is best "protected" by encouraging fair and open discussion of issues properly in the public arena.

The irony here is that in March 1950, the Ships Characteristics Board had defined Nautilus as an experimental test bed—as with *Albacore*—"to determine the capabilities and limitations of a high-speed nuclear powered attack submarine," specifically stating, "There will be no armament in the original ship." It was Rickover who pushed for the torpedo tubes.

The *Washington Post* was tipped off to the "Security and Policy Review" fiasco, and offered editorial comment: "In short, this appears to have been an attempt to censor an article not so much because of wrong facts as because of opinions and interpretations which officials found distasteful. This is very perilous ground, indeed. It is the sort of approach, if condoned, that could be used to muzzle all sorts of legitimate criticism; as such it ought to have the prompt attention of the president."

The Navy kept singing the same old, not necessarily inaccurate, song: Wallin, called to testify before the Senate Armed Services Committee, emphasized that Rickover was in a captain's billet, there were a number of other captains equally qualified to occupy that billet, and a flag officer was not needed. However, when the Senate Armed Services Committee held up the promotion of the thirty-nine captains who had been selected by the 1952 board and threatened to require the participation of civilians on selection boards, the Navy gave up; Secretary Robert B. Anderson convened a special continuation board, to consider engineering-duty-only officers facing forced retirement for retention on active duty for one year. The precept given the board specified that at least one had to have the nuclear qualifications that only Rickover possessed. Then, the next flag selection board, for 1953, was given similar guidance. Rickover was promoted.

Nautilus was under way for the first time in September 1954, as signaled by Commander Eugene Wilkinson in a most prosaic statement for such an historic moment: "Underway on nuclear power." The submarine—which bears a strong external resemblance to the Type XXI—is 323 feet long, 3,674 tons on the surface, and cost about $100 million (about two-thirds of which was for the power plant). Surface speed was eighteen knots, twenty-three knots submerged.

There was never much doubt that *Nautilus* would demonstrate the promised attributes of a "true submarine," but there was no concept of her exceptional capabilities until she actually entered service. *Nautilus* began making believers out of the submarine community on her shakedown cruise in May, when she steamed, entirely submerged, 1,381 miles from New London to San Juan, Puerto Rico—at an average speed of more than fifteen knots. Where the Guppies had been a revelation to ASW forces, the nuclear boats were a revolution. On her first exercise with an ASW Task Force, *Nautilus* overtook the force (moving at speeds to eighteen knots) and when attacked, outran the homing torpedoes. The Task Force claimed seven kills; post exercise evaluation revealed that in each case the submarine was as much as fifty miles away from the "attack." In another exercise, *Nautilus* proved to be of more value than all twenty-one participating snorkel submarines. At one point, she spent fifteen hours keeping station *under* the ASW formation—without being detected. By the time of her first refueling in 1957, after twenty-six months and almost seventy thousand operational miles, she had been "attacked" five thousand times, only three of which were judged successful. *Nautilus* performance soon put an end to diesel submarine construction for the U.S. Navy.

Nautilus was next refueled in 1959, after another twenty-six months and 93,000 miles. Retired, *Nautilus* is now on public display at the Submarine Force Museum in Groton, Connecticut.

There had been some improvements in guided and homing torpedoes immediately after the war, although the heavy warhead of the Mark XIV made it the weapon of choice for armored warships. The advent of *Nautlius* made it quite clear that greatly improved weapons were needed, the most critical element being high speed. A fast submarine could overtake its own Mark XIV while the torpedo was still accelerating.

The Mark 48 became the primary antisubmarine torpedo: with both active and passive sonar, originally powered by a turbine fueled with a combination of 90-percent hydrogen peroxide and diesel oil. The H_2O_2—as usual—was difficult to work with, and Otto W. Reitlinger of the Navy Research Lab developed a safer oxidant/fuel combination (called "Ottofuel II"). As it developed, the movement of the torpedo through the water made more noise than the engine, so a less-complicated piston engine was substituted. Also, true quiet was required only while the weapon was searching for a target; it could dash to the general area of the target at any speed, slow down to get a fix, then speed up for the attack. Nonetheless, the Mark 48 matched the low sound levels of the best of the postwar torpedoes while more than doubling top speed, to fifty-four knots. The British version, "Spearfish," retained the turbine and is reported to have reached seventy knots.

The second nuclear submarine, *Seawolf,* did not have as happy a career—although not for want of trying. *Seawolf* remained completely cut off from the atmosphere during a sixty-day, 13,700-mile test in 1958, and was the first nuclear submarine to take a President to sea (Eisenhower). There were problems with the liquid-sodium reactor. There were teething problems which delayed construction, and performance did not match that of *Nautilus;* when the heat-exchanger began to develop leaks, Rickover pulled the plug on liquid sodium, and in 1958–60 the reactor was replaced with the now-proven pressurized water reactor.

The first "real" nuclear submarine—that is, the first to realize the potential of both nuclear power and the high-speed *Albacore* hull—was *Skipjack,* commissioned April 1959, which easily exceeded thirty knots.

Rickover developed what he saw as a family of reactors, each sized to a different requirement: submarine, aircraft carrier, cruiser. One other development, of more limited utility, was the Natural Circulation Reactor (NCR). This design eliminates the need for noisy circulating pumps by letting natural convection do the job, but these reactors are necessarily taller than others and are useful only in large-diameter hulls such as the *Ohio*-class Trident submarine.

Rickover's passion for safety created a system where the crew was exposed to less radiation aboard ship than when ashore; heavy shielding encased the front and back of the reactor compartment (as well as the tunnel which allowed personnel to pass through the compartment). The ocean itself provided much of the shielding for the sides of the compartment.

Rickover also exercised iron control over the selection of officers for nuclear programs, which continued to fuel anti-Rickover sentiment (even passion) for the rest of his career. He asserted that such control was necessary because, initially, of the inherent hazards in nuclear power. Dr. Edward Teller proposed that nuclear submarines would be too hazardous to enter any populated harbor; they should be moored instead at artificial offshore islands created for the purpose. Rickover believed that he could guarantee the safety of his programs only if he controlled the selection and training of the officers. Over the years, the methods by which he made his selections came in for a great deal of criticism: Each candidate was subjected to an interview with Rickover, in which Rickover was likely to be blunt, demeaning, belittling, and harrassing. His defense: He wanted to see how they acted under pressure; if they couldn't handle it in a safe and artificial situation, how could he trust that they would stand up in a real emergency?

Rickover "retired" at the statutory age, sixty-two, in 1962, but remained on active duty through a series of two-year extensions for another twenty years. In 1973, he was promoted to the four-star rank of admiral, under Congressional prodding and as a final reward for his very real accomplishments. His eventual retirement was forced by a man half his age, Secretary of the Navy John F. Lehman, Jr., with the support of the seventy-year-old President, Ronald Reagan. It was clear to most observers—had been clear for some time—that Rickover had overstayed his welcome. Along the way, "his" nuclear navy had grown to encompass 169 submarines, four aircraft carriers, and nine cruisers/frigates, completed or authorized.

Nuclear power unleashed the designers' imaginations, and in the mid-1950s almost every type of Navy ship was considered for adoption as a submarine: tankers (31,200 tons submerged!), ammunition ships, amphibious landing ships, minesweepers, aircraft carriers. Several nuclear-powered submarine carriers were sketched out, ranging in size from 460 feet, seven thousand tons, to a clearly impractical 24,000-ton monster with a full carrier-type flight deck, 500 feet long and one hundred twenty feet wide. The Sea Dart jet-powered seaplane fighter was then under development, and there was some thought of adapting the A-4 Skyhawk to hydro-skis; a submarine carrier was just the thing to take an atomic-bomb-carrying plane to within quick striking distance of the enemy. But jet-powered fighter and attack seaplanes proved to be as dangerously impractical as the P6M minelayer and those projects were abandoned.

One unique type which was put into production was a nuclear radar-picket submarine. Toward the end of the war in the Pacific, American surface forces needed as much advance warning of the approach of enemy aircraft as possible; the radar picket destroyer provided the reach, but was itself vulnerable to attack, especially by the *kamikaze.* A radar picket submarine seemed to be an attractive alternative; ten fleet boats were converted to radar pickets and two were built from scratch after the war. The concept remained important (despite a nagging fear that even a submarine radar picket would soon enough reveal her presence to a searching enemy), and became compelling with the advent of the speed and staying power of the nuclear submarine. The first—and only—nuclear radar picket was *Triton,* at 447 feet and 5,450 tons the Navy's largest submarine to that time. As the U.S. Navy's only twin-reactor boat, *Triton* proved the type, but her mission was overtaken by the development of more effective early warning aircraft, operating from, and moving, with the carriers. Literally "flying combat information centers," they could not only detect enemy aircraft, but also control friendly aircraft to the attack.

Triton was deemed too large for conversion to an attack submarine, and in 1960 became the first nuclear boat to be put in reserve. She will always be remembered for her swan song—the first submerged circumnavigation of the globe. Under Commanding Officer Edward L. Beach, *Triton* followed the route taken by Magellan's five ships in 1519 when they began the first around-the-world voyage. Magellan was killed in the Philippines; only one of the ships and eighteen men completed the trip, which took three years. *Triton* handled the 36,014 miles in eighty-four days. She almost came to the surface once—off the Philippines—to transfer a sailor with acute appendicitis to a waiting ship, but just the top part of the sail was brought out of the water.

Triton performed no other memorable feats, but did serve as a prototype for such multi-reactor surface ships as the cruiser *Long Beach* and the carrier *Enterprise.*

The next major design was *Thresher,* the prototype for the modern attack submarine: fast, with a strong weapons suite and the latest sonar, and an ability to go to 1,300 feet, deeper than any other American submarine to date. Tests with *Albacore* notwithstanding, there was always the possibility that a fast submarine, temporarily out of control, might crash through the seven-hundred-foot test depth before she could recover. As a beneficial byproduct, a deeper-running boat would be less vulnerable to ASW weapons. A material was on hand—HY-80 steel—which would provide the necessary hull strength to reach 1,600 feet; however, existing fittings, bushings, and piping were not as robust, which determined the final design target of 1,300 feet.

There were some problems with *Thresher*—minor things mostly, but enough to send her back to the shipyard for changes after little more than a year in service. Then on April 10, 1963, after two years in commission and during post-shipyard sea trials, *Thresher* sank in 8,400 feet of water with the loss of 129 lives.

The post-accident investigation assumed—no proof could be assigned—that the accident might have been caused by the failure of a section of piping. Judging from sound recordings of the last moments of *Thresher,* she probably went into an uncontrolled dive, losing volume and gaining momentum with each foot.

The design had been widely duplicated; *Thresher* elements were in every submarine then under construction or development, including the ballistic-missile submarines deemed the most vital part of the nation's nuclear-deterrent forces. Was a design flaw responsible for the loss?

The probable scenario: A leak developed in some seawater piping; there was no central point from which the piping could be isolated; the flooding shorted out the electrical system, thus causing the reactor to shut down; when the captain tried to blow ballast, the sudden expansion of moisture-laden air formed an ice plug in the high-pressure lines. As it turned out, each component of the ballast-air system had been tested—but not the system as a whole.

Design flaw? Several. Loss of *Thresher* had vast political implications; any move by the Congress to put a safety "hold" on all submarine programs could have compromised the strategic-missile effort and given the Soviets time to catch up. A wide-ranging safety improvement program, SUBSAFE, was launched and applied, as possible, to all submarines. This required:

Any piping connected through the pressure hull to outside water pressure must be constructed as if it were part of the pressure hull.

Control valves, theretofore scattered about the boat in whatever position seemed convenient, were to be collected into a central manifold and the main control station.

Reactor shutdown would be under human control.

The diameter of high-pressure air lines was to be increased.

The *Thresher* class continued, but was redesignated *Permit* (the next boat in line).

The only other postwar American submarine loss—*Scorpion,* on May 24, 1968—was probably caused by an erratic onboard weapon. Sound recordings of *her* last moments were apparently obtained through the SOSUS secret seabed hydrophone array planted in mid-ocean to help track Soviet submarines.

The search for each of those boats brought out some of the finest underwater technology. The research bathyscaph *Trieste* located *Thresher* in June 1963, and recovered some bits of wreckage; *Scorpion* was found by the research ship *Mizar* in June 1968. And although no submarine could survive at the depth where *Thresher* went down, the loss spurred the development of the air-transportable Deep Submergence Rescue Vessel (DSRV), able to mate with and remove personnel from a submarine, twenty-two at a time, to a depth of six thousand feet.

In January 1968, en route from California to Hawaii, the nuclear-powered aircraft carrier *Enterprise* detected what turned out to be a November-class Soviet submarine. *Enterprise* increased speed; the November increased speed. *Enterprise* went up to her maximum thirty-one knots; the November kept station.

This was most troubling. The U.S. Navy assumed that the November class boats were equivalent to the *Nautilus*—fast, but not *that* fast. It was a big shock to discover that the Soviets appeared to have broken the thirty-knot barrier.

When the Soviets moved into nuclear power in the late 1950s, they gained a head start by following, stealing from, and copying American developments. The Soviets tested their first reactor—pressurized water—in the icebreaker *Lenin,* and had their first nuclear-powered submarine in service by April 1958.

The U.S. Navy was always spooked by Soviet claims: In 1948, a Soviet admiral apparently announced a fleet goal of 1,200 boats. A year after their first nuclear submarine went in service, Soviet Premier Nikita Khrushchev bragged to American President Eisenhower that *his* nuclear submarines were twice as fast. They certainly were building them faster. By 1963, five years into their nuclear program, the Soviets had twenty-four boats in three categories, all with the same basic reactor. Five years into the American program, the U.S. Navy only had eight boats in five different classes with six different reactors.

One thing the U.S. Navy did not know about the Soviet boats—indeed, one thing the Soviets themselves did not realize for many years—was the extremely hazardous design of the nuclear plant. The Soviets copied what they saw (and learned through spies), but not with the understanding of the problems of nuclear power and the care exercised by Rickover and the U.S. nuclear industry. Entire crews of Soviet boats may later have died from radiation poisoning.

However, the November surprise was real, not propaganda, and the Navy's answer was a new class of fast-attack submarines. The project would logically have been assigned to Electric Boat, but the classic "Electric Boat-Navy" issues had not gone away; only the specifics had changed. There were labor problems at the yard, and delivery delays, and some apparent dissembling by senior management (to the point that the General Manager of the yard fled the country under indictment). Just at that time, the Department of Defense had decided that Navy shipyards should get out of the design and construction business and concentrate only on overhauls, upgrades, and (potentially) battle damage. Therefore, Newport News Shipbuilding and Drydock Company, in Virginia, was given the lead to design *Los Angeles.* It was the first—and to date, the only—submarine designed by Newport News, although the yard continued to build submarines.

Rickover insisted that the solution to more speed for *Los Angeles* was brute force—a reactor about twice the size as any used up to that time, four times as large as some. As a result, *Los Angeles* was a boat designed to accommodate the power plant, not the other way around, and the compromises limited performance; for example, because the reactor was so large, hull weight had to be reduced, reducing planned operating depth. *Los Angeles* came in at 360 feet, 6,080 tons, with respectable but not spectacular performance.

Some innovations: a periscope with 24X magnification, almost four times the norm, and a periscope with a video camera system, whereby, for the first time, anyone in the crew could see what the captain saw through the periscope. This may have inhibited the captain, but opened a whole new range of possibilities for attack coordination. Sixty-two *Los Angeles*-class boats were commissioned over a twenty-year period; two sub-classes built within that run offered improvements over the earlier boats.

Los Angeles was that elusive "fleet boat" all navies of the world had been seeking for so long.

32 Unto the Present Time

The combination of the speed and the stealth and the weapon-carrying capacity of this ship really makes it the most powerful warship in the world; not just the most powerful submarine.
—Rear Admiral Bruce DeMars, following initial sea trials of the new Seawolf, *July 5, 1996*

It was easy to find compelling reasons to develop an atomic-bomb delivery system for a submarine, not the least of which would be an ability to loiter off a hostile coast unnoticed until given a signal to launch—or given a signal to retire should the political situation improve. The atomic weapons which demolished Hiroshima and Nagasaki to end the Pacific war were far too large, at ten thousand pounds, but efforts already were under way to create smaller warheads which could fit into a submarine-launched delivery vehicle.

And for that vehicle, the Germans had shown the way with the V-1 jet-powered unmanned flying bomb and the V-2 ballistic rocket. The V-1 was the most easily adapted to submarine launch and the first to be employed, as the non-nuclear "Loon." *Cusk* and *Carbonero* were converted as Loon launch boats; the first test flight was in February 1947. Loon was a Mach 0.6 vehicle: it was stored in a hangar and launched from a fifty-foot ramp mounted on the afterdeck, tracked by radar and command-controlled from the submarine.

However, the submarine was not exactly hidden once a decision was made to launch a missile, and this posed some security issues. The submarine had to surface, erect the ramp, mount the missile, fire, house the ramp, and submerge to periscope depth to control the missile's flight. The goal for all of that was five minutes. To enhance security, control of the missile could be handed off to another submarine running at periscope depth, perhaps eighty miles downrange, which could guide the Loon another fifty-five miles to the target. If the enemy didn't know that the missile was coming until it had arrived, the launching and handoff submarines might well remain undetected.

Loon was effective. In a 1948 fleet exercise, three were launched at a task group which not only knew the missiles were coming, but also from where—and one even had a radar pinger to help the gunners—yet none were intercepted.

The next-generation winged missile was the Mach 0.91 Regulus I, which could carry a three-thousand-pound nuclear warhead five hundred miles; as with Loon, the launching sub would command and track the missile, although now to a maximum radar range of 250 miles, where a second boat could take over. When Regulus was ready in 1953, Loon was retired. A planned Regulus II, twice the size, was overtaken by the ballistic missile Polaris (see below) and not put in service. *Grayback* and *Growler* were the first boats converted to handle Regulus, by using *Perch*-type troop compartments as hangars. In 1969, conversion had gone full circle when *Grayback* was turned into an amphibious transport. *Growler* is now on public display as part of the *Intrepid* naval museum complex in New York City.

The first guided-missile submarine built from the ground up was the nuclear-powered *Halibut;* she fired her first Regulus in 1960. *Halibut* was also the first submarine to be equipped with a vital element for accurate control of guided missiles—the Ships Inertial Navigation System (SINS). This used a complex set of inputs and algorhithms to constantly update what in the old days would have been the manual ded-reckoning track ("Ded," as in "deduced.") A special stabilized periscope was developed to obtain celestial fixes and refresh SINS with accurate navigational data.

When Regulus became obsolete, *Halibut* was modified for special deep-ocean intelligence missions, with underwater reconnaisance gear and thrusters to permit keeping an exact position. One known mission: In 1968, *Halibut* investigated the wreckage of a Soviet "Golf" nuclear submarine which had sunk in waters near Hawaii. Attempts to recover the Soviet boat itself were made in 1974 by the *Glomar Explorer*, the cover mission for which was deep-sea mineral exploration and recovery.

Adaptation of the V-2 type missile was more difficult. The major problem was size: 5.5-foot diameter for the body, almost twelve feet across the fins, forty-six feet in length. With a two-thousand-pound warhead, a V-2 weighed 12.46 tons, 80 percent of which was liquid oxygen/alcohol fuel. That highly volatile liquid fuel was another matter of great concern to the submariners who would have to coexist with the system for weeks, months on end.

The German barge-launcher offered some advantages over a submarine-based missile, as missile size was not driven by submarine size (and vice versa) and the hazards of liquid fuel were not as significant. A scale model of the barge-launcher was tested; the exhaust gas blew it apart.

In the 1950s, the Army, Air Force, and Navy were in competition for funds to develop guided missiles—and in competition for control of the programs as well. The Air Force wanted money for an Inter-Continental Ballistic Missile (ICBM); the Navy wanted money for an Intermediate Range Ballistic Missile (IRBM). The Department of Defense allowed funds for each, but reserved judgment on the matter of control. There was also an internal Navy competition between the Bureau of Ordnance, which handled weapons, and the Bureau of Aeronautics, which handled flying things. Chief of Naval Operations Arleigh Burke solved *that* issue by taking the program away from both of them and creating a Special Projects Office (SPO). The SPO, under the direction of Rear Admiral William F. Raborn, known as "Red" for a visible reason, provided the necessary force to push a series of increasingly effective missiles through development and introduce them into the fleet.

Navy ballistic missile development started with a joint Army-Navy effort to adapt the Army's Jupiter missile—a descendant of V-2—to submarines; the Army hoped this would lead to Army control of IRBM programs. Jupiter was larger than V-2, almost eight feet in diameter, ninety feet long, but in 1956 seemed to be the only available alternative. Plans were sketched for a submarine with four missiles in launching tubes which extended from the keel to the top of the sail.

In the meantime, research with solid fuel was rushing ahead, and the advent of the hydrogen bomb eliminated some of the problems: A smaller package simplified delivery vehicle design and greater yield reduced the need for pinpoint accuracy. By 1956, it was apparent that a solid-fueled missile small enough to fit inside the pressure hull could be ready by 1960. The Jupiter project was canceled, Regulus procurement was curtailed, and follow-on Regulus development was canceled.

Then, in August 1957, the Soviets launched their first ICBM, raising the very real threat that Soviet missiles could catch the Air Force Strategic Air Command unaware and on the ground. The more probable survivability of the submarine-launched missile force under surprise attack raised everyone's priorities: What had been a planned 1963 deployment was moved up three years.

The 1,200-mile A-1 Polaris missile was ready in 1960. To illustrate how far the engineering had come: A solid-fuel Jupiter-S, payload three thousand pounds, range 1,500 miles, was developed at the same time; it came in at ten feet by 41.25 feet, but weighed 162,000 pounds. The A-1 Polaris, with comparable payload and three hundred miles less range, was 4.6 feet by twenty-eight feet and weighed only 28,000 pounds.

To speed construction of the launch platform, the 252-foot *Skipjack*-class boat *Scorpion,* already under construction, was commandeered to become the first ballistic-missile submarine, *George Washington.* The hull was cut in half and a 130-foot-long midsection inserted, complete with sixteen vertical missile launching tubes and all related fire control and navigational systems. The name *Scorpion* was re-assigned to another submarine, which had an unhappy outcome—as noted above, it became one of only two American nuclear submarines to be lost to accident.

George Washington was commissioned in December 1959, fired the first test missile on July 20, 1960—and was on patrol in November. To launch a missile, compressed air was blown into the tube to equal outside pressure; the hatch was opened and a vestibule above an expendable diaphragm above the missile was flooded, while at the same time water was pumped out of a compensating tank. Air at 4,000 psi then blasted the missile from the tube through the diaphragm (water immediately filling in behind) and the rocket motor was ignited after the missile had cleared the surface.

A 1,200-mile range might seem impressive, but severely limited the waters from which a missile could reach Moscow—always the symbolic, if not the most strategic, target; the submarine would almost have to be operating in the Baltic. The Soviets knew this, and established an Anti-Ballistic Missile

(ABM) defense system in between, near Leningrad (the once and future St. Petersburg).

The U.S. Air Force had an "ABM" program of their own—they wanted to gain control of Polaris, to "eliminate duplication of effort." President Eisenhower cut through the internecine warfare by placing all strategic weapons under a joint Single Integrated Operational Plan (SIOP), to be developed by a Joint Strategic Target Planning System (JSTPS) at SAC headquarters. Around 1961, the commander of SAC, General Thomas Powers, set out to disprove the Navy's claims for Polaris survivability with a war game; despite the fact that the ground rules were front-loaded with some rather extraordinary assumptions about the size and capability of the Soviets, the three employed SSBNs survived long enough (in the six-week game) to launch thirty-six of their forty-eight missiles. General Powers decided that there wasn't enough evidence to support any conclusion!

The 1,500-mile A-2 version entered service in 1962; the 2,500-mile A-3 was ready in October 1963. A-3 had three warheads, to complicate ABM targeting and fire control; these were not independently aimed but would all land on the same target. A B-3 version, with Multiple Independently targeted Re-entry Vehicles (MIRV) was under development by 1961; renamed C-3, Poseidon, it went to sea in 1971. The programs to set missile guidance for the wide range of potential targets were simply overwhelming for the computer systems when Polaris first deployed; the boats carried a deck of punch-cards, each for a different target. Eventually, computer technology caught up with need and all programs could be carried in memory and even modified as necessary.

A major concern was the difficulty in getting a launch signal to a submarine operating underwater perhaps three thousand miles from a transmitter. Normal radio frequencies in UHF and VHF spectra do not penetrate water; the initial solutions included a floating whip antenna connected by wire, or a long trailing wire antenna floating on the surface. These were moderately secure from detection, but not completely, and kept the sumbarine tied to relatively shallow depth. The long-range solution was to develop a system using extremely low frequencies (ELF), which did penetrate the ocean albeit in such a large waveform that to transmit even one letter in Morse code could take minutes.

The Navy ran into a serious public relations problem with ELF (given the rather strange name, Project Sanguine—"the color of blood") in the mid-

1970s. The transmitter required a large antenna grid—large as in several hundred square miles of wires laid in the ground over a granite substrate. A forest area in northern Wisconsin met the physical requirements, but local residents were frightened by speculation that the radiated signal would cause a wide range of physiological problems in wandering cows and hiking humans. Navy assurances that an electric blanket radiated more energy were met with skepticism, and for some unaccountable reason, the Navy medical team which ran a series of experiments to prove the safety of the system did not get around to publishing their data until it was too late. Project Sanguine was canceled; technology marched ahead and a less intrusive system was developed and deployed.

A total of forty-one SSBNs in three classes (*George Washington, Lafayette,* and *James Madison*) were built between 1960 and 1968. To maximize the assets, two complete crews were assigned to each submarine, dubbed as "Blue" and "Gold," which exchanged duties after a minimal post-patrol turnover; the concept was borrowed from the British, who thus kept ships on indefinite assignment on Far East station in the 1920s. The in-port crew filled in their time with personal and professional matters, notably intensive training and educational programs. By mid-1972, the one thousandth SSBN patrol had been logged.

The SSBN had a presumed life expectancy of twenty years; the first replacements were needed in 1979. The next-generation missile started life as the Undersea Long-range Missile System (ULMS), but was given the more dynamic name Trident in 1972. It grew out of a Department of Defense study on strategic alternatives which also produced the B-1 bomber and a new silo-launched ICBM. Unlike the A-1 though C-3 missiles, which had been purposely designed to fit in the same launch tubes, Trident would be significantly larger and so would the launch submarine. Because larger size would increase the possibility of detection, there was at least one proposal, which did not go very far, to plant the missiles off enemy shores in submerged pods, loitering and ready. This was fraught with too much political jeopardy, however; the launching of nuclear missiles had to be under constant and inviolate human control.

The Trident submarine design team considered 117 alternative configurations, including missile capacity ranging from two to thirty-two. They eventually settled on twenty-four missiles in an 18,700 ton hull, just at the time the Salt I nuclear control treaty was being negotiated. Some in the Navy thought that Trident was being used as a bargaining chip, so submarine development was accelerated to get under a five-year window allowed in the treaty. Twenty-four *Ohio*-class boats were planned, but the end of the Cold War capped construction at eighteen.

The Trident C-4 missile, with twice the range of the C-3 Poseidon, first flew in January 1977. Polaris-Poseidon boats had necessarily been based in three locations nearer the anticpated targets—Holy Loch in Scotland, Rota in Spain, and Guam. Trident could hit Moscow from a submarine sitting in New York harbor—or anywhere in between. The overseas bases were closed.

Submarines have grown quieter as the designers have become more sophisticated. Some causes of noise were obvious, such as any opening, appendage, or seam on the hull, and the effects were minimized by redesign. Machinery makes noise; the British devised a method known as "rafting" whereby machinery was isolated from the hull on insulated platforms. As the boats became less noisy, unexpected sources of noise appeared, notably a wide range of otherwise overlooked equipment provided by various suppliers which was not quite up to specification.

Propeller noise was perhaps the single greatest problem. Reducing the speed of rotation reduced cavitation noise, but required a larger-diameter propeller. The longer blades were now rotating outside the cohesive wake of the hull, and each blade in turn whipped through the tubulence left by the sail and other appendages. That set up a series of intermittent vibrations, which caused especially troublesome low-frequency noise. As it turned out, carefully designed and manufactured scythe-shaped blades minimized the problem.

It also turned out that the Soviets were able to learn this secret, not through research and development but with the assistance of the Walker family spy ring in the mid-1980s. It seems that the Soviets were not even concerned with quieting until they learned the extent to which their submarines could be tracked with SOSUS and related systems. Once alerted, they were able to obtain the necessary blade-machining technology through back-door deals with Kongisberg Wappenfabrik of Norway, which developed the software, and Toshiba of Japan, which manufactured the milling machine.

By 1982, the declining performance of American submarines vis-à-vis assumed increased performance of the Soviets became a matter of great concern.

Thus began the push for the next-generation American submarine, anticipating the arrival of a number of new or improved weapons and technologies: a new reactor, pump-jet propulsion (pioneered by the British, it is somewhat less efficient than a propeller, but much quieter), new high-strength hull material, new sonar, new weapons.

The next-generation American submarine would the submarine of the 21st century, and the hull number, SSN-21, was picked as a symbolic reminder. The broad requirements for SSN-21, the fifth U.S. Navy submarine to bear the name *Seawolf*, were:

- The submarine must be fast—for quick transit to a crisis and for intercepting distant enemy forces.
- It must be able to go deep—to seek out the deep ocean sound channels the better to detect and track targets, to hide under thermal and salinity layers and avoid detection, to evade gravity weapons and escape seeking weapons.
- It must have enough power to recover from an uncommanded dive, even when underwater integrity is compromised by battle damage.
- It must have a large onboard weapons capacity—so that it can take on every ship in an enemy task group at once, or remain on station longer.
- It must have a variety of weapons—so that it can take on any adversary, on or under the sea or on land, under any conditions.
- It must be silent—to enhance its own listening abilities and limit detection by an enemy.
- It must have a variety of countermeasures to foil enemy detection and attack.
- It must have the most sensitive listening devices and the most powerful sonar—so that it can detect, identify, and track multiple threats.
- It must have a sophisticated, high-capacity command and control system so that it can deal with those threats.
- It must be versatile—easily accommodating special missions without compromising any capability.

- All onboard systems must be shockproof, rugged, and reliable—so that it can continue to perform any assigned mission under almost any conceivable conditions.

- And finally, after all those requirements have been met, it must have the reserve capacity to accommodate future, improved, and possibly larger and more power-hungry systems.

Seawolf meets those requirements, and more, in a 353-foot, eight-thousand-ton package; that's not much longer than *Los Angeles,* but the forty-foot diameter is ten feet greater. In Congressional testimony on March 20, 1990, Rickover's successor as Director of Naval Nuclear Propulsion, Admiral Bruce DeMars, noted that *Seawolf,* under way at "quiet speed," would be as quiet as *Los Angeles* sitting at the pier. "Quiet speed" may be above twenty knots.

One published estimate of *Seawolf*'s submerged speed is thirty-five knots; it could be—probably is—higher. The Navy has acknowledged that the boat has "exceeded" design specifications, but gives no details. There is a price, a big price, $2 billion a copy, which at times has raised opposition to the boiling point. Some would argue that because *Seawolf* is so expensive, perhaps some capabilities could be limited and a future submarine class could be built for much less.

Perhaps. If the price of a new reasonably capable submarine is to be substantially less than that of *Seawolf,* it must be built in reasonably large numbers. With the end of the Cold War, submarine production almost stopped dead—which raised the cost of building submarines. High production rates spread the cost of operating a shipyard across dozens of ships; with low production rates, an exceptionally high portion of submarine cost is overhead. The most cogent argument which affirmed funding for the third *Seawolf,* now under construction, was the very real need to "maintain the industrial base"—that is, keep enough skilled workers employed in building nuclear submarines today so that there will be some around when we need them tomorrow. Between 1900 and 1996, four Naval shipyards and thirteen private yards had built submarines for the U.S. Navy. In 1996, only one remained in the submarine business: Electric Boat. A nuclear shipyard is not something that quickly can be thrown up on the tidal marsh, like World War I's Hog Island. Nuclear submarines are not Liberty ships, assembled by unskilled workers following simple instructions.

But the question remains: Can a less-expensive submarine than *Seawolf* be built with sufficient capability to satisfy the needs of the Navy and protect the safety of the operators? Probably, and a great deal of time and effort is being devoted to that question. *Seawolf* was designed to counter an open-ocean threat from the Soviet submarine force, with an enhanced ability to operate in polar areas and the absolute latest in quieting technology. As the Soviet blue-water presence has diminished, the mission of the submarine force—*Seawolf* included—has been changing, with an increased emphasis on operations closer inshore. We are conditioned to think of submarines as machines for sinking ships, but Professor Papin's approach was to "touch the enemy and hurt them in sundry ways." The submarine offers significant advantages over airplanes or surface ships—for example, in covert electronic surveillance, deployment of special forces, minelaying in harbor approaches.

The threat is changing also, a shift from the Russian equivalent of *Los Angeles* to limited-capability (but nonetheless deadly) diesel submarines, bottom-planted surveillance systems, and mines. Yes, mines; the often-despised stepchild of Navy planners, the unglamorous and unchivalric weapon, the embodiment of "secret murder" hiding just out of sight. Three of the last four U.S. Navy ships damaged by an enemy were victims of mines. The fourth (*Stark*) was hit by a missile "inadvertently" fired from an Iraqi airplane. Gianibelli, Drebbel, Bushnell, Fulton, Colt, Maury—all must be smiling, from wherever.

The United States *is* developing a new class of nuclear attack submarines, usually referred to as the "NSSN" and shaped to the new requirements with major advances in detection and command and control systems; the baseline design of *Seawolf*, after all, was laid down in the early 1980s. The new attack submarine will have the sort of "fly-by-wire" control now used in aircraft, where electronic systems replace mechanical systems; in fact, the aerospace company Lockheed Martin is participating in the design (along with Newport News Shipbuilding Company, which hopes to be awarded some contracts for follow-on construction).

A modular isolated deck structure (dubbed "MIDS") will set preconstructed modules on special mounts, to give *Seawolf*-level acoustic stealth in a smaller hull. The new attack submarine should be five times quieter than a *Los Angeles*-class submarine. Because some computer and electronic technologies now become obsolete about every eighteen months, the systems will incorporate an open architecture to permit constant (and cost-effective) up-dating.

Computers are being employed, more than ever, in initial design and development—aerospace techniques applied to shipbuilding, resulting in fewer design documents, fewer specifications, and reduced "customization." One example: Where *Seawolf* has 98,000 unique parts, the new attack submarine will have only 10,500.

In keeping with the changing missions, the new attack submarine increasingly will function as a computer node: able to receive information from manned and unmanned aircraft, other ships, ground vehicles, sensors planted underwater and ashore, all to achieve what Admiral William Owens, former vice chairman of the Joint Chiefs of Staff, called "dominant battlefield awareness." In this regard, much of the submarine battlefield would now be ashore, not a thousand miles from land. A battlefield with targets for submarine-launched missiles—such as the *Tomahawks* sent against Iraq in 1991; a battlefield waiting for a clandestine incursion by the members of a submarine-launched team of SEALS or Special Forces who will already know what to expect before they paddle ashore.

Other relevant technology is on the horizon. Jules Verne's *Nautilus* generated electricity from seawater. Improbable? Daimler-Benz today is testing a prototype fuel cell in a passenger van which generates electricity from the chemical combination of hydrogen and oxygen. The "Necar II" gets the oxygen from the air, but must carry the hydrogen; the van can reach speeds of 70 mph and has a maximum range of 150 miles before the hydrogen runs out. Daimler-Benz engineers have proposed that hydrogen could be obtained from alcohol—more easily delivered and stored—but that technology does not yet exist. If they were working on submarine propulsion, they would find both hydrogen and oxygen ready at hand.

In the meantime, *Seawolf* appears poised to sail into the record books as the most effective submarine boat of all time. As DeMars noted, following *Seawolf*'s first sea trials, July 5, 1996: "The combination of the speed and the stealth and the weapon-carrying capacity of this ship really makes it the most powerful warship in the world; not just the most powerful submarine."

Epilogue

Well, enough. We leave the world of "history" and enter the world of current affairs; we leave a world where issues have already been decided and enter a world where the champions are still tilting at the lists. We withdraw gracefully, with just a final thought or two.

- The morality of "secret murder," an issue which occupied philosophers and politicians for the better part of five hundred years, was rendered moot by the widespread, indiscriminate warfare, especially as delivered from the skies, of the middle of this century. A cynic might presume that, if any vestige of chivalry remains, it may be limited to this: Don't machine-gun survivors in the water.
- The submarine has not fulfilled the prophecies of Fulton or Jules Verne; it has not brought an end to navies—let alone war itself—but it *has* greatly influenced the conduct of the major wars of this century, serves as the major deterrent to nuclear holocaust, and continues to influence the course of battles joined and unjoined.
- We have arrived at a point in history in which some forty-seven nations operate more than seven hundred submarines, almost three hundred of

them nuclear. Beyond the efforts under way in the United States and Russia, new designs are being pursued by Germany, Italy, Denmark, Norway, Sweden, and Japan. More than 118 submarines are known to be scheduled for construction over the next decade—fifteen for the United States, fifty-four in the Far East. Thailand and Singapore are joining the club; Finland, which provided "cover" for German developments between the wars, may be reinstating a submarine service. It is abundantly clear that the "submarine" is about to enjoy an official one hundredth birthday—dated from the acceptance of *Holland VI*—in the best of international health.

But wait a minute . . . *Singapore*? A nation the size of Philadelphia? The maritime powerhouse of *Thailand*? Indeed, they understand what submarine professionals have known almost forever: The "weapon of the weaker nation" delivers an inordinate return on investment. As Admiral Dewey mused in 1901, "The thought of those boats running around underwater would drive everyone crazy." They still do. In the 1982 Falklands war, two British ASW carriers, more than a dozen surface warships, five submarines (four of them nuclear), and a gaggle of patrolling aircraft were occupied—almost paralyzed—in trying to protect the rest of the force against an Argentine submarine threat which consisted of two badly maintained, poorly manned operational boats, one a World War II Guppy and the other an eight-year-old German Type 209. The Guppy spent a lot of time at sea without sighting any British ships and, when attacked by a helicopter, ran herself aground. The Type 209 claimed three attacks on British ships, none of which seemed to have been noticed by the targets.

However—be not deceived by that comic-opera flavor; the Falklands submarine war was deadly serious business. The British submarine *Conqueror* sank the World War II-vintage Argentine cruiser *Belgrano* (ex-USS *Phoenix*) with two World War II-vintage torpedoes; 368 sailors were killed.

The weapon of today's "weaker nation" will invariably be diesel-powered. Does this mean—especially with the end of the Cold War—that the nuclear submarine has been passed by? Can scarce national resources—U.S. and other—better be put to use in other systems? British journalist Dan van der Wat (*Stealth at Sea,* 1994) has begged the question; he points out that the "submersible diesel-electric torpedoboat" is "deadlier than ever," and suggests that the superpowers hold on to nuclear-powered submarines solely for reasons of prestige. In one of several arguments, he notes ". . . the British government proved in 1993 that it would rather sacrifice any item of public expenditure in its unprecedented debt crisis than this totem of status, by deciding to complete its Trident program regardless."

Well . . . should any navy which *can* build and operate the best elect instead to develop submarines with limited capabilities? Would any responsible naval planner (or politician) agree to such compromise? The issues today are much the same as on the eve of World War II, when the U.S. Navy's General Board wanted to build a number of smaller, less-expensive submarines of reduced capability. The submarine community *then* didn't want to go into combat with any boat which sacrificed advantage of speed, and range, and depth, and weapons, and habitability, just because the General Board wanted to be able to buy more and pay less—nor would any sensible sailor want to do so *today*.

According to U.S. Navy estimates, Russia put $7.2 billion into six new submarines produced in 1996. They were not diesel-powered. The next-generation Russian nuclear submarine, *Severodvinsk,* now under construction, may enter service in 1997. It is viewed as an advance over any *Los Angeles* boat.

Perhaps Russia is transfixed with totems of status.

Perhaps not.

Acknowledgments and a Note on Sources

For assistance with American sources: Commander Ed Johnson, USN, who reviewed the manuscript for technical accuracy and helped resolve naive ambiguities; unless otherwise noted, illustrations are from the collection of the author or U.S. Navy sources, with special thanks to Russ Egnor of the Office of Information and Chuck Haberlein of the Naval History Division.

For assistance with English sources: Lieutenant Commander Roger E. Arnold-Shrubb, JP, RN (Retired), the British Museum, and the staff of the British Submarine Museum at Gosport, who conducted a gracious tour.

For assistance with German sources: Dr. Eric Berryman of the U.S. Department of Defense.

For assistance with French sources: Christopher Beyeler of the *Service historique de la Marine* and Pierre Barre–the only quasi-French American I know who can penetrate Parisian traffic with aplomb.

For assistance with everything: the editors. Walter Boyne, who helped turn some ideas into a book, and Natalee Rosenstein, who helped turn meandering phrases into crisp prose.

Of the 425 books noted in the Introduction, I read perhaps sixty—listed in the following Bibliography—and looked into dozens more, which proved not to be of interest and are not therefore listed. Here are those works which were most helpful and, in the main, appear to be the more accurate:

For the early years: Alex Roland's 1978 *Underwater Warfare in the Age of Sail* is the best single treatment, and lays the groundwork for any philosophical discussion of clandestine warfare. *Gentleman's Magazine* is always cited in any discussion of early submarines; I assumed this would be difficult to find, and called on the British Museum for assistance—of course readily given. However, the same day I received copies of relevant pages from issues in 1747 and 1749, I found a complete set of *Gentleman's Magazine,* from 1731 into the 20th century, in my local University of Akron library!

David Bushnell's efforts with *Turtle* are most fully explored in the various volumes of *Naval Documents of the American Revolution;* the mystery of his later years is pieced together from several sources, including biographical treatments of Joel Barlow and Abraham Baldwin.

For Robert Fulton—three books are paramount. Parsons's 1922 study, *Robert Fulton and the Submarine,* Dickinson's 1923 biography, and Roland's work noted above.

The Colt story is told in Roland; the material on Lodner Phillips is largely from Patricia Harris's *The Great Lakes First Submarine: L.D. Phillips' "fool killer."* Additional information from Barber's *Lecture* and the U.S. Patent Office.

Perry's *Infernal Machines* provides some of the Civil War material, as does Friedman's *U.S. Submarines through 1945;* the Naval History Division's *Civil War Naval Chronology* was most helpful. The fate of *Hunley* is revealed in Cussler's *The Sea Hunters,* which also details his successful searches for the remains of *Pathfinder*—the first ship sunk in World War I—and Otto Hersing's *U-21,* the boat that did the sinking and later was scuttled en route to postwar internment at Scapa Flow. I have never met Cussler, but was in part responsible for his twenty-year passion for searching for shipwrecks. As a Special Assistant to the Secretary of the Navy in 1975–76, I officially encouraged a civilian-funded Bicentennial effort to locate the remains of John Paul Jones's *Bon Homme Richard,* sunk in the English Channel in 1779. Dr. Harold Edgerton volunteered the use of his newly developed side-scan sonar; Cussler provided the funding (from royalties on *Raise the Titanic*), and actively joined in the (unfortunately unsuccessful) searches of 1978 and '79—accompanied by my own former Bicentennial Special Assistant, Eric Berryman.

The best treatment of John Holland is Morris's 1966 biography of the inventor; additional material is from Simon Lake's *The Submarine in War and Peace* and the transcripts of several Congressional hearings.

For the World Wars, with special focus on German operations: Tarrant's *The U-Boat Offensive 1914–1945* and Terraine's *Business in Great Waters.* Hadley's *Count Not the Dead* gave insight into the propaganda wars. The British side is also represented in Compton-Hall's *Submarines and the War at Sea 1914–1918,* which provides colorful insights into submarine life and operations; Gray's *Damned Un-English Weapon* adds some material. The tale of the illegally procured American boats is told in Smith's *Britain's Clandestine Submarines 1914–1915.* The *Laconia* affair is fully covered in Frederick Grossmith's 1994 book—dedicated to the memory of his father, a *Laconia* survivor.

Principal American sources for World War I include Secretary of the Navy Daniels's *Our Navy at War,* DuPuy's *5 Days to War,* and my own *Age of the Battleship: 1890–1922;* for World War II, Boyne's *Clash of Titans,* Gallantin's *Submarine Admiral,* and Spector's *Eagle Against the Sun.* For the Japanese side: Boyd/Yoshida's *The Japanese Submarine Force and World War II.*

On Helmuth Walter's efforts with hydrogen-peroxide propulsion and streamlined hull shapes, the sources are fragmented and—as noted in the foregoing text—contradictory. My best source was Dr. Walter himself, although in 1966 when I consulted with him about the comic strip *Buz Sawyer,* I had no idea I would later be writing about his work.

The twin Friedman books, *U.S. Submarines through 1945* and *U.S. Submarines since 1945,* are invaluable sources, Alden's *The Fleet Submarine* is an excellent review, and Rockwell's *The Rickover Effect* gives an admitted partisan's spin to the development of the nuclear submarine. The Polmar/Allen 1982 biography, *Rickover,* is not as flattering, but wider in scope.

Finally, a rich resource, but not for the uninformed, is Sweeney's 1970 *Pictorial History of Oceanographic Submersibles.* The author assembled a wide array of materials with apparently no understanding of the significance or relationship of any. Three examples: He assumed that John Wilkens and the Bishop of Chester were two different men (they were not); he assumed that the "Rotterdam boat" and the submarines of de Son were two different submarines (they were not); and he completely misunderstood Holland's work (he did not know that *Fenian Ram* was a Holland boat, thought that *Holland VI* was a "duplicate *Plunger*" undertaken because of Holland's impatience with the Navy, thought the Navy's first Holland was *Holland IX,* thought the first British Holland was built in the United States . . .). Traveler, beware.

Bibliography

Alden, John D. *The Fleet Submarine in the U. S. Navy: A Design and Construction History.* Annapolis: Naval Institute Press, 1979.

Barber, Francis Morgan. *Lecture on Submarine Boats and Their Application to Torpedo Operations.* Newport, R.I.: U.S. Torpedo Station, 1875.

Boelcke, Willi A., Ed. *The Secret Conferences of Dr. Goebbels: The Nazi Propaganda War 1939–1943.* New York: E.P. Dutton & Co., 1970.

Boyd, Carl, and Yoshida, Akihiko. *The Japanese Submarine Force and World War II.* Annapolis: Naval Institute Press, 1995.

Boyne, Walter J. *Clash of Titans.* New York: Simon and Schuster, 1995.

Brittin, Burdick H. *International Law for Seagoing Officers.* Annapolis: United States Naval Institute, 1956.

Clancy, Tom. *Submarine.* New York: Berkley Books, 1993.

Compton-Hall, Richard. *Submarines and the War at Sea 1914–18.* London: MacMillan, 1991.

Cremer, Peter. *U-Boat Commander.* Annapolis: Naval Institute Press, 1984.

Cussler, Clive. *The Sea Hunters: True Adventures with Famous Shipwrecks.* New York: Simon & Schuster, 1996.

Daniels, Josephus. *Our Navy at War.* Washington, D.C.: Pictorial Bureau, Navy Department, 1922.

Davies, Roy. *Nautilus: The Story of Man Under the Sea.* Annapolis: Naval Institute Press, 1995.

Dickinson, H. W. *Robert Fulton, Engineer and Artist, His Life and Works.* London: John Lane, the Bodley Head, 1923.

DuPuy, R. Ernest. *5 Days to War: April 2–6, 1917.* Harrisburg, Pa.: Stackpole Books, 1967.

Fowler, William M., Jr. *Under Two Flags:*

The American Navy in the Civil War. New York: W.W. Norton & Company, 1990.

Friedman, Norman. *U.S. Submarines Since 1945.* Annapolis: Naval Institute Press, 1994.

Friedman, Norman. *U.S. Submarines through 1945.* Annapolis: Naval Institute Press, 1994.

Galantin, I.J. *Submarine Admiral.* Urbana: University of Illinois Press, 1995.

Gray, Edwin. *A Damned Un-English Weapon: The Story of British Submarine Warfare 1914–18.* London: Seeley, Service & Co. Ltd., 1972.

Grossmith, Frederick. *The Sinking of the Laconia.* Stamford, U.K.: Paul Watkins, 1994.

Hadley, Michael L. *Count Not the Dead.* Annapolis: Naval Institute Press, 1995.

Harris, Brayton. *The Age of the Battleship: 1890–1922.* New York: Franklin Watts, 1965.

Harris, Patricia A. Gruse. *Great Lakes' First Submarine: L.D. Phillips' "fool killer."* Michigan City, Ind.: Michigan City Historical Society, 1982.

Hinsley, F. H., and Stripp, Alan, Eds. *Code Breakers.* Oxford: Oxford University Press, 1993.

Hutcheon, Wallace. *Robert Fulton, Naval Warfare Genius.* Annapolis: Naval Institute Press, 1981.

Kemp, Paul. *Convoy Protection.* London: Arms and Armour Press, 1993.

Koppes, Clayton R., and Black, Gregory D. *Hollywood Goes to War.* New York: The Free Press, 1987.

Lake, Simon. *The Submarine in War and Peace.* Philadelphia: J.B. Lippincott Company, 1918.

Lipscomb, F.W. *Historic Submarines.* New York: Praeger Publishers, 1969.

Livezey, William E. *Mahan on Sea Power.* Norman: University of Oklahoma Press, 1947.

Manvell, Roger. *Films and the Second World War.* South Brunswick, N.J., and New York: A.S. Barnes and Company, 1974.

Marder, Arthur J. *From the Dreadnought to Scapa Flow; The Royal Navy in the Fisher Era, 1904–1914.* Volume I, *The Road to War, 1904–1914.* London: Oxford University Press, 1961.

Morris, Richard Knowles. *John P. Holland: 1841–1914; Inventor of the Modern Submarine.* Annapolis: United States Naval Institute, 1966.

Naval History Division, Navy Department. *Civil War Naval Chronology.* Volumes I–V. Washington, D.C.: U.S. Government Printing Office, 1961–1965.

Naval History Division, Navy Department. *Naval Documents of the American Revolution.* Volumes I–X. Washington, D.C.: Government Printing Office, 1964–1974.

Parsons, William Barclay. *Robert Fulton and the Submarine.* New York: Columbia University Press, 1922.

Perry, Milton F. *Infernal Machines: The Story of Confederate Submarine and Mine Warfare.* Baton Rouge: Louisiana State University Press, 1965.

Persico, Joseph E. *Nuremberg: Infamy on Trial.* New York: Viking Penguin, 1994.

Polmar, Norman, and Allen, Thomas B. *Rickover.* New York: Simon & Schuster, 1982.

Potter, E.B. *Nimitz.* Annapolis: Naval Institute Press, 1976.

Potter, E.B., and Nimitz, Chester W., Eds. *Sea Power: A Naval History.* Englewood Cliffs, N.J.: Prentice-Hall, 1960.

Puleston, W.D. *Mahan.* New Haven, Conn.: Yale University Press, 1939.

Read, James Morgan. *Atrocity Propaganda.*

New Haven, Conn.: Yale University Press, 1941.

Rockwell, Theodore. *The Rickover Effect.* Annapolis: Naval Institute Press, 1992.

Roland, Alex. *Underwater Warfare in the Age of Sail.* Bloomington & London: Indiana University Press, 1978.

Scharf, J. Thomas. *History of the Confederate States Navy.* Baltimore: The Fairfax Press, 1887.

Smith, Gaddis. *Britain's Clandestine Submarines 1914–1915.* New Haven, Conn.: Yale University Press, 1964.

Smith, H.A. *The Law and Custom of the Sea.* New York: Frederick A. Praeger, Second Edition, 1950.

Spector, Ronald H. *Eagle Against the Sun.* New York: The Free Press, 1985.

Stern, Robert C. *U-Boats in Action.* Carrollton, Texas: Squadron/Signal Publications, 1977.

Stern, Robert C. *Type VII U-boats.* Annapolis: Naval Institute Press, 1991.

Sweeney, James B. *A Pictorial History of Oceanographic Submersibles.* New York: Crown Publishers, 1970.

Tarrant, V.E. *The U-Boat Offensive, 1914–1945.* London: Arms and Armour Press, 1989.

Terraine, John. *Business in Great Waters: The U-Boat Wars, 1916–1945.* London: Leo Cooper, 1989.

Vat, Dan van der. *Stealth at Sea.* New York: Houghton Mifflin, 1994.

Verne, Jules. *Twenty Thousand Leagues Under the Sea.* New York: Bantam Books, 1962. Translation by Bonner, Anthony, of the 1870 edition.

Winterbotham, F.W. *The Ultra Secret.* New York: Dell Publishing Company, 1974.

Winters, John D. *The Civil War in Louisiana.* Baton Rouge: Louisiana State University Press, 1963.

Woodhouse, Henry. *Textbook of Naval Aeronautics.* Reprint of the 1917 Edition. Annapolis: Naval Institute Press, 1991.

Index